AF856831

Advances in Human Factors/Ergonomics, 10A

Social, Ergonomic and Stress Aspects of Work with Computers

Advances in Human Factors/Ergonomics

Series Editor: Gavriel Salvendy, Purdue University, West Lafayette, IN 47907, U.S.A.

Vol. 1.	Human–Computer Interaction (G. Salvendy, Editor)
Vol. 2.	Human–Computer Dialogue Design (R.W. Ehrich and R.C. Williges, Editors)
Vol. 3.	Expertise Transfer for Expert System Design (J.H. Boose)
Vol. 4.	Engineering Physiology: Physiologic Bases of Human Factors/Ergonomics (K.H.E. Kroemer, H.J. Kroemer and K.E. Kroemer-Elbert)
Vol. 5.	Human Factors Testing and Evaluation (D. Meister)
Vol. 6.	Applications of Fuzzy Set Theory in Human Factors (W. Karwowski and A. Mital, Editors)
Vol. 7.	Human Reliability: Analysis, Prediction, and Prevention of Human Errors (K.S. Park)
Vol. 8.	Human Aspects of Occupational Vibration (D.E. Wasserman)
Vol. 9.	Human Factors Research: Methods and Applications for Architects and Interior Designers (J.E. Harrigan)
Vol. 10A.	Social, Ergonomic and Stress Aspects of Work with Computers (G. Salvendy, S.L. Sauter and J.J. Hurrell, Jr., Editors)
Vol. 10B.	Cognitive Engineering in the Design of Human–Computer Interaction and Expert Systems (G. Salvendy, Editor)

Advances in Human Factors/Ergonomics, 10A

Social, Ergonomic and Stress Aspects of Work with Computers

Proceedings of the Second International Conference on Human–Computer Interaction, Honolulu, Hawaii, August 10–14, 1987, Volume I

Edited by

Gavriel Salvendy
Purdue University, West Lafayette, IN 47907, U.S.A.

Steven L. Sauter and **Joseph J. Hurrell, Jr.**
National Institute for Occupational Safety and Health, Cincinnati, OH 45226, U.S.A.

ELSEVIER

Amsterdam – Oxford – New York – Tokyo 1987

ELSEVIER SCIENCE PUBLISHERS B.V.
Sara Burgerhartstraat 25
P.O. Box 211, 1000 AE Amsterdam, The Netherlands

Distributors for the United States and Canada:

ELSEVIER SCIENCE PUBLISHING COMPANY INC.
52, Vanderbilt Avenue
New York, NY 10017, U.S.A.

ISBN 0-444-42847-X (Vol. 10A)
ISBN 0-444-42848-8 (Vol. 10B)
ISBN 0-444-42849-6 (Set)
ISBN 0-444-42396-6 (Series)

© Elsevier Science Publishers B.V., 1987

All rights reserved. No part of this publication may be reproduced, stored in a retrieval system or transmitted in any form or by any means, electronic, mechanical, photocopying, recording or otherwise, without the prior written permission of the publisher, Elsevier Science Publishers B.V./ Science & Technology Division, P.O. Box 330, 1000 AH Amsterdam, The Netherlands.

Special regulations for readers in the USA – This publication has been registered with the Copyright Clearance Center Inc. (CCC), Salem, Massachusetts. Information can be obtained from the CCC about conditions under which photocopies of parts of this publication may be made in the USA. All other copyright questions, including photocopying outside of the USA, should be referred to the copyright owner, Elsevier Science Publishers B.V., unless otherwise specified.

pp. 67–74, 211–218, 365–372: Work for the United States Government, not subject to copyright.
pp. 219–230: Not subject to copyright.

Printed in The Netherlands

FOREWORD

Computer technology has progressed to the point where computers are now indispensable in all areas of society. The interaction between human and computer has thus become more and more important. Conference Chairman Professor Gavriel Salvendy and all involved in organizing this Second International Conference on Human-Computer Interaction are most timely in their efforts to focus our attention on this vital relationship.

At the previous conference in 1984, the First USA-Japan Conference on Human-Computer Interaction, I had the opportunity to express my thoughts on our relationship with computer and communications (C&C) technology when I spoke on "Man-Machine Interaction in the C&C Age." My point at that time was that the progress of this technology must lead eventually to an improvement in the quality of human life. Our research and development efforts will gradually solve the remaining technological problems, but more important is to ensure harmony between people and C&C.

This harmony is still a thing of the future. In the present stage, we are still learning to adjust to C&C in order to take advantage of the benefits it offers. These efforts are paying off, and the systems are starting to take on the role of relieving humans of burdensome chores and assisting us to fulfill our creative potential.

The ideal state of human-computer interaction is still some way in the future. I see this as a state where C&C systems will assist humans to fully express themselves in every way, including many we cannot even imagine at present. We think, and these thoughts are translated into action and communication by our hands, mouths, eyes, and ears. It is the mind which integrates these various functions in daily life. Our thoughts, feelings, and wishes are expressed to a certain extent through the use of the various information media. C&C systems should integrate existing and future media to instantly and precisely convey and implement human thought processes. In this ideal state, C&C systems will interactively assist us to fully utilize our human creativity.

Much research and development remain to be done to ensure our smooth progress to this ideal interactive stage. A truly interdisciplinary effort will be required, not only between computer and communications experts, but also involving all the human sciences -- psychology, physiology, sociology, architecture, and the arts and culture in general. It is this which makes the present International Conference such a significant occasion. We expect far-reaching results to come from this gathering of experts in so many fields from so many parts of the world.

There is no question that we are moving step by step toward an information society based on C&C technology. My belief is that the human-computer interaction is the key factor which will permit C&C systems to operate for the full benefit and prosperity of humankind.

Koji Kobayashi
Chairman of the Board and
Chief Executive Officer
NEC Corporation

PREFACE

This book consists of 43 papers selected from the 752 submitted papers (from 31 countries around the world) to the conference of which 368 were accepted for presentation at the Second International Conference on Human-Computer Interaction held in Honolulu, Hawaii, 10-14 August 1987. The conference was organized under the auspices of the International Commission on Human Aspects in Computing, Geneva, Switzerland, whose aim is to provide international leadership in the subject area of the Commission. Under the auspices of the Commission, the Third International Conference on Human-Computer Interaction will be held in Boston, Massachusetts, in September 18-22, 1989. The select papers dealing with the Cognitive Engineering in the Design of Human-Computer Interaction and Expert Systems is presented in a companion volume edited by Gavriel Salvendy, the conference chair.

We wish to thank the International Commission on Human Aspects in Computing for sponsoring this conference and the board of the conference who made it all possible. The conference board members include the following:

Australia
Thomas J. Triggs
Brazil
Frida Marina Fischer
Canada
Neville D. Moray
East Germany
Winfried Hacker
England
Tom F. M. Stewart
Lillian G. Malt
Finland
Ilkka Kuorinka
France
Alain Wisner
Japan
Takao Ohkubo
Masamitsu Oshima
(honorary chair)
Hitoshi Watanabe
Korea
Ken S. Park
Poland
Leszek Pacholski
Sweden
Bengt Knave
Switzerland
Etienne Grandjean
Eberhard Ulich
Taiwan
Sheue-Ling Hwang
U.S.A.
Hubert E. Dunsmore
Ray Eberts
Gavriel Salvendy
Ben Shneiderman
Michael L. Smith
Robert C. Williges
U.S.S.R.
Boris F. Lomov
Vladimir M. Munipov
West Germany
Hans-Jorg Bullinger
Klaus-Peter Faehnrich

In most cases, the chapters in this book reflect the content and organization of conference sessions in which the papers were originally presented, though in some cases (e.g., Socio-Technical Aspects) chapters were formed from the combination of papers from several sessions with related content. It should be noted that the two sessions at the conference, corresponding to the chapters Organizational Demands and Cognitive Demands, were developed with support from the National Institute for Occupational Safety and Health (NIOSH). The contents of this book are reproduced herein as received from contributors. The opinions, findings, and conclusions expressed are not necessarily those of the National Institute for Occupational Safety and Health, nor does mention of company names or products constitute endorsement by the National Institute for Occupational Safety and Health. Regarding the session and chapter on cognitive demands, our interest was to call attention to the potential role of mental workload demand as a stress factor in computer-mediated work. Work with computers and the advancing information industry signifies the increasing cognitive nature of work. However, the cognitive demands construct, which includes aspects of attention, memory, decision-making, etc., has tended to escape recognition, characterization, and investigation in the stress literature. Many other unique contributions are to be found in these proceedings.

We want to especially thank Dr. Takao Ohkubo, the regional coordinator for the conference from Japan, who has put a lot of time and effort to make the Japanese participation most effective. The conference could not have been held and this book could not have been published without the diligent, hard and creative work of the conference administrator, Myra Leap. Our sincere thanks and appreciation goes to her.

Joseph J. Hurrell, Jr.
Steven L. Sauter
NIOSH
Cincinnati, Ohio 45226

Gavriel Salvendy
Purdue University
West Lafayette, Indiana 47907

May, 1987

CONTENTS

CONTENTS

Social, Ergonomic and Stress Aspects of Work with Computers,
edited by G. Salvendy, S.L. Sauter and J.J. Hurrell, Jr.
Elsevier Science Publishers B.V., Amsterdam, 1987 — Printed in The Netherlands

DIFFUSION AND IMPACT OF COMPUTERS IN ORGANIZATIONAL SETTINGS: EMPIRICAL FINDINGS FROM A HOSPITAL

JAMES G. ANDERSON[ab], STEPHEN J. JAY[bc], HARLAN M. SCHWEER[ab], and MARILYN M. ANDERSON[b]
[a]Department of Sociology & Anthropology, Purdue University, W. Lafayette, IN 47907, [b]Methodist Hospital of Indiana, 1701 N. Senate Blvd, Indianapolis, IN 46202; [c]Department of Medicine, Indiana University School of Medicine, Indianapolis, IN 46223, USA

ABSTRACT

This study examines the effects of a consultation network among physicians on the adoption and use of a computer-based hospital information system (HIS). The results indicate that the more prominent physicians who were frequently consulted by other physicians were the first to adopt the HIS in their clinical practice. Moreover, physicians adopted the HIS at approximately the same time as other physicians on the hospital service who have similar consultation patterns. These findings have important implications for future efforts to introduce computer-based technology into medical settings.

1. INTRODUCTION

There is evidence that informal communication among physicians plays a significant role in the adoption of medical innovations (Anderson et al., 1987; Coleman et al., 1966). However little research has been done on the structure and functioning of these informal communication networks. This study explores the structure of the consultation network that links physicians on a neurosciences medicine service to one another and to physicians on other hospital services. It also examines the effects of network structure on the diffusion process.

Three hypotheses concerning the nature of the influence exerted by the network structure on potential adopters of a medical innovation will be examined. First, research has shown that individuals whose pattern of relations with others integrates them into a system of potential adopters will be among the first to adopt an innovation (Rogers and Kincaid, 1981). Second, physicians who are structurally equivalent, that is who have similar patterns of communication with other potential adopters in the network, will have similar attitudes and practice characteristics. Third, physicians who are structurally equivalent will adopt an innovation at about the same time (Burt, 1982).

*This research was supported by Grant No. HS05398, NCHSR&HCTA.

2. RESEARCH METHODS

2.1 Practice Setting

This paper examines the consultation network for eight physicians on a neurosciences medicine service in a large private teaching hospital. Physicians on neurosciences medicine consult with one another and with 43 physicians on other hospital services. The hospital has over 1100 beds, a medical staff of 852 physicians, and a house staff of 136.

2.2 Medical Innovation

The medical innovation selected for study is a computer-based hospital information system (HIS) that was implemented throughout the hospital in 1978. The HIS permits physicians and other authorized hospital personnel to enter and retrieve medical orders, nursing notes, and other patient information using a light pen or keyboard at terminals. In order to use the system, physicians need to apply for a computer code and voluntarily undergo a short training session on the use of the HIS.

2.3 Data Collection

A questionnaire was used to obtain information on physician attitudes toward computer applications in medicine and self-reported use of the HIS. Data on age and the number of patients admitted to the hospital by each physician were obtained from the medical personnel system. The date when a physician applied for and received a code to use the HIS was obtained from the Department of Computer Services. A six week tape extract was created from the HIS purge tapes in order to identify physicians who served as consultants to each attending physician and the number of medical orders entered into the HIS by each physician. A consultation network for the neurosciences medicine service was constructed from these data.

2.4 Variables

Age is the physician's chronological age at the time of the study. Admissions is the mean monthly hospital admissions by the physician for the current fiscal year. A measure was created to describe each individual physician's position in the consultation network (Burt, 1982). Prominence is a measure of the extent to which physicians in the network consult a physician.

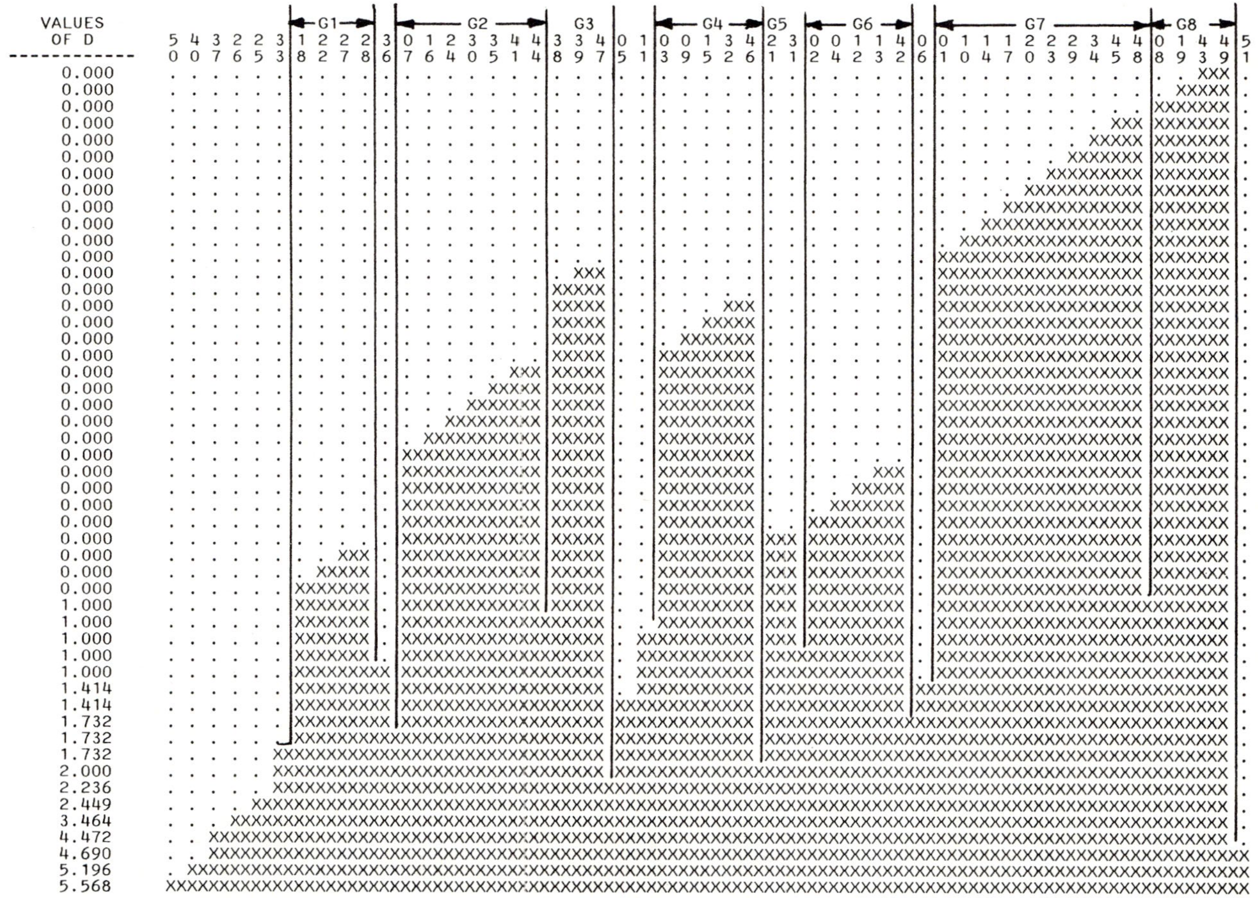

Fig. 1. Hierarchical cluster analysis indicating structurally equivalent groups of consultants for Neurosciences Medicine. The criterion distance, D, is the largest distance between any pair within a cluster.

TABLE 1

Structurally equivalent groups of physicians who are consulted by physicians on Neurosciences Medicine.

Physician	Specialty
EM 11	Emergency Medicine
UR 33	Urology
PSY 36	Psychiatry
<u>Grbup 1</u>	
EM 18	Emergency Medicine
FP 22	Family Practice
IM 27	Internal Medicine
OR 28	Orthopedics
<u>Group 2</u>	
IM 7	Internal Medicine
PSY 16	Psychiatry
AN 24	Anesthesiology
PSY 30	Psychiatry
OTO 35	Otolaryngology
IM 41	Internal Medicine
AN 44	Anesthesiology
<u>Group 3</u>	
CDM 38	Cardiovascular Dis. Med.
IM 39	Internal Medicine
IM 47	Internal Medicine
<u>Group 4</u>	
IM 3	Internal Medicine
FP 9	Family Practice
PSY 15	Psychiatry
IM 32	Internal Medicine
IM 46	Internal Medicine
<u>Group 5</u>	
NSS 21	Neurosciences Surgery
EM 31	Emergency Medicine
<u>Group 6</u>	
PSY 2	Psychiatry
EM 4	Emergency Medicine
NSS 12	Neurosciences Surgery
NSS 13	Neurosciences Surgery
IM 42	Internal Medicine
<u>Group 7</u>	
IM 1	Internal Medicine
OR 10	Orthopedics
EM 14	Emergency Medicine
PSY 17	Psychiatry
CDM 20	Cardiovascular Dis. Med.
FP 23	Family Practice
FP 29	Family Practice
PSY 34	Psychiatry
IM 45	Internal Medicine
IM 48	Internal Medicine
<u>Group 8</u>	
EM 8	Emergency Medicine
NSS 19	Neurosciences Surgery
NSS 43	Neurosciences Surgery
IM 49	Internal Medicine

A number of additional variables were computed to describe the adoption process. The first was the Year of Adoption, when the physician obtained a computer code. The second was the length of Delay in Adoption by the physician. A third was the average number of orders entered into the HIS per patient by the physician. A fourth variable measured the frequency of overall HIS Use by the physician. The last variable measured the physician's attitude toward using computers to assist with medical Decision Making.

2.5 <u>Analysis</u>

The consultation network consisting of the eight physicians on neurosciences medicine along with the 43 physicians that they consult on other hospital services was subjected to a hierarchical cluster analysis using the STRUCTURE computer program (Burt, 1982). Eight groups of structurally equivalent physicians on other hospital services who serve as consultants to physicians on neurosciences medicine were identified. In the second stage of the analysis, measures that describe the physician's position on the consultation network and practice characteristics were examined. Finally, observed response variables were compared to predicted response variables in order to examine the effects of network structure on adoption and use of the HIS.

3. RESULTS

Figure 1 presents the results of a hierarchical cluster analysis of Euclidean distances computed from the binary consultation matrix. The columns represent individual physicians. The rows indicate successive partitioning of physicians into structurally equivalent clusters (i.e., physicians with similar consultation patterns.

Eight groups of structurally equivalent physicians on other hospital services were identified and are listed in Table 1. Physicians in each group had the same pattern of consultations with one or more physicians on Neurosciences Medicine. Three physicians had unique consultation patterns (EM11, UR33, and PSY36).

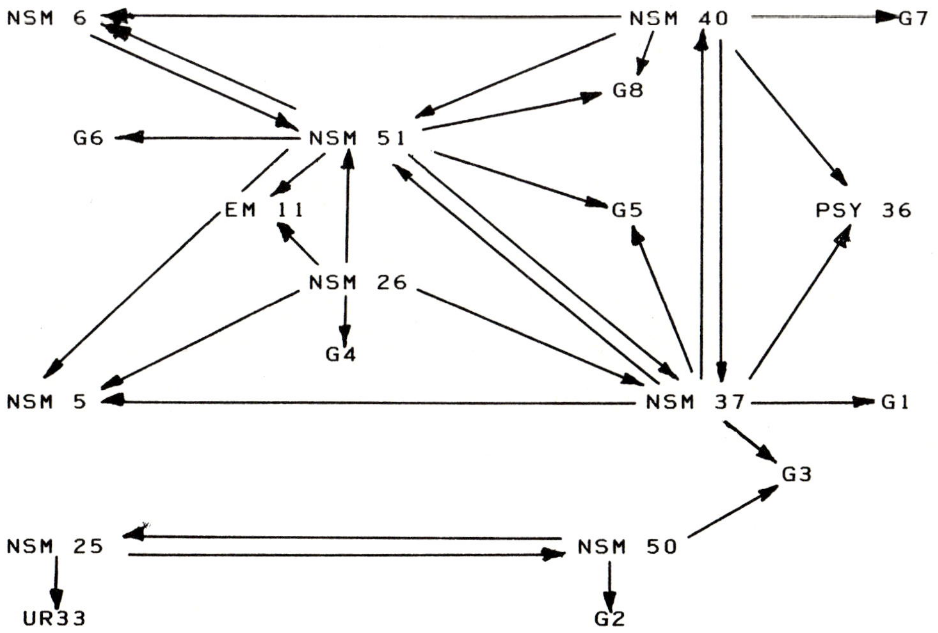

Fig 2. Consultation patterns among physicians on Neurosciences Medicine and with physicians on other hospital services

The structure of the consultation network is shown in Figure 2. There appear to be two subgroups on neurosciences medicine. Six physicians form one consultation group while two physicians (NSM25 and NSM50) only consult with one another and with physicians on other hospital services. All but two of the physician on neuroscience medicine (NSM5 and NSM6) have developed somewhat unique groups of consultants on other hospital services.

Table 2 provides information on each physician's position in the network structure and on practice characteristics. Prominence measures the extent to which a physician is consulted by other prominent physicians on the hospital service. It also indicates the degree to which each physician is integrated into the consultation network. As predicted, the four physicians with the highest prominence scores were the first to adopt the HIS in their practice. All of these physicians applied for a code and underwent HIS training within a year of the systems installation or their joining the hospital's medical staff. In contrast, three of the five less prominent physicians delayed from one to five years before adopting the HIS. At the same time, there is little difference between the more prominent and the less prominent physicians in use of the system and attitudes toward computer-assisted decision making.

TABLE 2

Measures of network structure and practice characteristics on Neurosciences Medicine

Physician	Prominence	Year of Adoption	Delay in Adoption	HIS Use[a]	Decision Making[a]	Order Entry
NSM 51	1.00	1980	0	--	--	1.57
NSM 5	0.93	1978	0	3.83	4.83	1.29
NSM 37	0.74	1978	0	1.33	3.00	4.00
NSM 6	0.74	1979	0	1.83	4.50	7.43
NSM 40	0.40	1983	5	1.17	4.33	8.14
NSM 26	0	1979	2	2.33	4.17	10.71
NSM 50	0	1978	0	0	3.83	1.71
NSM 25	0	1979	1	--	--	7.57

[a]Data were unavailable for two physicians

Table 3 shows the mean characteristics of the groups of physicians that comprise the consultation network. Coefficients of variation (i.e., the standard deviation divided by the mean) are also reported to indicate the homogeneity of each group.

TABLE 3

Mean characteristics of the groups of physicians that comprise the consultation network on Neuroscience Medicine

	Attribute						
Group	Age	Admissions	Year of Adoption	Delay in Adoption	HIS Use	Order Entry	Decision Making
1	43.50 (0.18)[a]	6.89 (0.76)	1978 (0.01)	1.00 (1.00)	2.58 (0.31)	3.12 (0.96)	4.17 (0.06)
2	44.42 (0.26)	8.08 (1.32)	1978 (0.01)	0.75 (0.67)	1.96 (0.79)	3.94 (1.28)	4.43 (0.07)
3	34.00 (0.13)	9.71 (0.66)	1978 (0.01)	0.50 (1.41)	2.39 (0.16)	2.07 (1.52)	4.39 (0.02)
4	40.00 (0.36)	5.17 (0.63)	1979 (0)	1.33 (0.87)	2.42 (0.42)	4.49 (1.13)	4.46 (0.10)
5	41.00 (0.28)	17.72 (1.41)	1979 (0.02)	0.50 (1.41)	3.25 (0.11)	7.34 (0)	4.33 (0.11)
6	43.60 (0.23)	9.74 (0.82)	1979 (0.03)	0.75 (1.28)	2.10 (0.36)	7.53 (0.83)	4.23 (0.14)
7	40.20 (0.09)	9.31 (0.92)	1980 (0.03)	1.71 (1.38)	1.61 (0.97)	8.37 (1.51)	4.06 (0.14)
8	40.00 (0.09)	13.57 (1.15)	1979 (0.02)	0.33 (1.73)	1.38 (0.81)	9.52 (0.87)	4.50 (0.30)
NSM	- 39.88 (0.18)	5.30 (0.68)	1979 (0.02)	1.13 (1.61)	1.75 (0.74)	0.95 (0.87)	4.11 (0.16)

[a]Coefficient of variation

It was hypothesized that structurally equivalent physicians would exhibit similar attitudes and practice characteristics. This is largely borne out by the data in Table 3. Physician groups are quite homogenous with respect to age, although four of the eight groups show greater variation on hospital admissions. In general, physicians in each group adopted the HIS at about the same time. However, there is greater variation within groups with respect to the time they delayed in adopting the computer system. Groups of physicians are quite homogeneous with respect to attitudes and overall HIS use but vary somewhat on direct order entry.

Finally, the influence of the network's structure on physicians was examined by comparing each physician's practice characteristics to the characteristics of other physicians on neurosciences medicine and on all hospital services who are structural equivalent to the physician. It was predicted that a physician would adopt the innovation at the same time that other physicians whom he perceives as similar to him or her adopt it. Moreover, it was hypothesized that physicians with similar interactional patterns with peers will exhibit similar practice patterns. The results are shown in Table 4.

TABLE 4

Effect of the consultation network on the adoption and use of the HIS by physicians on Neurosciences Medicine

	% physicians with no significant difference between observed and predicted responses:	
Characteristic	Neurosciences Medicine	All Hospital Services
Year of Adoption	75	38
Order Entry	71	0
HIS Use[a]	17	33
Decision Making[a]	50	33

[a]Data were unavailable for two physicians

Observed variables were compared to predicted values of variables that reflect the attitude and behavior of other physicians who are structurally equivalent to the physician using a t-test. As can be seen, 75 percent of the physicians adopted the HIS in the same year as their structurally equivalent peers on neurosciences medicine. Moreover, the number of orders entered by a physician per patient is quite similar to his or her structurally equivalent peers on the service. There is more variability in general frequency of use of the computer system, but attitudes are similar for half of the physicians. In general, structurally equivalent physicians on the same service have a much greater effect on adoption than physicians on other hospital services.

4. CONCLUSIONS

The results of this study provide empirical evidence that communication networks among physicians play a significant role in the adoption of medical innovations. This effect is strongest among physicians on the same hospital service. These findings have important implications for future efforts to introduce computer-based technology into medical settings. Future research will attempt to determine whether physician communication networks can be used to effectively introduce medical innovations.

5. REFERENCES

Anderson, J.G. and Jay, S.J. (Editors), Use and Impact of Computers in Clinical Medicine, Springer-Verlag, New York

Burt, R.S., 1982. Toward a Structural Theory of Action, Academic Press, New York.

Coleman, J.S., Katz, E., and Menzel, H., 1966. Medical Innovation: A Diffusion Study, Bobbs-Merrill, New York.

Rogers, E.M. and Kincaid, D.L., 1981. Communication Networks: Toward a New Paradigm for Research, Free Press, New York.

Social, Ergonomic and Stress Aspects of Work with Computers,
edited by G. Salvendy, S.L. Sauter and J.J. Hurrell, Jr.
Elsevier Science Publishers B.V., Amsterdam, 1987 — Printed in The Netherlands

ORGANIZATIONAL BARRIERS TO THE DIFFUSION OF COMPUTER TECHNOLOGY

H.P. LUNDSGAARDE and J.W. WILLIAMSON
Department of Anthropology, University of Kansas, Lawrence, KA 66045, and Health Services Research and Development Field Program, Veterans Administration Medical Center, Salt Lake City, UT 84148

ABSTRACT

The principal objective of this paper is to identify and document how inattention to contextual variables, and the low priority assigned to end-user involvement in the planning and decision-making phases of computerized expert and management information systems, results in costly delays, inefficient use of both human and computer resources, overall lag in the diffusion and cost-effective use of new technology, and the realization of improved health care outcomes.

Two interrelated sets of data are presented. First, we analyze the findings from the 1985 National Invitational Conference on Decision Support Technology sponsored by a large U.S. health care organization. These data represent the informed opinions and collective assessments of a nationally ranked panel of experts in medicine, computer science, and health services management. Second, we compare the findings from the DST conference with our own ongoing efforts to create new technology diffusion management strategies as well as new ways to implement quality assurance programs.

Utilizing our experiences with the creation and implementation of a DST Information Clearinghouse, we call attention to methods for identifying and solving developmental problems before they evolve into costly organizational mistakes. Such mistakes, as evident from the available data on the diffusion and adoption rate of large-scale computerized medical information systems, are neither readily admitted to by stakeholders nor are they easy to correct after the fact.

1. INTRODUCTION

This paper describes the initiation of a Health Systems Research and Development (HSR&D) program related to the development and implementation of Decision Support Technology in a large U.S. health care organization henceforth to be termed "The Organization." The analytic emphasis of this paper will be on the socio-organizational opportunities and barriers engendered by the DST initiative in "The Organization."

We define Health Services Research and Development (HSR&D) as the branch of applied health research that links clinical research and practice in the health care field by focusing on issues that range from health technology assessment and computers in medicine to direct medical practice evaluation and improvement.

We define Decision Support Technology (DST) as a general class of technologies that can interact with an information database to solve specific clinical and administrative problems. DSTs include technologies that may range from printed algorithms to computerized expert systems.

"The Organization" consists of a large network of acute care hospitals and clinics located throughout the United States that principally serve a population of adults and geriatric-age members entitled to pre-paid health care.

2. BACKGROUND

The research and development of DSTs has made remarkable progress over the last two decades. At present, hundreds of different DSTs currently exist and many are under development or in clinical use. This field is growing exponentially and will likely provide future generations of health care providers with powerful tools for assuring the quality and cost-effectiveness of health care.

Beginning in 1984, "The Organization's" Health Services Research and Development Service embarked on a program to develop a research agenda for the development and evaluation of these powerful DST resources for internal use. This direction seemed especially relevant in view of the policy commitment to development of a system-wide computerized medical information system in each of their hospitals. The plan for initiating this DST research program consisted of three events: The first event was a national Delphi study of "The Organization" itself to determine whether there was a perceived need for DSTs to facilitate decision making. Needs would be conceptualized in terms of improved outcomes related to both patient health and institutional cost-savings. The second event was signified by a national invitational consensus conference attended by leading DST developers, both in and out of "The Organization", and a select representative group of potential organization users of DSTs. The conference participants were asked to identify the potential costs and benefits of the proposed program and to identify the barriers to be overcome and resources to be utilized for the subsequent development of DSTs. Finally, the third event focused on analysis of the recommendations from conference participants to set a formal agenda for future HSR&D development in "The Organization."

3. METHODS

We will briefly describe and analyze the events relating to the DST development issue and the socio-organizational reactions that ensued. To assess organizational needs for DST, a two-iteration Delphi study was conducted. In the first iteration, a workbook was mailed to a representative group of insider decision makers (including medical, surgical, psychiatric, long term care, and support services, such as laboratory, pharmacy, and administrative services) together with a select sample of organization and non-organization DST developers. The participants were asked to list decisions within their own area, which if enhanced by such means as DSTs, would likely have substantial impact in terms of improved health outcomes for their patients or cost savings for "The Organization". Each nominated decision was thus given two priority weights ranging from 1 (low) to 10 (high) to indicate the likely quantitative impact of decision improvement on (a) patient health and (b) institutional cost-savings. In the second iteration of the study, the collated results were presented to the participants assembled face-to-face in a national invitational DST consensus conference. These results were collated, analyzed and openly discussed, using Nominal Group Technic (a group judgment method structured to reduce personal bias). A second workbook was completed and promptly distributed to the conference participants. The final judgments of the participants were subsequently aggregated and a mathematical consensus achieved regarding the final list of decisions and weights indicating likely impacts on health gain and cost savings.

Following the preliminary steps described above, participants in the national invitational consensus conference was asked to achieve the following tasks: (1) conceptualize a future DST research and development agenda, (2) estimate the costs and benefits related to such a program that might be accrued by the year 2000, and (3) identify projected resources and barriers to be recognized and managed to assure success of the overall DST program.

The event proved highly successful in the perception of both the participants and observers, which included "The Organization's" key national and local leaders, many different DST developers, and a distinguished group of outside observers.

4. FINDINGS

Conference participants supported a concerted, sustained, well-funded organization research agenda for internal DST development. Participants recommended that future research encompass the full spectrum of R & D stages ranging from needs assessment to post-implementation surveillance. Conference participants projected substantial gains in quality of health care, health outcomes, and cost savings with a cost-benefit ratio of nearly 3 to 1. They concluded that "The Organization" manifested distinct advantages and needs for the development and application of DSTs. Furthermore, the consensus of nationally recognized DST developers and potential DST users, both from within as well as from outside of "The Organization", estimated that if "The Organization" were to invest $15,000,000 annually in the development of DSTs, they might realize savings of as much as $517,000,000 by the year 2000 with concurrent improvement of 32% in patient health outcomes.

Conference participants identified four critical barriers to the development and implementation of DSTs in "The Organization". These barriers were, in order of importance: (1) lack of consistent and coordinated strategies for development and dissemination of DSTs, (2) lack of resources, encompassing personnel, finances, and technical support, (3) lack of attention to the selection and testing of DSTs, and (4) lack of attention to the attitudes and preferences of professionals who might be expected to use the new DSTs.

Despite and initial investment by "The Organization" of about $200,000 to study the feasibility of internal DST research development, and contrary to the evident health care and cost benefits to be realized by such development efforts, a subsequent policy decision was made to discontinue financial support for this HSR&D DST program development. The basic rationale for this decision was said to reside with the numerous new fiscal restraints placed upon "The Organization's" health care system, particularly in HSR&D, by new and changing economic cut-backs.

Four specific DST systems, developed by individual physicians who actively participated in the exploratory conferences, exemplify projects that not only promise significant improvements in health care but which face formidable barriers against implementation. These are: (1) a DST for Quality Assurance and Continuous Monitors for Preventing Tardive Dyskenesia, (2) a computer based Mental Status Examination that would result in an annual cost savings of $511,000 per organizational unit, (3) a Microcomputer-Assisted Oximeter Screening for Sleep Disorders estimated to eventually result in an annual cost savings of approximately $200,000 per organizational unit, and (4) an On-Line Computerized Initial Work-Up of Psychiatric Patients resulting in an annual cost savings of $170,000 per organizational unit. The developers of all four of these DSTs are affiliated with "The Organization." Although all four of these DST projects have been approved as relevant to organizational needs, funding and use remain uncertain despite their enormous cost-cutting and efficiency benefit potential. That the first of the above four was not funded is of interest since it follows close on the fact that "The Organization" lost two law suits amounting to $3,000,000 in damages paid to two former patients incorrectly placed on anti-psychotics that resulted in their development of tardive dyskenesia. The proposed DST might have clearly prevented these two unfortunate happenings. This DST proposal developed directly from "The Organization's" national DST conference. The proposed DST research agenda promised to produce many such proposals, but was vetoed. It would seem that, although policy makers within "The Organization" may have valued "golden eggs" they did not judge the golden goose to be feasible at this time.

Whereas we acknowledge, along with the experts who contributed to "The Organization"'s sponsored HSR&D project on DST, that limited and finite economic resources constrain any policymaker from meeting the demands of all developers or researchers, we also think that non-economic factors played a significant role in the decision-making processes. We hereafter refer to these factors as "barriers" to emphasize that policies created to develop, implement, or discontinue new technology can be predicted from knowledge of general innovation and diffusion theory (Barnett,1953; Hough, 1975; Rogers, 1983).

5. THEORETICAL PERSPECTIVES

Innovations, such as DSTs, have a definite growth cycle. They may begin and end as a small scale effort on the part of a solo researcher to solve a specific problem or begin as an experimental prototype and end as a larger and more complex system. Although innovations evolve in direct response to the various needs of these two groups, their respective priorities may be very different and at times in direct conflict. Many DST developers in the past, for example, have often failed to specify precisely when a particular system could be safely and effectively implemented in a clinical environment. As part of their HSR&D efforts to develop an DST information clearinghouse within "The Organization", their staff screened more than 9,000 citations and 1,300 relevant articles on DSTs and found that approximately ninety percent of all computerized medical expert systems have not been evaluated in clinical environments and that the diffusion of specific DSTs that have matured beyond an early developmental stage is the exception rather than the rule (Lundsgaarde, 1987). Simple attitudinal differences between developers and users about the potential benefits, costs, or relative risks associated with any new innovation can also contribute directly to the misuse or discontinuance of such innovations (Lundsgaarde et al., 1981).

Research on the diffusion of innovations shows that any new technology that does not directly consider the complex needs of future users of that technology may fail. It is obvious that computer technology already has been accepted as a necessary tool in both medical education and clinical practice (Association of American Medical Colleges, 1986). It is less obvious, however, how all current and future users of different computerized medical information systems, or even new DSTs, will come to realize the many benefits of this technology in daily practice (Lundsgaarde, 1987). Fineberg's (1985) survey of the effects of clinical evaluation on the adoption and diffusion of different medical technologies concludes with a number of important recommendations for strengthening the linkages between evaluation findings and practical applications. One of these recommendations urges developers and evaluators to focus attention on "...the diffusion of medical practice concepts and procedures so as to understand how to speed adoption of good practices and discourage use of those that are less effective or harmful." We add to this that DSTs will become both objects and tools in the growing area of Quality Assurance (Williamson, 1978).

Publications by the U.S. Congress Office of Technology Assessment, and a recent monograph on Assessing Medical Technologies (Institute of Medicine, 1985) provide a rich and varied reference source for formal evaluation strategies. They also contain many clues to the kinds of barriers that confront both researchers and developers of this new and exciting technology.

6. REFERENCES

Association of American Medical Colleges, 1986. Medical Education in the Information Age: Proceedings of the Symposium on Medical Informatics. AAMC Press, Washington, D.C.

Barnett, H.G., 1953. Innovation: The Basis of Culture Change. McGraw-Hill, New York.

Fineberg, H., 1985. Effects of Clinical Evaluation on the Diffusion of Medical Technology. In: Institute of Medicine., Assessing Medical Technologies. National Academy Press, Washington, D.C.

Hough, G.W., 1975. Technology Diffusion: Federal Programs & Procedures. Lomond Systems, Inc., MD.

Institute of Medicine, 1985. Assessing Medical Technologies. National Academy Press, Washington, D.C.

Lundsgaarde, H., Fischer, P., and Steele, E., 1981. Human problems in computerized medicine. Department of Anthropology Publications in Anthropology, 13, Lawrence, KS.

Lundsgaarde, H.P., 1987. Evaluating medical expert systems. Soc. Sci. Med., (in press).

Rogers, E.M., 1983. Diffusion of Innovations. Third revised edition. The Free Press, New York.

Williamson, J. W., 1978. Assessing and Improving Health Care Outcomes: The Health Accounting Approach to Quality Assurance. Ballinger Press, Cambridge, MA.

ith Computers, 17
rell, Jr.
1987 — Printed in The Netherlands

NICAL LABORATORY COMPUTER
S POST IMPLEMENTATION

IS DUCHON, Ph.D.[2]
tive Analysis and Information Systems,
ology and Laboratory Medicine
nt
nnati, OH 45221-130, USA

ects of a computer information
oratories, as reported by
months post implementation.
computer system were increases in
ned by technologists, and
ults reporting. Individual
ratories, differed in their
stem according to their perceptions of how the computer system affected their jobs. Technologists may differ in whether they view their jobs in terms of process (bench work) or outcomes (results reports).

The paper suggests how changes in processes and outcomes of work might affect responses to a computer system. It also draws some methodological implications concerning (1) the need for specific, context-dependent measures of relationship between computer system impact and job satisfaction, and (2) the value of combining qualitative and quantitative methods.

1 INTRODUCTION

An estimated 45% of computer-based information systems in medicine fail due to user resistance and sabotage, even though these systems are sound technologically (Dowling, 1980). However, few evaluations of computer systems in medicine go beyond such outcome measures as the system's effects on timeliness, completeness, error rate, retrievability, usage rates, and cost (Simborg et al., 1982). Studies of user satisfaction in medicine make little mention of the impact of computer systems on work or workers (Krobock, 1984). Some studies measure attitudes towards computers independently of working conditions and the computer's impact on work (Melhorn et al., 1979; Reznikoff et al., 1967). A few measure attitudes, job satisfaction and working conditions (Counte et al., 1985; Dlugacz et al., 1981; Dlugacz et al., 1982; Kjerulff et al., 1982;

This research was funded by the University Research Council of the University of Cincinnati.

Kjerulff et al., 1984) independently, using instruments that do not assess users' perceptions of specific ways in which a computer system affects their work. Such studies have found little or no correlation between job satisfaction and employees' acceptance and use of a medical information system (Brooks et al., 1977; Kjerulff et al., 1982; Kjerulff et al., 1984), leading some researchers to question whether attitudes are critical predictors of impact (Alexander et al., 1984).

This lack of relationship between attitudes toward the computer and perception of work conditions occurs despite the impact of a computer system on hospital employees' work patterns. Results documenting increased productivity (Flagle, 1974) or reduced paperwork (Schmitz, 1977) resulting from the introduction of a computer system suggest that a computer system affects the nature of the work performed by its users. Such a change in the nature of work would be expected to affect job satisfaction. Results indicating that work shift is a strong predictor of adaptation to a computer system also suggest that work enviroment might be significant (Kjerulff, 1982). The empirically-based Classic Diffusion Model (Rogers et al., 1971), too, leads one to expect that the perceived impact of a computer system would be influenced by its effect (actual and perceived) on work.

Computer systems do change the role of individual workers and alter their work behavior (Dowling, 1982). A few studies have shown that an individual's general receptivity to a medical computer system is influenced by that system's effect on (1) that person's status and prestige, (2) the nature of the work that that person performs, (3) exercise of that person's professional functions, and (4) perceived feedback and benefit in work (Alexander et al, 1983; Fischer et al., 1980; Lundsgaarde et al., 1981; Thoren et al, 1969; Young, 1981). Although the causal direction has not been established, these studies suggest that perceptions of changes in work are related to users' views of how a computer system benefits their professional roles.

2 LABORATORIES

Computers have been used in clinical laboratories more widely than in other areas of medical practice (Paplanus, 1985). The clinical laboratory represents a microcosm of automation needs throughout a medical center (Lincoln et al., 1980). The principle function of computers in laboratories is in data management, where they relieve the clerical burden of data acquisition and transcription. They also improve legibility, organization, and accuracy of laboratory results reports (Lewis, 1979). Thus, such computer systems affect both the work and the product of a clinical laboratory. However, the few evaluations of laboratory computer systems document primarily outcome measures (Brooks et al., 1977; Flagle, 1974). Although Fleck et al. (1974) report that a laboratory computer system was accepted by different groups within a laboratory because it provided different benefits for each of them, there has been little study of how these effects on process or work influence laboratory technologists' reactions to clinical laboratory information systems.

This paper describes results from one phase of a longitudinal study of the impact of a commercial clinical laboratory computer system on a midwestern university medical center. It explores the relationship between technologists' responses to the computer system and their perceptions of how that system affects their work.

3 SETTING

Research was conducted within the Department of Pathology and Laboratory Medicine at a 650 bed midwestern urban metropolitan university medical center. A commercial laboratory computer information system was installed in April 1985 for use by all nine laboratories within Laboratory Medicine. Data are presented from interviews immediately prior to installation, and from the first wave of survey questionnaires seven months later.

4 METHODS

Methods are detailed elsewhere (Kaplan, 1986; Kaplan, in press; Kaplan and Duchon, submitted for publication) and briefly summarized here.

4.1 Interviews and Observations

Researchers interviewed all laboratory directors and some head supervisory personnel prior to system installation. During the month prior to administering the survey, researchers were present in the laboratories to talk with laboratory staff and observe their activities. Starting three months after the computer system was installed, one author regularly has observed at weekly meetings where directors and head supervisors discussed laboratory management problems.

4.2 Survey Questionnaire

The survey instrument was composed of three parts. The first part consisted of measures, adapted from the Job Diagnostic Survey (Sims et al., 1976), to assesses the extent to which core job dimensions influence satisfaction with work (Hackman, 1983). The second part of the questionnaire used Likert-scale measures of expectations, concerns, and perceived changes that may be related to the use of the computer system. These measures were developed through the interviews and observations. Questions concerning expectations were adapted from Kjerulff et al. (1984). The survey instrument concluded with four open-ended questions to assess changes caused by the computer system and to elicit suggestions for improved system use.

All questionnaire items were pre-tested on a sample of laboratory personnel. After revision, the questionnaire was given to all 248 members of the laboratory staff seven months after computer system installation. Data from 119 (48%) questionnaires were analyzed. Most respondents were technologists, women with college degrees who worked first shift and who had not worked previously in a laboratory with a computer system.

5 ANALYSIS

5.1 Interviews and Open-Ended Questions

Prior to implementation, directors generally agreed that, although there would be more work for laboratory technologists, their jobs would not be changed by the computer system. This assessment was made in full knowledge that technologists would have to enter test results and some orders, and that they would reduce paper work and billing duties. Interviewees also expected that test results would be available more quickly. Reports were expected to have fewer transcription errors, be more legible, and provide useful cumulative information and patient profiles. These improvements were expected to reduce the number of unnecessary laboratory tests and stat (i.e., emergency) and duplicate test orders. The number of telephone calls also was expected to decrease.

Three themes predominated in answer to open-ended items on the questionnaire: (1) changes in technologists' work load, (2) improvements in results reporting, and (3) the need for physicians and nurses to use terminals rather than the telephone for results inquiry. Seven months after implementation, technologists expressed a general sense that their clerical duties and paperwork had increased, and productivity had suffered. However, they credited the computer system with making results available more quickly. They said that reports also were more complete, more accurate, easier to read, and provided a picture of "the whole patient." Nevertheless, they felt that doctors and nurses expected to get test results by calling the laboratories rather than through the computer system. Phone calls not only interrupt work, but respondents reported calls in which they felt blamed by others in the medical center for problems caused by the computer system.

When responses were grouped by laboratory, marked differences were evident between some of the laboratories (Kaplan, 1986). Laboratories differed in their assessment of changes in their workload, number of telephone calls, improvement in reporting, and attitudes expressed towards the computer system.

5.2 Scaled Response Questions

A factor analysis of Job Diagnostic Survey items led to the extraction of four factors that were used as job satisfaction variables: Skill Variety, Task Identity, Autonomy, and Feedback. Overall, no job satisfaction differences due to environmental or individual factors were found . Respondents reported the same levels of job satisfaction regardless of laboratory, education, shift, etc.

Five computer system variables were identified: External Communications, Service Outcomes, Personal Intentions, Personal Hassles, and Increased Blame. Data on these variables indicated that respondents generally were positive about the computer system. They reported that it had improved relations and aided communication between the laboratories and the rest of the medical center (External Communications), and that results reporting and service was better (Service Outcomes). Respondents also were very positive about their own continued use of the computer system (Personal Intentions). The respondents as a whole did not report an increase in irritants about their jobs such as the number of phone calls and the amount of work (Personal Hassles), nor did they report that laboratory staff was blamed more and cooperated with less by physicians and nurses (Increased Blame). Job satisfaction measures and computer system measures were found not to be correlated.

There are statistically significant differences among laboratories for three of the computer system variables: Personal Hassles, Personal Intentions, and External Communications.

Next, two new variables were created in order to test whether technologists differed according to the computer system's impact on process vs. product or outcome aspects of their jobs. One variable combined scores on External Communication and Service Outcomes, and the other combined scores on Personal Hassles and Increased Blame. Respondents tended to have high scores on one of these variables and low scores on the other. A favorability score for each respondent was computed by subtracting the sum of that person's scores on Personal Hassles and Increased Blame from her scores on External Communication and Service Outcomes. When the favorability score was regressed on Laboratories, statistically significant differences in favorability were found across laboratories. Thus, some laboratories, like some technologists, were process oriented while others were outcomes oriented.

It remains to account for these differences between individuals and between laboratories, and to explain why there was no correlation between job characteristic variables and computer system variables.

6 DISCUSSION

The five computer system variables reflect themes that are very similar to those found in the qualitative data from interviews and responses to open-ended questionnaires: changes in workload and changes in results reporting. This congruence of themes from independent sources of data strengthens confidence in the validity of the measures and the importance of differences between laboratories.

The repeated themes of system benefits through improved results reporting, and of disadvantages of increased workload, suggested that there might be a group of respondents corresponding to each of these two themes. One group would see their jobs in terms of producing results reports, the other in terms of the bench work necessary to producing those results reports (Kaplan, in press). The computer system could interfere with the bench work by increasing paperwork and requiring time for data entry, or the system could decrease interference by reducing telephone calls. Thus, the group seeing their jobs in terms of the bench work (i.e., the process of producing laboratory results) would respond to the computer system according to how it affected those items measured by the Personal Hassles and Increased Blame variables. The other group of respondents would see their jobs in terms of the results reports, (i.e., the product, rather than the process) of laboratory work. Their assessment of the computer system would be reflected in responses to External Communications and Service Outcomes items.

These kinds of differences in job definition would not be measured by the job satisfaction measures. Consequently, there would be no correlation between job satisfaction and supervisory status, education, previous employment, laboratory, and computer system measures.

Our results indicate that this analysis may be correct.

7 CONCLUSIONS

This study has methodological implications as well as implications for understanding how users respond to medical computer systems. Combining qualitative and quantitive methods proved especially valuable. In the absence of qualitative data, it would have been difficult to account for differences between laboratories and little would have been learned from the study. Even in the absence of quantitative data, differences between laboratories were striking because of the knowledge gained through field work and qualitative data analysis (Kaplan, 1986).

These findings also suggest the value of investigating specifically how a computer system affects users' jobs. Studies that independently assess satisfaction and attitudes ignore the specific impact of a computer system on work. Because standard job satisfaction measures do not assess job-specific factors, analysis needs to include context-specific assessment of a computer system's impacts on work, and of how users view their jobs. Such measures can be derived from knowledge of the setting gained by field experience, again suggesting the value of combining qualitative and quantitative methods. Additional work is needed that explores methods to productively provide the empirical data of specific systems and specific impacts that Alexander et al. (1983) recommend.

Lastly, the study suggests the need to move beyond outcome evaluations to investigation of how a computer system affects processes. By extending research to examine specific effects of individual computer systems in particular contexts, through a combination of qualitatative and quantitative methods, our general understanding of what affects the acceptance and use of medical computer systems can be improved.

8 REFERENCES

Alexander MJ, Siegel C, Dlugacz Y, and Fischer S, 1983. Post implementation changes in physicians' attitudes toward an automated drug review system. The Symposium on Computer Applications in Medical Care (SCAMC) 7. IEEE Computer Society Press, Silver Spring, Md., pp. 660-663.

Alexander MJ and Siegel C, 1984. Opinions and feelings: the validity of attitudes toward computers. SCAMC 8, IEEE Computer Society Press, Silver Spring, Md., pp. 540-542.

Brooks RC, Casey IJ, and Blackmon PW Jr., 1977. Evaluation of the Air Force Clinical Laboratory Automation Systems (AFCLAS) at Wright-Patterson USAF Medical Center. Vol. I: Summary. HDSN-77-4. (NTIS no. AD-A043 664). Vol. II: Analysis. HDSN-77-5. (NTIS no. AD-A043 665). Analytic Services, Arlington, Va.

Counte M et al., 1985. Long-term adaptation to the introduction of a medical information system. SCAMC 9, IEEE Computer Society Press, Silver Spring, Md., pp. 414-417.

Dlugacz YD, Siegel C, and Fischer S, 1981. Physician/Computer interaction. SCAMC 5, IEEE Computer Society Press, Silver Spring, Md., pp. 797-801.

Dlugacz YD, Siegel C, and Fischer S, 1982. Receptivity toward uses of a new computer application in medicine. SCAMC 6, IEEE Computer Society Press, Silver Spring, Md., pp. 384-391.

Dowling AF Jr., 1980. Do hospital staff interfere with computer system implementation? Health Care Management Review, 5: 23-32

Dowling AF Jr., 1982. Medically oriented computer-based information systems. Medical Care, 20: 253-254.

Fischer PJ et al., 1980. User reactions to PROMIS: Issues related to acceptability of medical innovations. SCAMC 4, IEEE Computer Society Press, Silver Spring, Md., pp. 1722-1730.

Flagle CD, 1974. Operations research with hospital computer systems. In: MF Collen (Ed), Hospital computer systems, John Wiley and Sons, New York, pp. 418-430.

Fleck A et al., 1974. Experience with the introduction and routine operation of a computer-based reporting system in a clinical biochemistry laboratory. International Journal of Bio-Medical Computing, 5: 189-202.

Hackman, JR, 1983. Work design. In: RM Steers and LM Porter (Eds), Motivation and work behavior, McGraw-Hill, New York, pp. 490-516.

Kaplan B, 1986. Impact of a clinical laboratory computer system: Users' perceptions. In: R Salamon, B Blum, and MJ Jørgensen (Eds), Medinfo 86: Fifth World Congress on Medical Informatics, North-Holland, Amsterdam/New York/Oxford/Tokyo, pp. 1057-1061.

Kaplan B, (in press). Initial impact of a clinical laboratory computer system: Themes common to expectations and actualities. Journal of Medical Systems.

Kaplan B and Duchon D (submitted for publication). Assessing relationships between a computer system and work in clinical laboratories. SCAMC 11, IEEE Computer Society Press, Silver Spring, Md.

Kjerulff KH et al., 1982. Predicting employee adaptation to the implementation of a medical information system. SCAMC 6, IEEE Computer Society Press, Silver Spring, Md., pp. 392-397.

Kjerulff KH and Counte MA, 1984. Measuring attitudes toward computers: Two approaches. SCAMC 8, IEEE Computer Society Press, Silver Spring, Md., pp. 529-535.

Krobock JR, 1984. A taxonomy: Hospital information systems evaluation methodologies. Journal of Medical Systems, 8: 419-429.

Lewis JW, 1979. Clinical laboratory information systems. Proceedings of the IEEE, 67: 1229-1300.

Lincoln TL and Korpman RA, 1980. Computers, health care, and medical information science. Scientific American, 210: 257-263.

Lundsgaarde HP, Fischer PJ, and Steele DJ, 1981. Human problems in computerized medicine. Publications in Anthropology, no. 13, University of Kansas Press, Lawrence, Kan.

Melhorn JM, Legler WK, and Clark GM, 1979. Current attitudes of medical personnel toward computers. Computers and Biomedical Research, 12: 327-334.

Paplanus SH, 1985. Clinical pathology. In: D Levinson (Ed), Computer applications in clinical practice: An overview, Macmillan, New York, pp. 118-122.

Reznikoff M, Holland CH, and Stroebel CF, 1967. Attitudes toward computers among employees of a psychiatric hospital. Mental Hygiene, 51: 419-425.

Rogers EM and Shoemaker FF, 1971. Communication of innovations: A cross-cultural approach, The Free Press, New York.

Schmitz HH, 1977. Productivity effectiveness: It can be done in the health care field. Proceedings, Society of Management Information Systems 9, pp. 253-258.

Simborg DW and Whiting-O'Keefe QE, 1982. Evaluation methodology for ambulatory care information systems. Medical Care, 20: 255-265.

Sims H, Szilagy A, and Keller R, 1976. The measurement of job characteristics. Academy of Management Journal, 19: 195-220.

Thoren BJ, Smith DP, and Gould LC, 1969. Attitude study, training help employees adapt to use of computers. Hospitals, Journal of the American Hospital Association, 48(March 1): 61-64.

Young DW, 1981. Doctors' attitudes to a computer-based clinical information system. Methods of Informatics in Medicine, 20(April): 196-199.

Social, Ergonomic and Stress Aspects of Work with Computers,
edited by G. Salvendy, S.L. Sauter and J.J. Hurrell, Jr.
Elsevier Science Publishers B.V., Amsterdam, 1987 — Printed in The Netherlands

COMPUTERIZATION OF PSYCHIATRIC CASE REGISTERS: PUBLIC PERIL VS. HAZARDS TO CONFIDENTIALITY

MICHAEL RAHAV Ph.D
Nathan Kline Institute for Psychiatric Research, Orangeburg, NY 10962

ABSTRACT

This paper studies the consequences of computerizing psychiatric patients' records on the confidentiality and the rehabilitation of former mental patients. The paper focuses on one country, Israel, and its experience in recently automating its National Psychiatric Case Register. Analysis of the Israeli experience reveals that fear and suspicion of the mentally ill, and the desire of various agencies to identify the mentally ill and screen for them, were main motives in the computerization initiative. The paper concedes that in certain instances, such knowledge may, indeed, prevent tragedies and save lives. The paper concludes, however, that the damage to the medical ethics of confidentiality and to the population of former mental patients is very severe. In the long run, using computerized psychiatric records for screening gun applicants or army recruits may backfire, hurting not only the mentally ill but an entire society.

INTRODUCTION

The aim of this paper is to examine some of the consequences of using computer technology to maintain mental health records. Specifically, the paper aims at studying: a) stigmatization or labeling effects computers may have for people with records of psychiatric treatment and, b) the opportunities computers provide for protecting the public from mental or former mental patients who may be dangerous to themselves or others.

The analysis presented here is based on the experience of one country, Israel, of computerizing all records of psychiatric patients.

Records on all people admitted and discharged from any psychiatric hospital in Israel have been accumulated ever since the State of Israel was founded in 1948 (Rahav et al 1981). These records were maintained on cards, forming a national psychiatric case register (PCR). This PCR constituted a one way avenue of information: a great deal of data were input but yielded very little output. Following eight years of costly and often frustrating attempts, a fully automated PCR finally started operating in 1983 (Rahav, et. al. 1986).

USES OF INFORMATION OF THE ISRAELI PCR

The enormous and costly task of computerizing national PCR would not have been possible had there not been certain government agencies that needed information from this PCR and urged its establishment. Were these agencies interested in information for planning, budgeting, quality assurance and research? Apparently not. The agencies involved were in charge of "sensitive" domains in community life that needed the information from the PCR for screening applicants for handguns, new recruits for the Israeli Army, or drivers for special automobiles (Rahav, 1983).

The computerization of the PCR in 1983 made screening of large number of individuals, needed by the agencies above, practical and efficient. This was especially the case in Israel because of the Person Number System (U.S., 1980). Every resident of Israel is identified by a unique ID number from birth on across any files or data banks. Screening individuals for psychiatric history entails therefore looking for matched person numbers.

The newly available psychiatric information brought waves of requests from various government agencies to screen certain groups of people whose mental health was of concern to them. Before the computerization of psychiatric information, the question of these people's mental status had never come up. Now that the mental health information is easily available, it became of utmost importance and urgency. The newly available information seems to beget new needs for information that did not exist prior to its inception. Let us review further the forces and motivations that let to the automation of Israel's PCR, which may explain the priority certain government agencies placed on obtaining people's psychiatric records.

HOW AND WHY PSYCHIATRIC RECORDS BECOME COMPUTERIZED?

The question of why and how a computerized information system on the mentally ill comes into being is hardly ever a trivial one. By allocating the resources needed to establish such a system, and insisting on its creation despite high costs and failures, society reveals the importance and priority it attributes to information on the mentally ill. Why all this interest in psychiatric information in particular? As suggested earlier it was, probably, not in order to increase research capabilities nor was it to improve treatment of mental patients. The public pressure to have psychiatric information easily accessible, I believe, reflects public concern about the mentally ill and the public's wish to know who the mentally ill are. This wish is anchored in specific laws entitling certain agencies to obtain confidential information on the one hand, and requiring those very ones who are obligated to safeguard the confidentiality of this information, to violate this obligation on the other.

It was the insistence of government agencies like the Firearms Register to obtain access to psychiatric records that was the driving force behind the establishment of an automated psychiatric information system. As long as information was on cards, screening was not practical. This meant that people with a history of treated mental illness might have gone undetected, obtaining handguns, being recruited into the army, or being licensed to drive buses. For policy makers who believe that mental and former mental patients are violent and dangerous, this was an unacceptable prospect. It was fear, suspicion and mistrust of the mentally ill that made government agencies persist and demand the automation of psychiatric information. Indeed, fear and mistrust of the mentally ill have been shown to be pervasive in Israel (Rahav, et. al., 1984). They have probably played a more important role in bringing about the automation of psychiatric data than did research, evaluation and planning needs.

CONSEQUENCES OF COMPUTERIZING ISRAEL'S PCR

The computerization of Israel's PCR has made it possible to immediately and quickly identify every resident of Israel who has ever been mentally ill (if we accept history of psychiatric hospitalization as a definition of mental illness). The computerization of psychiatric data which has made screening feasible, has also introduced a dilemma between the right of society to protect itself from people who may constitute a danger to themselves or others, and the right of individuals to privacy and confidentiality. For many policy makers, psychiatric information is viewed as a means of controlling the mentally ill; that is, knowing who they are, where they are, and consequently, having the ability to circumscribe their activities. Their constant nightmare is the image of a mental patient, armed with a gun, going on a shooting rampage on a busy street. The potential for danger inherent (according to these beliefs) in mental or former mental patients carrying guns, serving in the army, or driving buses is so severe that this potential should be prevented at all costs.

Is the assumption about the danger emanating from the mentally ill a true one? Evidence needed to answer this question is sparse and inconsistent (Rabkin, 1979). While the peril to society from mental patients is, to a large extent, unknown, the damage to mental patients and former patients is awesome and real. Making psychiatric records available for screening purposes invades the privacy of the patient. It accesses data that were collected under an oath of secrecy and confidentiality. These are data that were elicited from patients who believed they were protected from any indiscrete use of the intimate and personal information they shared with their therapist. For many of these patients the data are used in a punitive manner, where persons are penalized for having a record of treatment in the past, without any consideration of their present state of mental health.

By allowing various agencies to access routinly its PCR, Israel has turned it into a national labeling device. Entering personal records into the computerized PCR initiates a labeling process that is, probably, more powerful than the ones described by Scheff (1966) and Goffman (1963). According to Scheff it is the reaction to aberrant behavior, which, if not reacted to might have passed unnoticed, that shapes and stabilizes a possibly transitory act into institutional "mental illness." But societal reaction is human. People may forget or forgive and accept a person who once might have violated certain behavioral codes. The computer does not. That person's behavior is on record and it is used every time a former patient attempts to assume a "sensitive" enough task. It is exactly these "sensitive" tasks that entail responsible and trustworthy behavior which former mental patients are now being denied.

They are too "crazy" for those roles which imply full membership in community life. The stigmatizing potential of the computer is so effective that it does not allow patients to flee their past and forget their status as mental mental patients (Rahav, 1985). It forces former mental patients to continue to be "mental patients".

Using psychiatric records for screening people may have the exact opposite effect of that intended. As people learn that seeking help may be known to various authorities and may bring severe punitive repercussions, they will decline to look for help. People who could have benefited from psychiatric help will keep living in the community unaided, likely to get involved in the very tragedies and mishaps we are trying to prevent.

The damage to psychiatry itself is also of the utmost consideration. Psychiatric therapy is based on gaining patients' trust and cooperation through the promise of confidentiality. Sharing patients' records with certain agencies is equivalent to hanging a sign on the clinic's door: Everything you say in this room may be used against you."

What makes the dilemma between public peril and protective privilege so accute is that in certain instances, severely disturbed patients should be known to the authorities. Identifying such patients through screening and denying them certain priveleges such as bearing handguns, may indeed save lives and prevent tragedies. Rare episodes such as the shooting at President Reagan by John Hinkely, or the murders committed by "Son of Sam", turn this dilemma into a choice between saving lives and protecting the public on the one hand, and safeguarding and protecting the confidentiality of psychiatric treatment records, on the other. The computer did not create this dilemma, but it certainly exacerbated it. It is only with the understanding of the potential for abuse computers may have, and better understanding of what mental illness is and what is is not, that we may continue to use computers in maintaining psychiatric records.

REFERENCES

Goffman, E., 1963. Stigma. Prentice-Hall, Englewood Cliff, N.J.
Rabkin, J.G., 1979. Criminal behavior of discharged mental patients: A critical appraisal of the research. Psychological Bulletin, 86: 1-27.
Rahav, M., 1983. The involvement of the mentally ill in "sensitive" activities: Depriving individuals of some certain civil rights, legally. Paper presented at the International Congress on Psychiatry, Law and Ethics, Haifa, Israel.
Rahav, M., 1985. Labeling the mentally ill through psychiatric records: The Israeli case. Israel Journal of Psychiatry and related sciences, 3: 221-231.
Rahav, M., Popper, M., Nahon, D., 1981. The Psychiatric Case Register of Israel: Initial results. Israel Journal of Psychiatry and Related Sciences, 18: 251-267.
Rahav, M., Popper, M., Nahon, D., 1986. The Psychiatric Case Register as an administrative screening device: The example of Israel. In ten Horn G.H.M.M., Giel, R., Gulbinat, W.H., Henderson, J.H. (Eds.).
Psychiatric Case Registers in Public Health, Elsevier, Amsterdam.
Rahav, M., Struening, E.L., Andrews, H., 1984. Opinions on Mental Illness in Israel. Social Science and Medicine, 19:1151-1158.
Scheff, T., 1966. Being Mentally Ill. Aldine, Chicago.
U.S. Department of Health & Human Services, 1980. The Person Number System of Sweden, Norway, Denmark and Israel. DHHS, Publication No. (PHS) 80-1358, Series 2, No. 84, Hyattsville, MD.

CONTENTS

Social, Ergonomic and Stress Aspects of Work with Computers,
edited by G. Salvendy, S.L. Sauter and J.J. Hurrell, Jr.
Elsevier Science Publishers B.V., Amsterdam, 1987 — Printed in The Netherlands

ELECTRONIC MONITORING AND JOB STRESS

MICHAEL J. SMITH, PASCALE CARAYON AND KATHLEEN MIEZIO

Department of Industrial Engineering, University of Wisconsin, Madison, Wisconsin, U.S.A.

ABSTRACT

Serious concerns about the potential health effects of electronic monitoring of worker performance have been voiced by organized labor, government agencies and researchers around the United States. These groups believe that electronic monitoring can lead to job stress and result in diminished worker health (mental and physical), satisfaction, and the motivation to work productively. On the other hand, management groups claim that electronic monitoring is an effective tool for proper utilization of human resources as it provides the statistical base of management information necessary for improved production and quality control.

This paper will evaluate the potential for electronic monitoring to produce job stress. A more thorough review of monitoring influences on production, stress and motivation can be found in Smith et al, 1986. Various aspects of job design such as job demands, socialization, supervision and task control are examined in light of influences by electronic monitoring. The relationships between these job design considerations and occupational stress are discussed. The paper concludes with suggestions for the design of electronic monitoring systems that could minimize worker stress and enhance the productive capacity of the work system.

INTRODUCTION

Workplace electronic monitoring of employee performance is becoming more widespread, and with this increased application a number of important issues have been raised. Some deal with personal privacy considerations, while others center on the potential influences on worker stress and health. Sides have been drawn with unions and employee advocacy groups arguing against the use of electronic performance monitoring and going as far as recommending legislation to prohibit such use, and employers and their associations maintaining that electronic monitoring is vital to doing business and is a management right under federal labor legislation. This last contention by management has spurred congressional interest in changing labor laws to redefine labor and management responsibilities and rights. Advocates claim that such drastic action is required to provide the necessary legislation foundation for new organizational structures that will be needed in the next decade. Technological changes will tranform the workplace, forcing new roles for labor and management, and "fuzzying" the lines of responsibility between the two groups. At the heart of this issue is electronic monitoring which provides opportunities to share information at various levels of an organization or to limit such sharing, and which reduces the requirements for direct human supervision thereby influencing the organizational structure dramatically. Because of its central role in technological change and personal privacy issues, electronic monitoring will be one of the first arenas in which efforts to change federal legislation on privacy and labor laws will take place.

Such attempts must take into consideration the significance of electronic monitoring for improving working conditions and job design which may be eroded by technological advances, and at the same time needed production improvements which can maintain the competitive advantage of U.S. products. A major concern of labor unions is the potential for more stressful working conditions because of monitoring which management readily accepts as a cost of production improvement. This position states that management gets all the benefits while workers pay with diminished health. However, there is reason to believe that proper application of job design principles can develop monitoring systems that can provide benefits to both labor and management. This can be achieved by establishing monitoring systems that provide for more employee participation and control over work processes and which are coupled with employee motivation programs that reward participation, performance and commitment. Such systems can provide meaningful and satisfying feedback to employees, better means for employee evaluation, and production process information necessary to effectively manage production resources.

It is important to note that current electronic monitoring systems do not incorporate necessary features that will improve job design to benefit employees, and therefore the experience gained from these systems has little benefit in defining future directions and needs to achieve systems that provide concurrent benefits to labor and management. Currently, there is no research evidence that demonstrates that electronic monitoring increases individual worker production, total workplace productivity, enhances the ability to manage and control workers, or improves the evaluation of worker performance. On-the-contrary, current electronic monitoring systems are so limited in their capabilities that they have fueled the objections of worker advocacy groups to electronic monitoring of worker performance. These systems rely primarily on second-by-second counts of the quantity of worker output without any consideration of the quality of the product, differences in providing specific products that change performance requirements and exceptional circumstances that throw-off timetables. Reliance on the number of units per hour puts inordinate emphasis on speed and can increase work pressure which is stressful, diminish job satisfaction, and decrease worker pride. Such systems are doomed to failure in the long run as they build "sweat-shops" full of unhappy employees who frequently move from employer to employer.

Those few electronic monitoring systems that do examine quality as well as quantity typically do the measurements in such an objectionable, "sneaky" way that the approach acts as a demotivator to workers rather than as a source of helpful information. Unannounced eavesdropping on telephone solicitations or monitoring the number of corrections made in a typed document are examples of current systems. A major concern in devising new electronic monitoring systems is ensuring that the information gathered represents the full gamut of employee performance including quantity, quality, service and initiative. Current systems do not gather all of this information and technological and worker evaluation advances are needed to be able to address all of them. But even if they can be measured, such as with the quality examples above, if the information is not applied in an appropriate way, then it will be harmful to workers and the production process. It is often said that of today's electronic monitoring systems are not, but the way in which they can be abused is "evil." Future systems that monitor more explicit information have the potential to be "more evil" than current systems if they are abused. The purpose of this paper is to define the elements in job design that electronic monitoring has the potential to influence in a way that can make work more stressful. The recognition of these potential stressors can serve as the impetus for designing more healthful electronic monitoring systems.

MONITORING AND STRESS

"Big Brother" is watching you. You cannot escape his scrutiny. This message is one that resounds throughout the union halls of America. There is a general "paranoia" that electronic monitoring is the backbone of an impersonalized workplace in which every "sneeze" or every "breath" of a worker will be recorded and analyzed; and more importantly will be used in some way to punish workers, or to change their behavior. These fears have not been generated in a vacuum. Employer practices in using electronic monitoring have shown some of the fears to be well founded. Horror stories of monitoring misuse have been reported in the Washington Post (Perl, 1984), Dunn's Business Month (Hershman & Rozen, 1984), and ABC-TV Nightly News (Jennings, 1984). Such stories include descriptions of the overwhelming effects of monitoring on worker psychological functioning, effects such as fear, anxiety, hatred and loss of self image. In addition, influences on health due to stress are reported. Westin (1984) indicates that management must realize that monitoring can cause increased production standards which employees feel are unfair. This can lead to morale and commitment problems, create physical and psychological stress and increases in undesired behavior such as absenteeism, turnover and lowered productivity.

While these effects all seem reasonable and have firm theoretical backing, currently there is insufficient research literature to support the contention that electronic monitoring leads to stress and diminished health. Results from interviews with monitored workers provide preliminary evidence for such a linkage, but are not sufficient to make such a claim (Smith et al, 1986). In addition, the excesses of some employers in applying monitoring which result in these horror stories most likely are not the norm for industry. While this cannot be established without more comprehensive surveys of employment practices regarding electronic monitoring, Westin (1984) provides the best available data on such practices from 110 companies. His data indicates more moderate application of electronic monitoring in personnel and administrative functions than are reported in the abuse stories. However, Westin's data base is too limited to be able to reflect general trends.

One approach to examining stress issues is to look at monitoring concerns from a "worst case" perspective. This allows for the establishment of the most damaging applications which then provides the basis for the limits of acceptable practice as well as direction for establishing policy considerations. Actually, employees are not informed as to the precise nature of the behavioral information which may be collected from a computer system and how this information may be used. For example, a telephone monitoring capability was included with a new telephone system for customer service operators. The monitoring system which distributed calls, timed the calls and allowed for unintrusive eavesdropping was instituted under the guise of workflow scheduling in order to determine the peak periods when extra operators are needed. But instead of using the monitoring system for that purpose only, management established standards (seconds per transaction) and used this information for discipline. Abusive use of electronic monitoring systems by misinformed management can lead to workers' stress.

A major concern in electronic monitoring is the influences that it can have on worker self-image and on feelings of self-worth. In one sense, monitoring should enhance feelings of worth if the results of worker efforts are positive and the workers get feedback to that effect. Likewise, management interest in the worker as a valuable resource can be demonstrated by the attention provided by monitoring. However, both effects may be seen differently by workers if poor performance can lead to some form of punishment or reprimand. This fear of evaluation can produce anxiety and heightened sensitivity to adverse feedback that may damage self-esteem and self-image. Both the type of technology and the context of the monitoring determine its influences on stress, not the monitoring process itself.

Tied to fear of reprimand is the pressure to perform above average that monitoring can induce. Some managers feel that this is a desirable effect since it implies high production. But occupational stress research indicates that such work pressure is not conducive to good performance and brings about adverse health consequences (Cooper & Marshall, 1976; Smith, 1987).

In fact, there are a range of stressful working conditions that may be the outcome of electronic monitoring of employee performance. These include heightened work pressure, routinized work activities, paced work, potential for increased work standards and workload, variable work standards, lack of control over the tasks, lack of decision latitude, reduced peer social support, reduced supervisory support and fear of job loss. The following is a summary of the stress implications of various job conditions all of which could be influenced by electronic monitoring of worker performance.

Two organizational factors have been shown to be of special significance for increased job stress and decreased worker health. These are: (1) job involvement or participation; and (2) organizational support, as reflected by supervisory style, support from managers and career development. Lack of participation in work activities has been demonstrated to result in an increase in negative psychological mood (Margolis et al, 1974; Caplan et al, 1975; Smith et al, 1981). In terms of organizational support, it has been shown that close supervision and a supervisory style characterized by constant negative performance feedback are related to high levels of stress and poorer worker health (Caplan et al, 1975; Smith et al, 1981). The implication of these findings is that excessive, impersonal electronic monitoring of employee performance that produces close supervision and constant negative performance feedback can promote worker stress.

It has also been demonstrated that workers' feelings of lack of involvement are related to stress and potentially to health complaints. (WHO, 1984; Rutenfranz et al, 1977; Karasek, 1981; Caplan et al, 1975; Gardell, 1976; Margolis et al, 1974; Haynes & Feinleib, 1980; Colligan et al, 1977). Electronic monitoring has the propensity to reduce worker feelings of job involvement and may in this way increase worker distress. The chances to participate and be involved in the job process may be diminished in work systems that are driven by employee performance monitoring. As monitoring is used to reduce worker involvement and participation in their work tasks, the more stress can occur. Thus, a central element in current monitoring theory, that is, to keep close track of employee actions to be able to detect undesirable patterns and then make modifications in employee actions, will produce an organizational environment that is conducive to stress.

Career development is another factor that has been related to stress. Concern over chances for promotion has been shown to be a significant stressor for office workers while being passed over for promotion has been related to increases in both job stress and ill-health (Smith et al, 1981; Arthur & Gunderson, 1965). In this regard, monitoring can have both beneficial and negative effects. If monitoring provides for more objective employee evaluations and employee promotions are tied into the evaluation process, then monitoring may have a positive impact on workers. However, if the monitoring is perceived as unfair and not representative of performance, then this could produce a stressful influence.

Security is the other side of career development. The threat of job loss is a very potent stressor. It has been tied to serious health disorders such as ulcers, colitis, severe emotional stress and patchy baldness as well as to increased muscular and emotional complaints. In the past, monitoring has been used for employee dismissal due to unsatisfactory performance and such use can be very stressful.

Stress-related work-task factors such as task variety, take clarity (confusion), challenge, complexity, utilization of skills and abilities, and activity level may be indirectly influenced by electronic monitoring. All of these factors have been related to increased stress and negative psychological states such as boredom, confusion and frustration and have also been related to increased risk of health disorders. In particular, monitoring may reduce task complexity, variety, challenge and skills use due to the need for management to simplify work tasks and break them down into measurable, simple units that can be easily measured and evaluated.

The lack of discretionary control over work activities has also been shown to be related to increased risk of coronary heart disease (Karasek, 1981). Monitoring can greatly reduce the amount of discretionary control and participation that a worker has unless specific actions are taken by management to include these elements in the use of the monitoring process; current systems do not provide for these.

Mental workload factors, such as quantitative underload/overload, and workpace, can cause stress. In this case, a balance is required to avoid negative health consequences. Quantitative overload has been shown to be a significant stressor for various occupations including scientists, machine operators, and data-entry clerks. The impact of work overload varies from psychological disturbances to increased disease risk. Monitoring is often accompanied by the establishment for work standards to assess employee performance. These standards are often not based on scientific grounds, but on computer system capabilities (Smith et al, 1981). If the standards are excessive they will produce stress. On the other hand, if a scientific basis is used to set the standards, this influence of monitoring could reduce stress if prior workload requirements were excessive.

Workpace is a very important workload factor (Smith et al, 1981; Cakir et al, 1980). The speed or rate of work has been implicated as a significant issue in factory-worker ill-health. Recent technology, such as computers, that can operate at high speeds on a continuous basis has increased the pacing impact on office workers. Research suggests that pacing produced by computerization may have an even greater stress effect than factory pacing (Smith et al, 1981). Monitoring, and the establishment of excessive work standards has the potential to increase work pace,to simulate machine pacing, and thus to producing stress.

Time pressure, such as having to meet deadlines, is a stressor that may interact with both work hours and workpace (Friedman et al, 1958). Studies have shown increases in stress levels as difficult deadlines draw near. Monitoring may produce such deadline pressure on a "constant" basis that may be more damaging than occasional deadline pressure.

Although there is insufficient research literature to support the contention that electronic monitoring increases stress and diminishes health, the above review of possible relationships to known stressors shows that electronic monitoring has the potential to be stressful. Indeed, electronic monitoring may create adverse working conditions such as paced work, lack of involvement, reduced task variety and task clarity, reduced peer social support, reduced supervisory support, fear of job loss, routinized work activities, and lack of control over tasks. The results of our interviews with monitored workers provide some preliminary evidence for such a linkage between stress and electronic monitoring (Smith et al, 1986).

CONCLUSION

The literature evaluations and our interview experience with monitored workers (Smith et al, 1986) provide the basis for some general conclusions about the potential influences of electronic monitoring on job design, stress and worker motivation.

Monitoring has the capability to change job design substantially. All theories of human performance at work from Taylor (1947) to Hackman (1976) discuss the significance of having up-to-date, accurate information about individual performance so that aspects of workplace redesign or personal motivation can be applied to increase productivity. While the theories may differ on the use of the information, they agree on the need for such information. Thus it is apparent that electronic monitoring can play a significant role in providing this information which can be applied in accordance with whichever theory has been chosen. However, it is also clear that to be effective monitoring must provide more significant information than just the quantity of individual output; it must also provide insight into the quality of output and the individual contributions of workers, such as initiative, which are not adequately reflected in simple measures of quantity and quality.

Monitoring has direct effects on employee perceptions of stress and has the potential to influence worker motivation and behavior. If monitoring is used in an evaluative way by management, then it can create distress, influence individual motives and goals adversely, and could have both positive and negative effects on performance. If monitoring is used to provide accurate, meaningful feedback in a timely way and is non-evaluative, it provides the basis for enhanced goal setting and motivation and provides the cues necessary to improve performance. Such monitoring also can provide the basis for establishing an equitable and reasonable work standard that can influence performance motivation under incentive conditions, as well as worker perceptions of fairness and job satisfaction under non-incentive conditions.

A major fear of organized labor is that continuous monitoring will be used to establish unfair production standards which will overtax worker capabilities. To counteract such outcomes it is likely that employees will limit their performance as they have under incentive systems so that work standards will not increase. This will defeat the positive benefits of monitoring in enhancing performance motivation, and therefore the most effective uses of monitoring will be for non-evaluative purposes.

Monitoring can provide the necessary feedback climate that the research literature extols as necessary for the best possible employee performance. But monitoring also changes the nature of jobs, and most importantly, the way workers interact on the job. Socialization and talking are reduced in most automated worksettings; and monitoring adds the concern about being watched on a frequent basis. Thus, the workers are afraid to socialize. In our interviews (Smith et al, 1986) all of the workers expressed the desire to work hard and to produce quality products. This is reflected by the fact that the vast majority indicated that they produced more than the established work standard at their worksites even though they did not receive extra payment for this. They all also commented on the importance of work as a place to socialize and to interact with others. It seems reasonable that such socialization can be accommodated in automated work systems and that monitoring can be established in such a way that it will not diminish social interaction.

Monitoring also influences the way in which workers are supervised. Smith et al (1981) indicated that computer automation with monitoring produced more coercive, stricter, number-counting supervisory style which replaced a more helpful, less performance oriented supervisory approach. Our interviews (Smith et al, 1986) support the belief that most employees feel that supervision changes with monitoring, and changes are for the worse. Again, this suggests that monitoring should not be used as part of an individual evaluation approach, but should be used to feed back information directly to individuals about their performance in a non-evaluative way so that they can make necessary corrections in their behavior.

Monitoring can influence motivation by the type of information fedback to individuals. Value judgement information such as that in annual performance evaluations generally is not perceived as important, accurate or worthwhile by workers. In fact workers misjudge what they are told about their performance. Performance feedback on a continuous basis can help in setting individual performance goals, can act as a direct performance motivator and can help direct behavior that is tied to financial incentives. Thus, information and cues presented act as primary sources for motivating worker behavior, increasing satisfaction and reducing stress.

Whether monitoring can influence the health of workers is not clear. There is no research about such health effects, and only a limited amount of research on the stress effects of monitoring. Monitoring has been shown to be associated with worker stress (Smith et al, 1981; Smith et al, 1986) and to have the necessary influences on job design to be a potent stressor (Smith et al, 1986). Thus, if not used properly electronic monitoring has the potential to be a serious threat to worker health as mediated through increased job stress and reduced job satisfaction.

REFERENCES

Arthur, R.J., & Gunderson, E.K., "Promotion and mental illness in the Navy", Journal of Occupational Medicine, 1965, 7, 452-456.

Cakir, A., Reuter, H., Von Schmude, L., & Armbruster, A., Visual Display Terminals, New York: John Wiley & Sons, 1980.

Caplan, R.D., Cobb, S., French, J.R.P., Harrison, R.V., & Pinneau, S.R., Job Demands and Worker Health, Washington, D.C.: U.S. Government Printing Office, 1975.

Colligan, M.J., Smith, M.J., & Hurell, J.J., "Occupational incidents rates of mental health disorders', Journal of Human Stress, 1977, 3, 34-39.

Cooper, C.L. & Marshall, J. "Occupational Sources of Stress: a review of the literature relating to coronary heart disease and mental ill health", Journal of Occupational Psychology, 1976, 49, 11-28.

Friedman, M., Rosenman, R.H., & Carroll, V., "Changes in the Serum Cholesterol And Blood Clotting Time in Men Subjeted to Cyclic Variation of Occupational Stress," Circulation, 1958, 852-861.

Gardell, B., "Technology alienation and mental health", Acta Sociological, 1976, 19, 83-94.

Hackman, J.R., & Oldham, G.R., "Motivation through the design of work: Test of a theory", Organizational Behavior and Human Performance, 1976, 16, 250-279.

Haynes, S.G., Feinleib, M., "Women, work and coronary heart disease: Prospective findings from the Framingham heart study", American Journal of Public Health, 1980, 70, 133-141.

Hershman, A. & Rozen, M., "Corporate Big Brother is watching you", Dunn's Business Month, January, 1984.

Karasek, R.A., "Job decision latitude, job design, and coronary heart disease", in G. Salvendy and M.J. Smith (Eds.)., Machine Pacing and Occupational Stress, London: Taylor and Francis, 1981, 45-56.

Margolis, B., Kores, W.M., & Quinn, R., "Job stress: An unlisted occupational hazard", Journal of Occupational Medicine, 1974, 16, 654-661.

Perl, P., "Rising performance standards keep some in agency on edge", Washington Post, September 3, 1984.

Rutenfranz, J., Colquhoun, W., Knauth, P. and Ghata, J., "Biomedical and psychosocial aspects of shift work", Scandinavian Journal of Work Environment and Health, 1977, 3, 165-182.

Smith, M.J., "Occupational Stress" in G. Salvendy (Ed.) Handbook of Ergonomics/Human Factors, New York: John Wiley, 1987.

Smith, M.J., Cohen, B.G., & Stammerjohn, L.W., Jr., "An investigation of health complaints and job stress in Video Display Operations", Human Factors, 1981, 23, 387-400.

Smith, M.J., Carayon, P. and Miezio, K. Motivational Behavioral and Psychological Implications of Electronic Monitoring of worker performance, Washington, D.C.: U.S. Congress, Office of Technology Assessment, 1986.

Taylor, F.W., "The Principles of Scientific Management", 1911, in F.W. Taylor, Scientific Management, London: Harper & Row, 1947.

Westin, A., "Privacy issues in the monitoring of employee work on VDT's in the office environment: practices, interests, and policy choices", Office of Technology Assessment, Washington, D.C., 1984.

WHO Psychosocial Factors and Health: Monitoring the Psychosocial Work Environment and Worker's Health, Geneva: World Health Organization, 1984.

Social, Ergonomic and Stress Aspects of Work with Computers,
edited by G. Salvendy, S.L. Sauter and J.J. Hurrell, Jr.
Elsevier Science Publishers B.V., Amsterdam, 1987 — Printed in The Netherlands

A CONCEPT OF CONTROL: IMPLICATIONS FOR STRESS AND PERFORMANCE IN HUMAN-COMPUTER INTERACTION

MICHAEL FRESE
Dept. of Psychology, Ludwig-Maximilians-Universität, Leopoldstr. 13, D-8000 München 40, Federal Republic of Germany

ABSTRACT

Control at work has an influence both on stress-effects and on performance. A conceptualization is presented that relates to the action process and its implications for stress-effects and performance are described. Empirical data shows consistent stress and performance effects of control over the introductory process of new technology, of control over the training process, and of control at work on transfer of training.

1 INTRODUCTION

The issue of control (or decision latitude) has become an important topic in stress research (e.g. Frese, 1977, Karasek, 1979, Karasek et al., 1981). In this article I would like to pursue the theme of control a little further with regard to stress and performance in human-computer interaction, presenting first theoretical conceptualization of control, discussing the function of control for stress-effects and performance, and presenting some data on the importance of control.

2 THE THEORY OF CONTROL

Experiencing control means to have an impact on the conditions and on one's activities in correspondence with some higher order goal (Frese, in press).

A person exerts control when she has influence over her actions and over the conditions under which she acts. An action consists of a sequence (this sequence is variable, of course): goal development and goal decision, plan development and decision, execution of the action and use of feedback (Frese & Sabini; 1985, Norman, 1986). Having influence means to be able to decide what goals, what plans, what kind of feedback a person is using under what conditions. If the environment does not provide the freedom to decide, the person does not have any control. This concept is summarized in Table 1. Decision possibilities can appear with regard to sequencing, timeframe, and content. Decision points with regard to sequence may for example mean that people are able to determine which tasks they do first and which one second, in which sequence plans are being formed and executed, or in which sequence they call on signals to inform them of the success of their activities. Timeframe refers to two sets of decision possibilities: First, the decision, <u>when</u> a task is tackled or a plan is performed; second, deciding on <u>how long</u> it will take to work on a task or on a plan. Similarly, the timing of the signal alludes to when it will appear and how long it is displayed. Content refers to the substance of the decisions with regard to task, plan, signal and conditions: What particular task is done, what plan is formed, what kind of signals does one choose to use and what conditions exist for work.

TABLE 1

Aspects of External Control

Action sequence	Decision possibilities: Sequence	Timeframe	Content
Tasks (Goals)			
Plans			
Feedback (Signals)			
Conditions			

Four issues are worth emphasizing in this context. <u>First</u>, the decisions must refer to a goal (or in the case of a task, to a superordinate goal). One decides with some goal in mind. As long as something is not related to a goal, non-control does not matter.

<u>Second</u>, freedom to decide has a positive quality only when the decisions do not involve high risks. If all the alternatives involve high risks, then controllability may lead to an aversive situation. This is related to the issue of goal again, because we usually do not develop goals that are very risky.

<u>Third</u>, there is a hierarchy of goals within any one person at any one time (there may be, of course, multiple and conflicting goals). The higher up in this hierarchy a particular goal is, the

more important and central the decision become, (e.g. life or career plan decisions). Thus, control and non-control must be weighted by the importance of the goal.

Fourth, aside from the above points about the significance of goals and goal hierarchy, it is important to ask the question, how much of the time a person is exposed to non-control or control situations. The theory predicts that exposition time is an important variable (Frese, 1984). If one is constantly under conditions of non-control (little decision making power) even in small matters, there is an impact even if the goals are not very important. The longer, one is exposed to these conditions, the stronger is their impact on experienced control, even if higher order goals (like life goals) are still to be influenced by the individual.

There are certain prerequisites of control: skills, functionality, transparency and predictability:

Skills: The external conditions represent only potential decision points with regard to some goal. To realize this potential, one needs to have knowledge and skills as internal requirements.

Functionality: It refers to whether a system (e.g. a computer program) allows and enhances the completion of a task. One issue of functionality is, for example, whether the system models real world tasks. Thus, a statistics program should calculate correctly and do what it is supposed to do. A spell program should have enough words in the dictionary. Without functionality there is no control, because the decisions are not meaningfully related to the goal (task) any more. However, a high functionality does not necessarily imply, that there are decision points available.

Transparency: Transparency implies that the user can develop an internal model of the functions of the system (Maass, 1983). Thus, the system should not confuse the user by giving different commands to do the same thing under different modes or by giving explanations that are inconsistent or only half true. Under conditions of non-transparency, the user cannot make adequate decisions referring to his goal. Transparency is not identical to control because it is possible to develop a system that is completely transparent, that nevertheless offers only little control (e.g. an expert system that perfectly explains why it is doing what, but that does not allow users' decisions).

Predictability: There is some overlap with the concept of transparency. If a system is not predictable, it is most likely not transparent. However, transparency refers to the present, predictability to the future. If a system's behavior cannot be foreseen it is not predictable. In the context of predictability, two issues have been discussed: predictability of what (pwhat) and predictability of when (pwhen) (Miller, 1981). Pwhat implies that one knows what happens when a certain command is typed into the machine. Pwhen implies that one knows when a certain command is executed and completed. The relationship between predictability and control is complex. It is possible that a system is predictable but not controllable (e.g. everything is determined by the system but there is a signal that tells what will happen next). It is also possible (although unlikely) to conceive of a system that is controllable but not predictable (e.g. a system in which one does not know what will happen but that allows to change things once they have happened). For practical purposes, lack of predictability make decisions meaningless, because when the states of a system cannot be foreseen, one cannot make adequate decision.

3 CONTROL AND STRESS-EFFECTS AND PERFORMANCE

Control may have a direct influence on the stressors or it may function as a moderator of the relationship between stressors and stress-reactions. The direct effect on stressors is apparent when people can change those working conditions that they find stressful. The moderator effect has been consistently shown in experimental research but also in field studies on the relationship between stress and health. Stressors given under conditions of non-control lead to helplessness and physiological effects (Seligman, 1975; Weiss, 1977). Field studies have shown similarly, that the combination of stress and control can lead to heart attack (Karasek et al., 1981) and psychosomatic complaints (Semmer & Frese, 1987).

Performance effects existed when there was only potential control (Glass & Singer, 1972). In this series of experiments the subjects had a button that could turn off a loud noise (the stressor). In one condition, the subjects were asked not use this button (and all of them complied). This condition produced less performance decrements in various tasks than not having such a control button.

The performance effects because of non-control may be due to the following factors: (1) Helplessness: It leads to passivity and to a reduction of active and exploratory strategies to solve tasks and problems at hand. (2) Reactance effects: In the attempt to regain control, one does not concentrate on the task and may even actively sabotage those people (or systems) who are seen to be the agents responsible for reducing control (Wicklund, 1974). (3) Diffusion of responsibility: In non-control situations the person is not responsible for the results and, therefore, is less active to bring them about. (4) Reduction in learning: When one is not in control, feedback is not perceived as being brought about by one's own actions; therefore, one does not learn from feedback. (5) Fitting: When there is control, it is possible to fit the working conditions to one's psychophysiological prerequisites thus making it easier to work.

4 EMPIRICAL FINDINGS

4.1 Changing technology and control

We were interested whether new (computer) technology would change control at work. Since we had data on German blue collar workers in 1979 (before large scale computer technology was introduced in w.- Germany), we made a second wave of questionnaires to the same subjects in 1985. One third of them were now working with new technology. Among these the following groups of people could be differentiated: (1) Those who worked with the computer but did not have any influence on the programing (N=18), (2) those who had some influence on programing (e.g. they collaborated with the programers) (N=17), (3) those who programed themselves (N=10), and (4) finally people who worked with robots (N=9). Control at work was ascertained with a questionnaire on how much influence a worker had on sequencing of tasks, on choice of tasks that he had to do, etc. (Semmer, 1984).

Table 2 shows the results; group 3 has the highest control at work, group 4 the lowest. However, this does not test the question whether or not these blue collar workers gained or lost control because of the introduction of computers. Interestingly, an analysis of covariance (with earlier control as covariate) showed no gain or loss at all. Apparently, those blue collar workers with high control in their former jobs were also the ones who got the jobs in which they programed themselves or had at least some impact on programing; similarly those with low control remained in jobs with low control after the introduction of new technology (Frese & Zapf, 1987).

TABLE 2
Different groups of (computer) workers and control at work (means)

no influence	influence on progr.	program themselves	robots
(1)	(2)	(3)	(4)
5.09	5.19	6.74	4.30

Control is not only important on the level of work but also on the organizational level (the issue of participation). This is of particular importance when new technology is introduced. We were interested in whether it made a difference for workers to have a "say-so" in the process of introducing new technology. A scale on this issue shows: (1) Those who have high control over technological changes also report that their work situation has improved over the last 6 years ($r=.31$, $p<.05$, $N=53$). (2) Control over technological changes is negatively related to psychological dysfunctioning, with anxiety ($r=-.24$, $p<.05$, $N=56$) and marginally with psychosomatic complaints ($r=-.21$, $p=.06$, $N=55$). (3) There is a high correlation with job satisfaction ($r=.41$, $p<.01$, $N=56$). (4) Finally, people who perceive themselves to have control over technological changes report better social support from their supervisor ($r=.30$, $p<.05$, $N=56$). Thus, to have control over technological changes may have positive effects on workers' psychological functioning, on job satisfaction, on a decrease of anxiety, and on a seeing the job to be better.

4.2 Control in the training process

In an experiment on training we pursued the question of control as well. In one experimental group the training was highly structured and gave a step-by-step procedure with little chance to develop one's own mental model. In the other group, the subjects (all of them complete computer novices) were asked to develop their own hypotheses about the program that they were about to learn and they were encouraged to explore. Thus, the latter group had control over the development of their mental model and had more infleunce over how to proceed in their exploratory behaviors. This group proved to be superior in various performance variables after the training, e.g. they need less time to correct errors when writing a text and a smaller number of keystrokes; they were also better in a transfer task (Frese et al., 1987).

4.3 Control and the transfer from training to everyday work

Finally, we wanted to know, whether control at work helped the transfer process. The hypothesis was: Even when people learned a lot in the training, they would only apply this knowledge if they had a high level of control at work; otherwise reactance and helplessness effects would decrease their motivation to use the new system in their everyday work. In a study of engineers, using the same scale of control at work as in the studies above, this turned out to be true (v. Papstein, 1987). The correlation between system knowledge (ascertained at the end of the training) and hours of use of the new system in their work was .66 ($N=14$, $p<.01$) for those with a high level of control at work, while this correlation was .03 ($N=11$, n.s.) for those with low control. The difference between these two correlations is also significant ($p<.05$). If these results can be reproduced in other studies, they suggest that it is useless to buy expensive computers and deliver expensive trainings, if the company does not provide a sufficiently high degree of control to their workers.

5 CONCLUSION

The concept of control developed in this article can be used on different levels: It is possible to analyse working conditions with the dimensions of Table 1 (e.g. how much control does a person have over the introductory process, over the timing, or over sequencing of work tasks); it is also useful to analyse computer programs to ascertain whether or not they provide decision points along the lines of Table 1. For example, our results show that direct manipulation interfaces lead to superior performance in the long-range training process (Frese, Schulte-Göcking & Altmann, 1987). It may be useful to study whether this is an effect related to the higher degree of control that direct manipulation provides or whether it relates to other variables.

Control at work has been shown to have an impact on stress-effects. In a series of studies, we have additionally shown that it also has an impact on performance, e.g. in training and in transfer of training. Our theory and the results suggest that we should consider the concept of control to be one of the central variables in the use of computers at the work place, in the development of training concepts, and in the design of computer systems.

6 REFERENCES

Frese, M. (1977). Psychische Störungen bei Arbeitern: Zum Einfluß von gesellschaftlicher Stellung und Arbeitsplatzmerkmalen

(Beiträge zur klinischen Psychologie). Salzburg: Müller.
Frese, M. (1984). Transitions in jobs, occupational socialization and strain. In V. Allen & E. V. D. Vliert (Eds.), Role transitions: Explorations and explanations (pp. 239 - 253). New York: Plenum Press.
Frese, M. (in press). A theory of control and complexity: Implications for software design and integration of the computer system into the work place. In M. Frese, E. Ulich, & W. Dzida (Eds.), Psychological issues of human computer interaction at the work place. Amsterdam: North-Holland.
Frese, M., Albrecht, K., Altmann, A., Lang, J., Papstein, P. v., Peyerl, R., Prümper, J., Schulte-Göcking, H., Wankmüller, I., & Wendel, R. (1987). The effects of an active development of the mental model in the training process: Experimental results on a word processing system. Manuscript
Frese, M., & Sabini, J. (Eds.) (1985). Goal directed behavior: The concept of action in psychology. Hillsdale: Erlbaum.
Frese, M., Schulte-Göcking, H., & Altmann, A. (1987). Lernprozesse in Abhängigkeit von der Trainingsmethode, von Personenmerkmalen und von der Benutzeroberfläche (direkte Manipulation vs. konventionelle Interaktion). In Schönpflug, W. & Wittstock, M.(Eds.)Software - Ergonomie. Stuttgart: Teubner.
Frese, M., & Zapf, D. (1987). Die Einführung von neuen Techniken am Arbeitsplatz verändert Qualifikationsanforderungen, Handlungsspielraum und Stressoren kaum: Ergebnisse einer Längsschnittuntersuchung. Zeitschrift für Arbeitswissenschaft.
Glass, D. C., & Singer, J. E. (1972). Experiments on noise and social stressors. New York: Academic.
Karasek, R. A. (1979). Job demands, job decision latitude and mental strain: Implications for job redesign. Administrative Science Quarterly, 24, 285 - 308.
Karasek, R. A., Baker, D., Marxner, F., Ahlbom, A., & Theorell, T. (1981). Job design latitude, job demands, and cardiovascular disease: A prospective study of Swedish men. American Journal of Public Health, 71, 634 - 705.
Maass, S. (1983). Why systems transparency? In T. R. G. Green, S. J. Payne, & G. C. van der Veer (Eds.), The psychology of computer use (pp. 19 - 28). London: Academic Press.
Miller, S. (1981). Predictability and human stress: Toward a clarification of evidence and theory. In L. Berkowitz (Ed.), Advances in experimental social psychology, Vol. 14. (pp. 203-256). New York:Academic.
Norman, D. A. (1986). Cognitive engineering. In D. A. Norman & S. W. Draper (Eds.), User centered system design. Hillsdale: Erlbaum.
Papstein, v. P. (1987). Transfer von Training. München: Unpubl. Master Thesis.
Seligman, M. E. P. (1975). Helplessness: On depression, development and death. San Francisco: Freeman.
Semmer, N. (1984). Stressbezogene Tätigkeitsanalyse: Psychologische Untersuchungen zur Analyse von Stress am Arbeitsplatz. Weinheim: Beltz.
Semmer, N., & Frese, M. (1987). Control at work as moderator of the effect of stress at work on psychosomatic complaints: A longitudinal study with obejctive measurement. München: Manuscript.
Weiss, J. (1977). Psychosomatic disorders. In J. D. Maser & M. E. P. Seligman (Eds.), Psychopathology: Experimental models (pp. 232-269). San Francisco: Freeman.
Wicklund, R. A. (1974). Freedom and reactance. Hillsdale: Erlbaum.

Social, Ergonomic and Stress Aspects of Work with Computers,
edited by G. Salvendy, S.L. Sauter and J.J. Hurrell, Jr.
Elsevier Science Publishers B.V., Amsterdam, 1987 — Printed in The Netherlands

PSYCHOSOCIAL FACTORS IN THE INTRODUCTION OF COMPUTERIZED TECHNOLOGY: EXPERIENCES FROM THE FINNISH ENGINEERING INDUSTRY

P. SEPPÄLÄ, E. TUOMINEN and P. KOSKINEN
Department of Physiology, Institute of Occupational Health,
Topeliuksenkatu 41 a A, SF-00250 Helsinki (Finland)

ABSTRACT
The paper discusses psychosocial issues, such as organization of work, job content, qualifications, autonomy, variation of work, social contacts etc., related to the introduction of computerized technology (FMS) into Finnish engineering companies. According to the findings on the four engineering firms, the majority of the FMS operators interviewed reported mainly positive changes in their work. The qualifications demanded and opportunities to learn new things had increased. The work was more interesting, responsible and variable. On the other hand, many workers also felt that the work was more routinized and opportunities to control one´s work pace had decreased.

1 INTRODUCTION

The flexible manufacturing system (FMS) is one of the manifestations of technological advance in the engineering industries made possible by the rapid progress in information technology and numerical control (NC).

The definition of FMS allows a diversity of sizes in installations, ranging from the small production cell with one or two machining centers to interconnected complex systems including automatic warehousing and transportation systems (Ebel, 1985).

The wider use of computer-controlled flexible manufacturing systems can be expected to have far reaching implications in terms of manpower, division of labor, job structure, work content and training requirements (Ebel, 1985; Bullinger et al., 1985). However, no universal trend can be discerned when flexible manufacturing systems and computer numerical control in general are introduced into the engineering industries (Sorge et al., 1985). Depending on plant size, batch size, and the culture of an enterprise, different approaches in work organization, division of labor and training have been adopted (Sorge et al., 1985; Köhler et al., 1983; Brödner, 1985).

The psychosocial problems, e.g. time pressure, monotony, lack of autonomy, social isolation, decreased qualifications, etc. that have been found in previous studies on factory automation may, or may not emerge depending on the decisions made and solutions adopted (Lennerlöf, 1984; Friedrich and Bengtsson 1986; Friedrich and Svensson, 1986).

The study presented in this paper is part of a larger project entitled "Information technology and the working environment". In one of the subprojects, flexible manufacturing systems implemented so far in Finland, and those to be implemented in the near future, will be studied.

The aim of the study is to analyze psychosocial issues, such as work organization, job content, qaulification, autonomy, variation of work, social contacts etc., related to the introduction of computerized technology into the Finnish engineering companies.

2 MATERIAL AND METHODS

So far we have visited five enterprises that have introduced, or are at the stage of introducing, new flexible manufacturing technology. The factories under study are specialized production units of bigger companies. Flexible manufacturing systems are used for machining parts of diesel engines, transmission systems, axles, rock drills and elevators.

This paper is based on the material collected from four FMS, which have been in use for at least half a year at the time of the study.

The FM systems studied have been built up from CNC-machining centers equipped with automatic material storage and transportation systems. In addition, some jobs at single CNC-machining centers and robotized cells have been analyzed.

The research methods used are interviews, job analyses, observations and questionnaires.

3 RESULTS

3.1 General findings from the production organizations

The typical production organization of a conventional engineering enterprise is schematically presented in Fig. 1. The functions needed in the production process from customer order to shipping of the products are separated and hierarchically organized. The jobs and task structures follow the same specialization and hierarchical organization.

In all of our cases before, or parallelly to the introduction of flexible technology changes in the organization of production flow, job structures and responsibilities have been carried out. The functional organization described above (Fig. 1) has been replaced by the system in which the manufacturing is organized into the product shops, each manufacturing a particular logical entity of a finished product (e.g. transmission system of a tractor, break system for the rear axle of a truck).

FMS is usually the key machine group in processing the parts needed for the product manufactured in a certain product shop. Consequently, the high

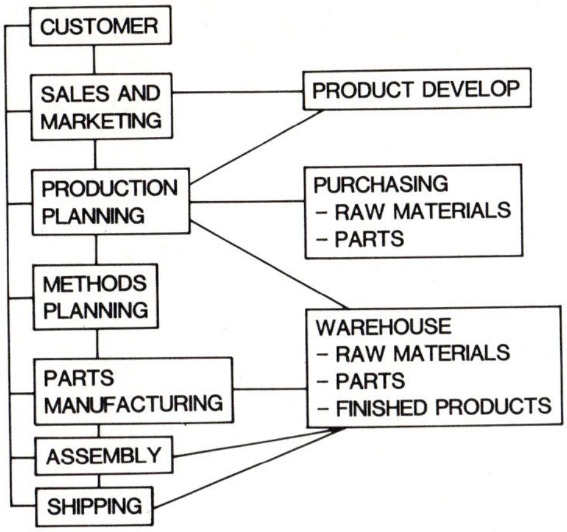

Fig. 1. Model of the typical production organization of a conventional engineering enterprise.

availability and running time of the FMS is crucial for the smooth production process.

In product shop organization, the previously centralized functions like production planning and control, ordering of materials and tool service are often distributed into the product shops. The manager of the shop or a supervisor is usually in charge of these functions. The traditional role of the supervisors has also been changed. Instead of giving detailed instructions and orders to the operators, the supervisor is more a co-ordinator of the whole production line.

More responsibility concerning the details of performing the tasks at the work stations has been given to the operators. The operators of the crew of a FMS usually perform activities such as the changes needed in machining programs, presetting of tools, replacing broken tools, minor maintenance, inspection of the processed workpieces immediately after the machining, etc.

3.2 Nature of work and work content

The main task of the FMS crews is to keep the system running continuously and as long as possible, even through the night. In one of the systems studied this is achieved by using human operators also in the night shift. The other three systems are operated unmanned by night. This is made possible by the automatic pallet storage system and automatic tool monitoring system. If disturbances occur, the system will stop automatically.

The duties of the FMS operators in a normal work shift are: 1) fixing the work pieces into the fixtures on the pallets and removing them at the pallet handling station, 2) sending the pallet into the FMS, 3) cleaning the fixtures and work pieces, 4) finishing the work pieces after processing, 5) inspection and measurement of the processed work pieces, 6) replacing broken tools, and correcting minor disturbances and programming needed in this context.

In all of the four systems studied the major programming is done in the office by a programmer. In two cases also presetting of the tools and tool service in general is the responsibility of a separate organizational unit. In two cases the FMS operators themselves take care of the presetting of tools and ordering of the reserve materials needed.

In one of the systems the skills of all crew members were not quite homogenous. That is, all crew members were not able to operate the machines, replace the tools or do the programming. Consequently, the tasks in the FMS were polarized so that part of the workers did only the manual material handling, finishing and measurements, whereas another part of the workers did the more skilled and demanding tasks. This was apparently one reason for the relatively low total running time of the system.

3.3 Work tasks as experienced by the operators

(i) Qualifications demanded by the jobs. According to the majority of the FMS as well as other operators experience, training and professional skills of the operators corresponded well to the demands of the work tasks in general. They also felt that they had many opportunities to use and further develop their knowledge and skills in their work.

Some psychosocially important job characteristics as experienced by FMS and other operators are shown in Table 1. The qualifications needed most in computerized machining in general are good memory, keen concentration, alertness, precision and reasoning.

The majority of workers felt that in addition to the basic training of a metal worker, one to four years of experiences is necessary before one is able to manage all the tasks needed for operating computerized machine tools.

(ii) Variation of work. About one third of the workers assessed their work to be rather variable, even though nearly all respondents felt that they had to repeat the same tasks or subtasks throughout the working day. In general, feelings of monotony were not experienced very often.

(iii) Autonomy. The majority of the respondents considered their work rather independent or autonomous, and most workers felt that they were free to leave the work stations for a while if necessary. As mentioned, it is possible to run the FM systems also unmanned even for several hours.

Otherwise most operators felt that they had rather small possibilities to influence the working methods or control their work pace. This is determined mainly by the processing capacity of the machines or the production schedules.

(iv) Social contacts. Opportunities for social contacts and discussions with other workers were experienced to be good. No one felt isolated during the work day. Most operators had contacts with at least 3-6 persons during the shift.

(v) Stress and strain. The work was not experienced especially stressful. Furthermore, nearly half of the respondents were rather or very satisfied with their jobs and only one person was very unsatisfied. However, about 20-50 % of the operators felt that environmental conditions, bad working postures and work movements, as well as confinement to the work station were strenuous.

TABLE 1

Job characteristics as experienced by the FMS and other operators (n = 28).

	Low score %	High score %
Qualifications demanded		
- good memory	0	93
- keen concentration	11	64
- alertness	4	79
- precision	0	93
- decision making ability	7	46
Variation of work		
- monotony	32	18
- repetition of same sub-tasks	4	86
Autonomy		
- independence in work	4	61
- freedom to leave the work station	0	75
- freedom to take a rest pause	4	32
Social contacts		
- opportunities to talk with fellow workers	0	89
- feeling of isolation	86	0
General job satisfaction	4	43

TABLE 2

Impact of the introduction of FMS on job characteristics as experienced by the FMS operators (n=19).

	Decreased %	No change %	Increased %
Working instructions	10	16	74
Freedom to control one´s work pace	63	11	26
How interesting the work is	16	10	74
Work pace	5	42	53
Opportunities to use one´s abilities	21	16	63
Monotony of work	42	26	32
Status of work	0	32	68
Difficulty of work	0	11	89
Pleasantness of work	26	26	48
Responsibility	5	21	74
Dependence on fellow workers	5	37	58
Skills required of a metal worker	17	11	72
Variation of work	42	5	53
Routinized tasks	6	33	61
Opportunities to learn new things	0	11	89
Demands for knowledge and skills	5	21	74
Outside control of one´s work	5	21	74
Confinement to the work station	0	37	63

3.4 Changes in job characteristics related to the new technique

During the interview, the FMS operators were presented with a list of job characteristics related to qualifications, autonomy, skills, etc. in work. They were asked to evaluate to what degree each aspect has decreased, increased or remained unchanged compared to their previous job (Table 2). From the psychosocial point of view, these changes can be considered as positive or negative.

From the psychological point of view the following changes reported by the majority of the FMS operators can be considered positive:

- the work was more interesting
- opportunities to use one´s skills and abilities had increased
- there were better opportunities to learn new things
- demands for knowledge and skills had increased
- the professional skills and knowledge demanded from a metal worker had increased.

On the other hand, many operators felt that some negative changes had taken place. These were:

- increased outside control of one´s work
- routinized work tasks
- decreased opportunity to determine one´s work pace.

All in all, the changes experienced were related to the nature of the former job of a respondent.

4 CONCLUSIONS AND DISCUSSION

When looking at the implementation of flexible manufacturing systems into the Finnish engineering companies and their consequences on the jobs and work characteristics, we can note that the introduction of new computerized technology is not the only change in the companies. Simultaneous to the technological changes, new organizations and management philosophies have been adopted.

The introduction of product shops and manufacturing cells, as well as the just-on-time (JOT) principle in production planning and control have had an impact on all the jobs. The main consequences of the new manufacturing systems have been increased decentralization of responsibility as regards production flow and the quality of the product, as well as increased appreciation of qualified, multi-skilled and highly motivated workers.

Despite these general phenomena, there are some differences between companies as to the participation and training of workers, and division of labor between organizational units. For example, the degree to which the FMS crew itself can take care of the presetting of tools and participate in work planning varies from case to case. These differences account for the variation found in the opinions of the workers revealed in the interviews.

There are also differences in the products and in the technical solutions for handling the work pieces in the system. Thus, in some cases operators are more exposed to inconveniencing environmental factors (oil, dust, etc.) and repetitive manual work than in other cases.

When evaluating the results of the introduction of flexible manufacturing systems into the FInnish companies considered in this paper, we can conclude that the process of technological change has been carried out rather smoothly in general. The problems encountered are mostly related to the history and technological advancement, as well as the training policy of the companies.

In one of our cases, where the change seems to have succeeded especially well, the company has used numerically controlled machine tools already for several years. The workers recruited to the FM systems were experienced operators of CNC machining centers, so that the step to the more advanced technology was not as radical as in the cases where the FMS operators had to start from the very basics of numerical control. In addition, the management had adopted a more "human centered" than "technocentric" approach in job design (Brödner, 1985).

As to the psychosocial aspects of the jobs we found that the majority of the FMS operators interviewed reported mainly positive changes in their work. The qualifications demanded and opportunities to learn new things had increased. The work was more interesting, responsible and variable. On the other hand, some workers also felt that the work was more routinized and opportunities to control one´s pace had decreased. In addition, a considerable amount of manual material handling is needed when loading and unloading the fixtures used in otherwise highly automated machining and material handling systems. So, the picture often given by mass media of a factory without manual work is in most cases only a myth, at least in small and medium size industries.

As a final remark we can say that the systems studied had been in use for a relatively short time, and are still undergoing changes and development. For example, industrial robots will be used increasingly in some tasks which are conducted manually at present, and more tasks experienced as very new and challenging now may become more routinized in time. It is therefore perhaps too early to say what the FMS jobs will be in the long run.

REFERENCES

Brödner, P., 1985. Skill based production- the superior concept to the "unmanned factory". In: H.-J. Bullinger and H.J. Warnecke, Toward the Factory of the Future, Springer-Verlag, Berlin, pp. 500-505.

Bullinger, H.-J., H.J. Warnecke, and H.-P. Lentes, 1985. Toward the factory of the future. In: H.-J. Bullinger and H.J. Warnecke (Eds.), Toward the Factory of the Future, Springer-Verlag, Berlin, pp. XXIX-LIV.

Ebel, K.-H., 1985. Social and labour implications of flexible manufacturing systems. Int. Labour Review, 124: 133-145.

Friedrich, P. and Bengtsson, I., 1986. Automatiserad balkborrning. Arbetsmiljökonsekvenser vid datorstödd automatisering inom verkstadsindustrin- fallstudie 1. Undersökningsrapport 1986: 8, Arbetarskyddsstyrelsen, Stockholm.

Friedrich, P. and Svensson, G., 1986. Automatiserad bearbetning av kringutrustning till gnistbearbetningsmaskiner. Arbetsmiljökonsekvenser vid datorstödd automatisering inom verkstadsindustrin- fallstudie 3. Undersökningsrapport 1986:10, Arbetarskyddsstyrelsen, Stockholm.

Köhler, C., Schultz-Wild, R., and Lutz, B., 1983. Flexible manufacturing systems - manpower problems and policies. Paper prepared for the 1983 World Congress on the Human Aspects of Automation, August 8-11, 1983 - Ann Arbor, Michigan, U.S.A.

Lennerlöf, L., 1984. Automatisering och människor. Liber Förlag, Stockholm.

Sorge, A., Hartman, G., Warner, M.. and Nicholas, I., 1985. Microelectronics and Manpower in Manufacturing. Applications of Computer Numerical Control in Great Britain and West Germany. Gower Publishing Company Limited, Aldershot, Great Britain.

Social, Ergonomic and Stress Aspects of Work with Computers, 59
edited by G. Salvendy, S.L. Sauter and J.J. Hurrell, Jr.
Elsevier Science Publishers B.V., Amsterdam, 1987 — Printed in The Netherlands

JOB PERFORMANCE AND QWL IN A DATA ENTRY TASK

Paul T. Cornell
Steelcase, Inc.; P.O. Box 1967; TC-1W-04; Grand Rapids, Michigan, 49506 (USA)

ABSTRACT

Seven bank proofing departments were examined using a four factor model of the workplace. The factors included the physical environment, job content, organizational/social concerns, and characteristics of the workers. The effects of these on job performance and quality of worklife (QWL) were assessed in a multidimensional manner. After stepwise regression reduced the number of variables to 17 predictors and five criterion variables, a canonical correlation was performed. Two variates were found, one essentially QWL, the other performance. The model received support in that numerous interactions were noted and variables which contributed to one outcome did not necessarily contribute to the other. The four factor approach seems to have merit as a conceptual framework for viewing intervention strategies.

As the U.S. economy evolves from an industrial to information base, the nature of human factors input required to design the "workplace" has changed dramatically. As Bleecker (1986) discusses, the changes brought about by this shift are not only manifested in our work tools and technology, but in the resources we use, the products produced, and the way work is organized.

In office environments, the transition to automation has not been without problems. Concerns with these workplaces fall into three categories of ergonomics, job content and management. In this context, ergonomics relates to the physical environment, e.g., furniture design, lighting, acoustics, etc. These issues are often the subject of research activities and are included in many state legislative proposals. In terms of job content, some feel that automation will result in less job control and greater fragmentation, with computers assigned the "interesting" job functions and workers left with the mundane tasks that cannot be automated (Zuboff, 1982; Driscoll, 1982). In terms of management, perhaps the biggest concerns are monitoring performance and controlling/pacing operators (Gregory & Nussbaum, 1982; Ellis, 1982).

All three of these areas, ergonomics, job content, and management, are not new fields of study, as they have traditionally been part of the field of industrial/organizational psychology (e.g., Landy & Trumbo, 1976; Gilmer & Deci, 1977). What is new is the increased attention these topics are receiving within the human factors community (e.g., Hendrick & Brown, 1984). This broadening perspective has given rise to a new focus within the profession, popularized in the term "macroergonomics" (Dray, 1984; Hendrick, 1986).

History of Job Design Practice

Davis and Taylor (1979) summarize the history of job design theory and practice as containing four themes: task and job rationalization (e.g., scientific management, ergonomics), job content (e.g., job enrichment, job characteristics), role content (e.g., industrial democracy), and organizational units (e.g., sociotechnical systems theory). Each approach was developed in response to an existing condition in the workplace. Each has had its successes and failures.

In viewing Davis and Taylor's timeline, it is interesting to note that each successive theory took more into account than the one preceeding. This is not to say that time proven methods, such as MTM (Konz, 1983), are no longer practiced. But in general theories became broader in their perspective.

Two recent studies epitomize this perspective, Rousseau (1978) and Campion and Thayer (1985). Rousseau used measures of departmental, positional, job, and individual characteristics to predict employee attitudes and behavior. Using a canonical correlation, she was able to look at the effects of multiple independent variables on multiple dependent variables. Campion and Thayer used an interdisciplinary approach combining industrial engineering, organizational psychology, biomechanics, and experimental psychology. They also conducted a canonical correlation analysis and were able to relate different outcomes to different job design approaches. These two studies demonstrated the benefits of a multidimensional perspective; the conclusions drawn would have differed had not the variables been viewed simultaneously.

Four Factor Model

A four factor model is proposed as a means for integrating the various job design theories and tools (Figure 1). One factor, the physical environment, concerns the physical context of work, e.g., the furnishings, lighting, and tool design. The second factor is job content and focuses on the characteristics of the job. Variables studied here include autonomy, skill variety, task significance, and feedback. A third area deals with the organizational and social context of the work environment. Management style, leadership behavior, span of control, and centralization are examples of this factor. The fourth category includes worker characteristics. Education, age, skill, needs, and physique are but a few of the areas contained in this factor. These exist and are influenced by a larger context of social-political-economic variables that are typically beyond the control of the practicioner.

It was hypothesized that these four factors act independently and collectively to determine job performance and QWL. It was also hypothesized that the workers are the key ingredient--the success or failure of any intervention or change will depend upon the workers. The study explored the model and these two assertions.

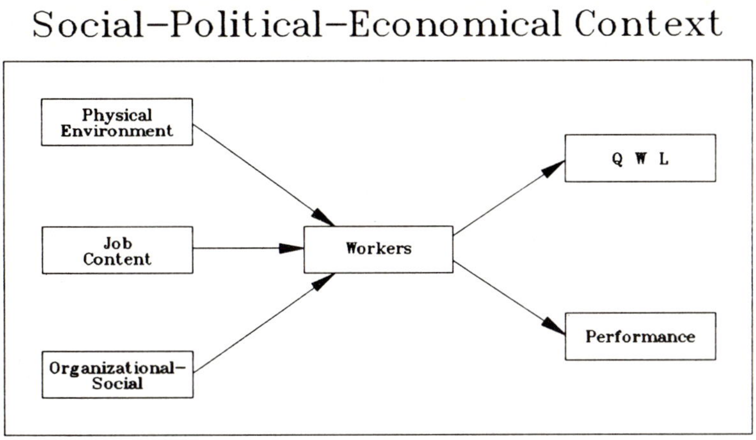

Figure 1. Four Factor Model of the Workplace

METHODS

Participants

Data were collected in the proof departments of seven U.S. banks. (Proof encoding is a data entry task, performed at a dedicated workstation, which requires operators to read the amount on a check, enter the data on a keyboard, and drop the check into a feeder.) The banks were located in Akron, Baltimore, Chicago, Columbus, Los Angeles, San Diego, and San Jose. The participation of employees was voluntary and they were not required to take part. A total of 163 operators participated.

Apparatus

Data were gathered via three instruments: a worker questionnaire, a management interview, and observation.

Observational Checklist. Document encoding machines were evaluated using guidelines from the ergonomic and industrial engineering literature (Cakir, Hart & Stewart, 1980; Grandjean, 1980; Niebel, 1982). Due to the redundancy of recommendations, five variables were developed which grouped the specific items from each reference. The summary variables were: movement accommodation, physical accommodation, vision, chair design, and noise. Scores rating the physical environment were determined by the total number of criteria passed. None of the criteria were prioritized or weighted. Noise levels were measured using a General Radio Company Type 1565-A sound level meter. This handheld device measures levels from 50 to 140 dB on scale A, B or C.

Questionnaire. The first part of the questionnaire consisted of biographical data and asked about age, height, marital status, gender, years of education, working hours, thruput (i.e., documents encoded per hour), tenure as an operator, tenure at the bank, and the types of encoding machines used. The second part consisted of the 30 items of the Job Characteristics Inventory (Sims et al., 1976). Using a five-point scale, participants were asked to rate their job on autonomy, skill variety, feedback, and task identity. The third part contained portions of the Job Diagnostic Survey (Hackman & Oldham, 1980). This assessed context satisfaction, general satisfaction, motivation, growth satisfaction, and growth need strength (GNS). The last page contained a physical stress measure based upon Patchen's (1970) work. The five items queried participants on the frequency with which they experienced trouble getting to sleep, headaches, upset stomach, eye discomfort, and backaches. These five were summed to provide an overall stress measure. A response scale of zero (almost never) to five (several times a week) was used.

Interview. The management interview was used to gather data on the organization and the department. In addition to absenteeism and turnover, Pugh et al.'s (1968) measures of formalization and centralization were used. With the formalization measure, the manager noted which of the 14 forms existed in her department. The score on this variable was the total number of forms, 0 to 14. For centralization, a scale of 0 to 5 was first modified to reflect the organization of the particular department (the higher the scale value, the higher the job level). Managers then used this scale to note the level responsible for each of 25 decisions. These values were summed to derive the centralization score. Size was defined as the number of full- and parttime proof operators. Hierarchy was the number of job levels in the department.

Procedures

Prior to administering the questionnaire, participants were given a verbal overview of the study. They were informed it was voluntary and that responses would be kept confidential. The survey took 20 to 30 minutes to complete.

The interview took place in the manager's office. These were first-line supervisors who had day-to-day responsibility for running the proof department. All but one were former operators. The manager herself completed the formalization and centralization questions. The interview lasted from 30 minutes to an hour.

The observation took place during work periods. The encoding machines were evaluated on each item on the checklist. Noise levels were measured while the machines were in full operation. Ambient measures were taken in several locations throughout the room. Measures were also taken behind operators at head level.

RESULTS

Variable Set Reduction and Interactions

Of the 163 completed questionnaires, 40 were unusable. Due to the sample size, it was necessary to reduce the variable set prior to the canonical correlation. Thus, variables exhibiting multicollinearity or nonvariance were omitted. Of the original 30, 15 independent and five dependent variables were retained. Interaction terms of theoretical interest were added to the set using crossproduct terms (Table 1), bringing the total to 64 variables.

Table 1. Variable Set Used in Stepwise Regressions

Factor (# of variables)	Interacting With (# of interaction terms)
Physical Environment (2) chair design available kneeroom	Workers (6) age, height, fulltime
Job Content (4) skill variety autonomy feedback task identity	Workers (20) age, education, experience, context satisfaction, GNS Organizational/Social (8) span of control, department size
Organizational/Social (2) span of control department size	Workers (10) age, education, fulltime, context satisfaction, GNS
Workers (7) age height years of experience years of education fulltime (1 if fulltime, 0 if parttime) context satisfaction growth need strength (GNS)	
Quality of Worklife (4) general satisfaction motivation growth satisfaction physical stress	
Performance (1) thruput	
15 independent variables 5 dependent variables	44 interaction terms

Stepwise Regression

Stepwise regression was used to reduce the variable set further. The analyses were performed using BMDP2R with an F-to-enter and F-to-remove of 2.79 ($p \leq 0.10$). This high alpha value was selected to reduce the likelihood that a significant variable would be eliminated prematurely. Five analyses were conducted.

Five variables were significant in accounting for general satisfaction. These were context satisfaction, kneeroom x fulltime (the interaction of kneeroom and fulltime), skill variety x context satisfaction, department size x fulltime, and feedback x span of control. The r-square was 0.51.

Four variables entered the equation for motivation, including skill variety x context satisfaction, task identity x age, feedback x GNS, and span of control x fulltime. R-square with motivation was 0.24.

For growth satisfaction, the r-square was 0.67 and four variables were significant. These were GNS, height, autonomy x context satisfaction, and skill variety x context satisfaction.

Only three variables entered the equation for stress and the r-square only reached 0.09. The variables were context satisfaction, chair design x age, and span of control x fulltime.

Seven variables were significant in predicting variance in thruput. Included were education, chair design x age, task identity x experience, autonomy x GNS, span of control x fulltime, department size x context satisfaction, and autonomy x span of control. The r-square was 0.56.

A total of 17 variables were significant in one or more regression equations. These are shown in Table 2.

Canonical Correlation

The five outcome variables comprised the criterion set and the 17 significant variables from the regression analyses made up the predictor set. Two canonical correlations were significant at the 0.001 level with correlations of 0.87 and 0.77. The variable loadings are shown in Table 2. These are the correlations of the original variable with each canonical variate, and show the degree to which each variable is related to the variate. The redundancy coefficients were 0.29 and 0.12, indicating that 41.1% of the variance in the group of five outcome variables was accounted for by the two canonical variates.

Table 2. Canonical Variable Loadings

Outcome Variables	Canonical Variate 1	Canonical Variate 2
general satisfaction	0.81	-0.09
motivation	0.48	0.01
growth satisfaction	0.96	0.11
stress	-0.24	0.23
thruput	-0.22	0.97
Predictor Variables		
height	-0.14	-0.01
education	-0.44	-0.15
context satisfaction	0.86	0.02
GNS	-0.22	0.07
chair design x age	-0.05	-0.11
kneeroom x fulltime	0.02	0.61
skill variety x context	0.85	0.11
autonomy x context	0.87	0.23
autonomy x GNS	0.27	0.39
task identity x age	0.24	0.33
task identity x experience	-0.01	0.55
feedback x GNS	0.32	0.09
span of control x fulltime	-0.10	0.66
department size x fulltime	-0.10	0.42
department size x context	0.60	-0.16
autonomy x span of control	0.20	0.69
feedback x span of control	0.25	0.46

DISCUSSION

The results of Table 2 lend support to the hypothesis that the variables impact performance and QWL independently and collectively. Only four variables entered the regression equations as main effects, the other 13 entered as interaction terms. Of the 13, 11 interactions included a worker characteristic variable, supporting the hypothesis that workers moderate the effects of the other variables. The content and form of the model, as shown in Figure 1, seem justified by these data. More research is required regarding the specific relationships of the model before it becomes a useful tool.

Canonical variate 1 related to QWL and variate 2 to performance. The predictor variables generally loaded into one variate and not the other. Thruput correlated -0.12 to -0.23 with the QWL variables, so it is not surprising to see the negative loadings of performance with variate 1. This suggests that what improves QWL may not lead to higher performance. To change agents, this information has to be handled delicately. One cannot emphasize one outcome to the exclusion of the other; yet, depending on the problem areas, one would employ different interventions to improve the situation.

Hendrick, Dray, and others have called for human factors professionals to take a broader approach in how they design the office workplace. While the lessons learned from designing manufacturing and military workplaces are still valid, there are some different factors at play in the office. Many of the significant variables found in this study are not even considered in studies of other environments. Our basic assumptions about motivation, satisfaction, QWL, and performance need to be modified as we move our research focus from these other workplaces to the office.

REFERENCES

Bleecker, S. Taking the factory out of the office. Computerworld, June 16, 1986.

Cakir, A., Hart, D. & Stewart, T. Visual display terminals. New York: Wiley & Sons, 1980.

Campion, M. & Thayer, P. Development and field evaluation of an interdisciplinary measure of job design. Journal of Applied Psychology, 1985, 70, 29-43.

Davis, L. & Taylor, J. (Eds.) Design of jobs (2nd ed.). Santa Monica: Goodyear, 1979.

Dray, Susan. Personal communication, August, 1984.

Driscoll, J. Office automation: The dynamics of a technological boondoogle. In R. Landau, J. Bair & J. Siegman (Eds.), Emerging office systems. Norwood: Ablex Publishing Corporation, 1982.

Ellis, W. Statements of concern. Office: Technology and People, 1982, 1, 17-18.

Gilmer, B. & Deci, E. Industrial and organizational psychology (4th ed.). New York: McGraw-Hill, 1977.

Grandjean, E. Fitting the task to the man, an ergonomic approach. London: Taylor & Francis, Ltd., 1980.

Gregory, J. & Nussbaum, K. Race against time: Automation of the office. Office: Technology and People, 1982, 1, 197-236.

Hackman, J. & Oldham, G. Work redesign. Reading: Addison-Wesley, 1980.

Hendrick, H. Macroergonomics: The systems approach to VDU systems. Proceedings of the Conference on Work with Display Units, 1986, 86-89.

Hendrick, H. & Brown, O. (Eds.) Human factors in organizational design and management. Amsterdam: North-Holland, 1984.

Konz, S. Work design: Industrial ergonomics (2nd ed.). Columbus: Grid Publishing, 1983.

Landy, F. & Trumbo, D. Psychology of work behavior. Homewood: Dorsey, 1976.

Niebel, B. Motion and time study (7th ed.). Homewood: Irwin, 1982.

Patchen, M. Participation, achievement and involvement on the job. Englewood Cliffs: Prentice-Hall, 1970.

Pugh, D., Hickson, D., Hinings, C. & Turner, C. Dimensions of organizational structure. Administrative Science Quarterly, 1968, 13, 65-105.

Rousseau, D. Characteristics of departments, positions, and individuals: Contexts for attitudes and behavior. Administrative Science Quarterly, 1978, 23, 521-540.

Sims, H., Szilagyi, A. & Keller, R. The measurement of job characteristics. Academy of Management Review, 1976, 19, 195-212.

Zuboff, S. New worlds of computer-mediated work. Harvard Business Review, 1982, Sept-Oct, 142-152.

Social, Ergonomic and Stress Aspects of Work with Computers,
edited by G. Salvendy, S.L. Sauter and J.J. Hurrell, Jr.
Elsevier Science Publishers B.V., Amsterdam, 1987 — Printed in The Netherlands

RESPONSES OF TYPE A AND TYPE B INDIVIDUALS PERFORMING A SUPERVISORY CONTROL SIMULATION

S. G. HART, V. BATTISTE
NASA-Ames Research Center, Moffett Field, CA, U.S.A.

M. A. CHESNEY, M. M. WARD, and M. McELROY
SRI International, Inc., Menlo Park, CA, U.S.A.

ABSTRACT

A supervisory-control simulation, presented with different levels of complexity and time pressure, was used to examine candidate behavioral, subjective, and physiological measures of mental workload. The predicted relationships were found among physiological and subjective workload measures, but their correlations with performance were low. Significant Difficulty X Personality Type interactions were found for heart rate, systolic, and diastolic blood pressure: Type A men exhibited greater physiological responsiveness to difficulty manipulations than Type B men, suggesting that behavioral characteristics are salient when interpreting physiological (but not subjective and behavioral) workload measures.

1 INTRODUCTION

Computer aiding and automation are being introduced into advanced systems to enhance their capabilities and allow operators to perform additional tasks in environments where they could not function unaided. This is gradually changing the nature of human operators' activities from direct mechanical control to indirect supervisory control (e.g., scheduling, monitoring, decision making, and planning). In general, this has succeeded in reducing physical workload, but at the cost of an increase in mental workload. This tradeoff between physical and mental workload has been difficult to quantify, however, because many traditional measures of workload are insensitive to mental activities. For example, performance-based measures are inappropriate when operators' overt responses are infrequent and their consequences indirect. Subjective ratings may be more appropriate as they can reflect an operator's global assessment of the personal consequences of performing a task.

However, substantial variability in ratings given by different individuals in response to apparently identical situations is common. Thus, considerable research has been undertaken to identify the sources of variability and to minimize their effects on ratings. To some extent, rating variability reflects differences in the strategies people adopt and the effort they exert; each one actually experiences somewhat different levels of workload. In addition, individual differences in workload definitions produce rating variability. One approach that accommodates such differences when deriving a measure of overall workload (Hart, Battiste, and Lester, 1984) is by giving more weight to the subscales that each rater identifies as most relevant. Compared to global workload ratings, this method reduces between-rater variability by 25% or more.

Several measures of cardiovascular activity have been proposed as measures of mental workload: heart rate, heart rate variability, and blood pressure. Heart rate reflects an overall increase in arousal, but it is relatively insensitive to subtle workload variations, especially when stress and anxiety do not contribute to task demands (Hart and Hauser, in press). Heart rate variability (or sinus arrhythmia), is more sensitive to variations in mental workload, decreasing as task load is increased (Mulder, 1979). Blood pressure has been used to assess workload less often because noninvasive measures can be obtained only intermittently; inflating the blood pressure cuff, detecting systolic and diastolic levels, and deflating the cuff to re-establish normal blood flow is time consuming.

Just as behavior and subjective experiences differ, individuals' metabolic adjustments in response to task demands vary as well. For example, different personality traits may introduce variability in physiological workload measures. Classifying people as having either Type A or B behavioral characteristics may provide information about why, when confronted with the same task demands, their cardiovascular responses differ; Type As may exhibit increased cardiovascular responsiveness to environmental stimuli whereas Type Bs may not (Chesney and Rosenman, 1983). In addition, these behavior patterns may influence how people perceive, perform, and evaluate tasks; Type As may adopt different personal performance criteria, exert different levels of effort, and report different subjective experiences than Type Bs. The Type A behavior pattern is characterized by a habitual sense of time urgency, competitiveness, aggressiveness, and ever-increasing expectations of personal productivity, characteristics relatively absent in the Type B behavior pattern. Type As tend to work faster (Damos and Bloem, 1985) than Type Bs, even when no time limits are imposed, and report boredom and show performance decrements when forced to slow their pace (Glass, 1977). Finally, Type As tend to compete with (unrealistically high) self-imposed performance standards. For example, Damos and Bloem (1985) found that Type As were less satisfied with their performance than Type Bs, although they performed better.

This study was designed to evaluate the effects of experimentally controlled variations in time pressure and task complexity on males classified as either Type A or Type B to determine whether these behavior patterns are, in fact, associated with differential responsiveness. Four levels of supervisory-control task difficulty were selected from those used in an earlier study (Hart, et al, 1984), to provide a suitable range of task demands. We hypothesized that Type As would exhibit momentary elevations in cardiovascular measures in response to task-related stressors while Type Bs would not. Furthermore, we anticipated that Type As would report lower effort and fatigue and would be more critical of their performance achievements than Type Bs. This would identify the Type A/B behavior patterns as salient sources of individual differences in subjective and cardiovascular measures of workload and could be used to predict how specific individuals might respond to different tasks.

2 METHOD

2.1 *Subjects*

Thirty five males (aged 17 to 35 years) served as paid participants. None had previously performed the experimental task. Two were eliminated because their data were incomplete, two did not meet the age (18 years) or resting blood pressure criteria, and three were classified as a mixture of Type A and B behavior patterns. The data from 14 Type As and 14 Type Bs were analyzed.

2.2 *Equipment*

The simulation was programmed on a Digital Equipment Corporation PDP-11/40 computer and an Evans and Sutherland Picture System I. Subjects were seated in front of a 25.6 cm Xytron monitor in a secluded area. They interacted with the system by positioning a stylus on a magnetic response pad, entering selections by depressing its tip. The display area was projected onto a 5.1 cm area on the response pad so subjects could reach every function with minimal hand movements. Blood pressure and heart rate were assessed automatically by the Vita-Stat 900D automatic blood pressure machine using a cuff placed on the subject's left arm.

2.3 *Experimental Task*

The simulation consists of task elements that must be performed and control functions that subjects may invoke to perform the task elements. (Fig. 1) Each of the five "tasks" is represented by a box which is filled periodically with new groups of task elements (represented by symbols). The control functions are represented by selectable areas on the display. Each actuation imposes that function once on the selected task. The initial decision to perform a task is made by selecting it and opening its box (OPEN). After a box is opened, task elements stream out in a vertical line Each time the "PERFORM" function is selected, the topmost element in the stream disappears and five points are added to the score. By selectively opening and closing boxes, subjects control the elements available for performance. Several tasks can be completed in parallel or one can be finished before beginning the next. Various penalties are imposed if subjects delay performing specific activities and they can "shed" tasks (if they are behind) or request additional tasks (if they are ready to perform them ahead of schedule). Subjects select tasks and control functions based on their assessment of the situation, task urgency and difficulty, and the reward or penalty for performing or failing to perform them. The optimal strategy differs for each scenario. The goal was to score as many points as possible (up to a maximum of 500) during each six-min scenario.

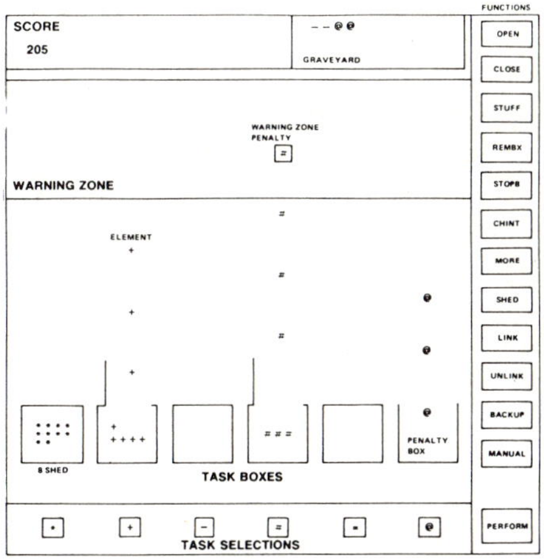

Fig. 1: Experimental Display

2.4 *Experimental Variables*

The number of task elements that could be performed was the same (100) for each scenario, but their arrival times and movement rates were varied between scenarios. Complexity was determined by the number of Elements/Task (04 or 10) and the task-arrival schedule (massed or staggered). Although, 04-element tasks could be completed more quickly than 10-element tasks, additional time was required to switch among tasks to obtain the same points and successive groups arrived more often. Groups of elements arrived simultaneously with the simpler massed schedule. Arrival times were staggered in the more complex schedule. Time pressure was manipulated by varying element-movement rate. The basic rate (1.25 cm/sec) inhibited skilled operators from performing elements as quickly as they could issue commands. An acceleration function was imposed in half of the scenarios (1.52 cm/sec/sec); the longer elements remained unperformed, the faster they moved. Accelerated elements could be performed more rapidly, but they were available for performance for only a short time. Two groups of 10-element tasks appeared in each box at the same time without acceleration for the LOW workload scenario or according to a staggered schedule with acceleration for the M-LOW workload scenario. Five groups of 04-element tasks appeared in each box at the same time without acceleration for the M-HIGH workload scenario and according to a staggered schedule with acceleration for the HIGH workload scenario. The scenarios were designed to impose attainable, but different, requirements; trained operators could complete one group of task elements before the next arrived, but the mental and physical effort required to do so varied.

2.5 *Rating Scales*

Subjects rated their experiences immediately after each scenario using the NASA Bipolar Rating Scale (Hart, et al, 1984) , which consists of nine workload-related subscales. Individual biases about the importance of each dimension are assessed in a pretest where subjects identify the more important dimension in 36 pairwise comparisons of the nine factors. The frequency with which each dimension is selected serves as its weight. The ratings for each scenario, each multiplied by its weight, are averaged to produce a workload score that ranges from 0 to 100.

2.6 *A/B Interview*

A 10-15 min Type A Structured Interview was given by a trained interviewer (Chesney, Eagleston, and Rosenman, 1980). It was taped for later assessment by the interviewer and a second rater, with a third rater to resolve disagreements. It provided a structured situation where the interviewer sought to create a sense of time urgency and provide an opportunity for subjects to compete with the interviewer or themselves through confrontations and challenges. The assessment represented a global subjective evaluation of the presence or absence of specific characteristics associated with the Type A behavior pattern; 14 subjects were rated as Type A, 14 as Type B, and 3 as a mixture of A and B patterns.

2.7 *Procedure*

A one-hour, coached training session familiarized subjects with the tasks, equipment, and procedures. Four practice and four experimental scenarios followed, presented in a different order to each subject. Each scenario was described and a schedule of task arrival times was provided before it began. The structured interview was completed last. The experiment lasted approximately 5 hour. Thirty measurements of blood pressure and heart rate were obtained; three readings were taken (at two-min intervals) before and after the experiment and during each of the four practice and four experimental scenarios.

3 RESULTS AND DISCUSSION

Mixed-design analyses of variance were performed on the scores, workload ratings, heart rate, and systolic and diastolic blood pressure. The between-groups factor was behavior type (A/B). The within-groups factors were Practice/Experimental and Difficulty (LOW, M-LOW, M-HIGH, HIGH). Additional two-way analyses of variance for repeated measures were performed on cardiovascular measures for Type A and Type B subjects individually.

3.1 *Performance*

Scores improved significantly ($F(1,26) = 38.97$, $p<.001$) between practice and experimental scenarios. However, Type Bs improved less (from 309 to 376) than Type As (from 326 to 412). Scores differed significantly ($F(1,26) = 58.78$, $p<.001$) among the four scenarios (Fig. 2); the detrimental effects of schedule complexity and time pressure were exaggerated when more groups of elements had to be performed to achieve the same score. These results are similar to those in the earlier study (Hart et al., 1984). Type As' scores were higher (369) than Type Bs (337), although the difference was not significant. There were no significant interactions.

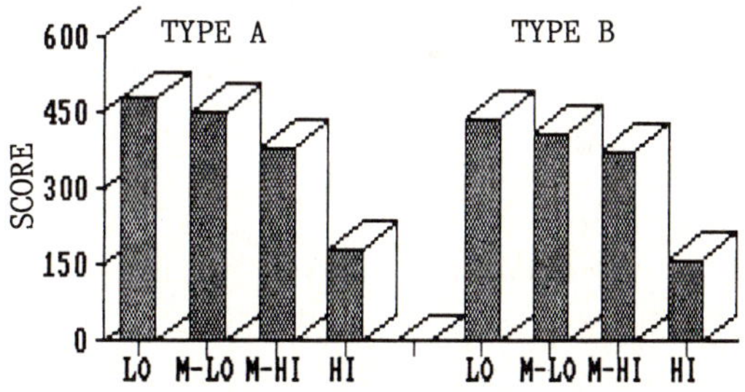

Fig. 2: Average scores.

3.2 *Subjective Ratings*

Ratings on the nine subscales multiplied by weights that reflected their subjective importance, were combined and averaged. Workload ratings were not significantly different between Practice and Experimental sessions, unlike the performance scores. Although there was a significant difference among ratings due to the experimental manipulations (F $(3,78) = 37.09$, $p<.001$) the pattern of results was somewhat different than it was for scores; the complexity and time pressure manipulations had a greater impact on subjective workload ratings than did the number of elements/task. (Fig. 3) Again these results are similar to the previous study. There was no significant difference in workload ratings between Type As and Bs, nor any significant interactions.

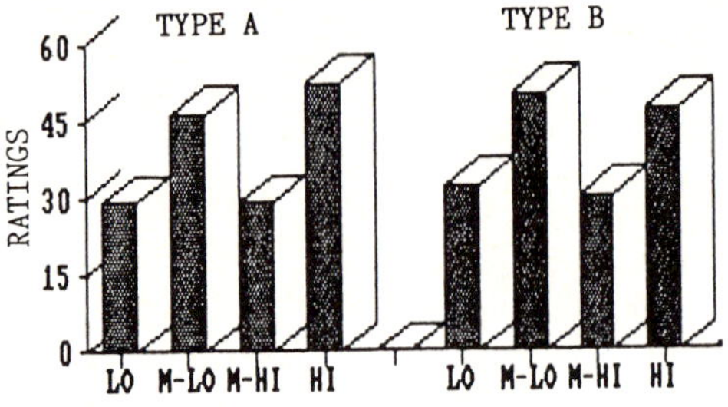

Fig. 3: Workload Ratings

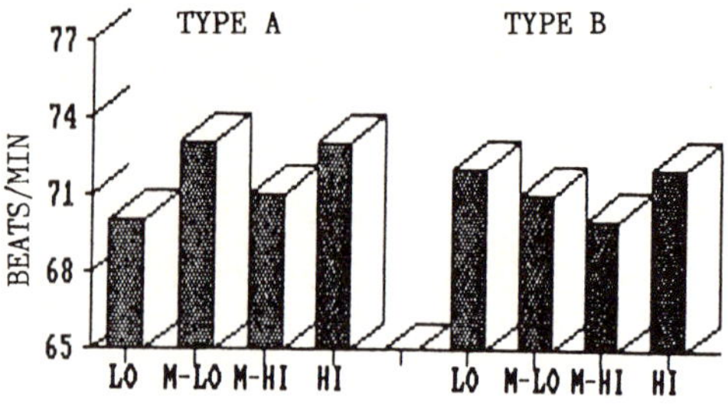

Fig. 4: Heart Rates

3.3 ***Cardiovascular Measures***

The three values recorded during each measurement interval were not significantly different for either heart rate, systolic or diastolic blood pressure. Thus, analyses were performed on the mean of the three samples for each measure.

(i) ***Heart Rate.*** Pre-test (65 beats/min) and post-test (64 beats/min) heart rate levels were virtually identical, and there was no Type A/B difference. Heart rates increased to 71 beats/min, on the average, when the supervisory control simulation was introduced, although there was no difference between Practice and Experimental scenarios. Heart rate was somewhat responsive to the experimental manipulations (F (3,78) = 2.89, p<.04), and there was a significant Type A/B x Difficulty interaction (F(3,78) = 2.78, p<.05). When Type A and Type B heart rates were analyzed individually, the effects of the experimental manipulations were significant for Type As (F(3,39) = 4.25, p<.05) but not for Type Bs (F(3,39) = 1.28, p>.05), as predicted (Fig. 4). Furthermore, there was a significant correlation between heart rates and workload ratings for As (r = 0.65), the correlation was not significant for Type Bs (r = -0.37).

(ii) ***Blood Pressure.*** Subjects were selected on the basis of having normal blood pressure; their systolic blood pressure averaged 123 and 125 and their diastolic blood pressure averaged 78 and 84 in the pre- and post-tests, respectively (represented on the abscissas of Fig. 5a and 5b) However, systolic and diastolic blood pressure levels for Type As and Bs reacted differently to the experimental manipulations.

Systolic blood pressure increased slightly between the pre- and post-test (F(1,13) = 5.77, p<.05) for both Type A and Bs. Systolic levels increased significantly (F(1,13) = 20.48, p<.001) when the experimental task was introduced (from 123 to 140). The increase was not significant for Type Bs (20) or for Type As (14). There was no significant difference between Type As and Bs, between Practice and Experimental sessions, nor among Difficulty levels. However, there was a significant Type A/B x Difficulty interaction (F(3,78) = 4.85, p<.003), as predicted. When analyzed separately, variations in systolic blood pressure arousal related to experimental manipulations were significant for Type As (F(3,39) = 4.48, p<.01), but not for Type Bs (F(3,39) = 2.07,

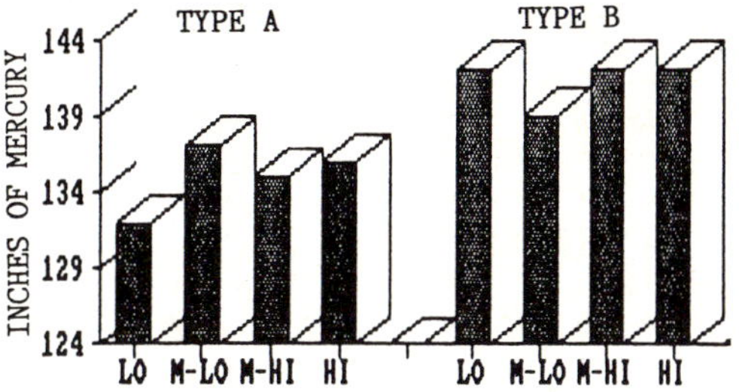

Fig. 5a: Systolic blood pressure.

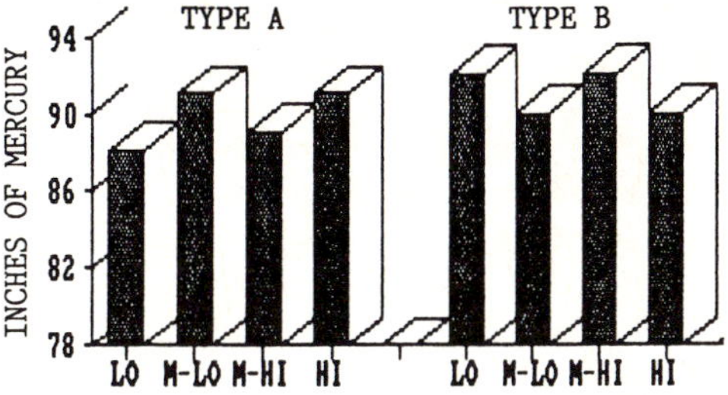

Fig. 5b: Diastolic blood pressure

$p > .05$). As for heart rate, the pattern of systolic blood pressure changes and workload ratings were similar for Type As, although nonsignificant (r = 0.54), and unrelated for Type Bs (r = -0.29).

Diastolic blood pressure was not significantly different between Type As (79) and Bs (77) in the pretest. There was a significant increase between pre-test (78) and post-test (84) measurements ($F(1,26) = 16.31$, $p < .01$), however, indicating a slower return to pre-experimental baseline levels. Diastolic blood pressure increased (to 90) with the introduction of the experimental trials for both Type As and Bs. Again, there were no significant Type A/B, Practice/Experimental, or Difficulty effects. However, the predicted Type A/B x Difficulty interaction was again found ($F(3,78) = 4.45$, $p < .006$). When analyzed separately, variations in diastolic blood pressure arousal related to experimental manipulations was significant for Type As ($F(3,39) = 3.31$, $p < .05$), but not for Type Bs ($F(3,39) = 1.62$, $p > .05$). As found for heart rate and systolic blood pressure, the pattern of diastolic blood pressure changes and workload ratings were similar for Type As, again nonsignificantly (r = 0.68), and unrelated for Type Bs (r = 0.18). There was a significant correlation between diastolic blood pressure and heart rate (r = 0.74) and between diastolic and systolic blood pressure (r = 0.67) for Type As. For Type Bs, only the relationship between systolic and diastolic levels was significant (r = .68)

4 CONCLUSIONS

The experimental manipulations generated significant differences in performance and subjective evaluations for all subjects. In addition, for Type As, but not Type Bs, significant changes in heart rate and blood pressure were observed in response to the experimental manipulations. These results provide insight into one possible source of variability among workload assessment measures obtained from different subjects under apparently similar circumstances. At least for the physiological measures included in this study, identifying the behavior pattern of the subject enhanced the interpretability of the assessment technique; measures taken from Type As were significantly more responsive to experimentally manipulated workload levels than were measures taken from Type Bs. In addition, these data provide experimental support for the suggestion that the cardiovascular responsiveness of Type A personalities may be a potent variable in the development of subsequent cardiovascular disease, even for individuals who exhibit otherwise normal resting blood pressure levels.

Although subjective ratings continued to be the most sensitive measure of workload, the evaluations given by the two groups may not have been equally representative of their experiences. Type A subjects evaluated the quality of their performance more stringently and they appeared to be less sensitive to the effort they exerted to achieve their higher scores; their physical and mental effort subscale ratings were lower and they reported similar levels of fatigue, even though they appeared to be working harder than Type Bs, from the higher number of function-key actuations that they made., This suggests that the two groups are not equally accurate in evaluating specific task- and operator-related aspects of their experiences. The cardiovascular measures provided converging information about the influence of the experimental tasks only for the subjects who were most responsive to them (i. e., Type As).

Thus, the behavioral characteristics that are relevant in classifying individuals as Type A or B appear to predict how each group will react to different experimental manipulations. Although their subscale ratings provided some indication that Type A and B subjects differed in their personal performance criteria, sensitivity to the challenge of complex tasks or effort required, and acceptance of objectively higher levels of time pressure and task difficulty, the differences were not sufficiently great to have achieved statistical significance in this study. However, it does suggest that the distinction between Type A and Type B behavioral characteristics can provide useful information about a potential source of individual differences in workload experience and evaluation to consider when interpreting their subjective workload ratings.

5 REFERENCES

Chesney, M. A., Eagleston, J. R. and Rosenman, R. H., 1980. The Type A structured interview: A behavioral assessment in the rough. *J. Behav. Assess.* *2*(4), pp. 255-272.

Chesney, M. A., and Rosenman, R. H., 1983. Specificity in stress models: Examples drawn from type A behavior. In C. L. Cooper (Ed.), *Stress Research* (pp. 21-34). NY: John Wiley and Sons.

Damos, D. L., and Bloem, K. A.. 1985. Type A behavior pattern, multiple-task performance, and subjective estimation of mental workload. *Bull. of Psycho nomic Soc.*, *23*, pp. 55-56.

Glass, D. C., 1977. Stress behavior patterns and coronary disease. *American Scientist*, *65*, pp. 177-187.

Hart, S. G., Battiste, V., and Lester, P. T., 1984. POPCORN: A supervisory control simulation for workload and performance research. In *Proc. of 20th Ann. Conf. on Man. Cntrl* (NASA CP-2341) (pp. 431-454). Washington, D. C.: Natl. Aero. and Space Admin.

Hart, S. G. Hauser, J. R. and Lester, P. T., 1984. Inflight evaluation of four measures of pilot workload. In *Proc. of 28th Ann. Mtg. of Hum. Fac. Soc.* (pp. 945-949). Santa Monica, CA: Human Factors Society.

Mulder, G., 1979. Sinusarrhythmia and mental workload. In N. Moray (Ed.), *Mental workload: Its Theory and Measurement* (pp. 327-344) NY: Plenum Press.

Social, Ergonomic and Stress Aspects of Work with Computers,
edited by G. Salvendy, S.L. Sauter and J.J. Hurrell, Jr.
Elsevier Science Publishers B.V., Amsterdam, 1987 — Printed in The Netherlands

PERSONALIZED RECOGNITION OF MENTAL WORKLOAD IN HUMAN - COMPUTER SYSTEM

M. NOWAKOWSKI
Instytut Organizacji i Zarządzania, Politechnika Wrocławska, 50-370 Wrocław, Wybrzeże Wyspiańskiego 27 (Poland)

ABSTRACT

A knowledge based adaptive system is capable of controling the task allocation and load leveling functions within the human - computer systems. This approach uses on-line mental workload assessment procedures. Upon detection of an abnormal level of mental workload the adaptive system redefines and reallocates the task or subcomponents of an overall task. The determining of acceptable workload levels for mental tasks is not trivial. The difficulty lies in fact that not every human responds to a given mental stressor and when there is a response, there is no uniform pattern. The main goal of the paper is outline a method of recognition of workload in respect of individual response pattern. The proposed approach utilizes concepts of fuzzy measure and fuzzy integral and classification methods. A method of personalized recognition of mental workload may be employed in knowledge - based adaptive systems and they both may be useful in creating human - computer symbiotic systems that utilize the full capabilities of all the sub-systems involved.

1. INTRODUCTION

The allocation of tasks between human and computer is necessary step in the design of any highly automated system. One approach that may be employed when both the human and the computer are capable of performing a particular set of tasks is a dynamic approach. Responsibility for performing each task goes to the entity with the resources available at that time. A dynamic approach uses the systems resources more fully, provides a relatively constant or acceptable workload level for the human, and allows the human to have a flexible role in the system. (Greenstein et. al., 1986).

Chignell and Hancock (1986) proposed a method for building intelligent person - system interfaces which are responsive to changes in task demand. This approach uses on-line mental workload (MWL) assessment procedures. Comparison of available perceptual and cognitive capacity with the current level of MWL demanded of the operator by the task, activates a compensatory procedure that shifts the processing load toward or away from the human in order to optimize overall system performance.

A knowledge - based adaptive mechanism (KBAM) is capable of controlling the task allocation and load leveling functions within human - machine system (Chignell and Hancock, 1986). Load leveling is required because of fluctuations in the perceived or actual dificulty of the task over time and it is particularly important when the mismatch is sufficient to produce inadmissible overload or underload. Any KBAM which acts as buffer, task generator or load leveler between the task and the operator, must be capable of assessing the MWL imposed within a short time frome and in sufficiently accurate fashion. Task reallocation can be accomplished through user generated commands instructing the computer to assume control of certain tasks, or by the intervention of KBAM system.

Dialogue - based communication may be counter to the goal of workload reduction. Not only is the total workload increased, but it is likely that engaging in dialogue will detract from the human's ability to perform the primary task. An alternative mode of communication which does not increase the human's workload is model based communication (Greenstein et. al., 1986). In the case of KBAM systems, upon detection of an abnormal level of MWL the adaptive mechanism redefines and reallocates the task or subcomponents of an overall task, unless speccifically directed to do otherwise by the user.

This procedure is depended on the derivation of the error signal (Chignell and Hancock, 1986) which is input into compensatory system. The error is based on fuzzy set - theoretic formulation and multivariate profile of physiological responses and performance measures are mapped on to fuzzy sets. The determining of acceptable workload levels for mental tasks is not trivial. In contradiction to physical tasks, mental ones do not have established workloads limits. The difficulty lies in fact that not every human responds to a given mental stressor and when there is a response, there is no uniform pattern. We need to develop levels of acceptable workload that would be unique for an individual.

The main goal of the paper is outline a method of MWL recognition in respect of individual response pattern. The proposed approach utilizes concepts of fuzzy measure and fuzzy integral and classification methods. A system of personalized recognition of MWL may be employed in knowledge - based adaptive system and they both may be useful in creating human - computer symbiotic systems for knowledge acquisition and action that utilize the full capabilities of all the sub-systems involved.

2. MODELLING AND CLASSIFICATION OF MWL

Mital and Karwowski (1986) proposed a modelling approach to quantify mental stresses in they are qualitative and to integrate them with other physical and mental stresses. This approach utilizing fuzzy set theory is oriented towards developing upper limits of workload levels or acceptable response levels. They have defined the set "strain" within the boundaries "yes" and "no" such that they indicate the state of person either under strain or within the compensatory zone respectively. A person whose responses are at the lower end of compensatory zone will be identified by response membership value equal to O (no), while some one with indication of severe fatique symptoms will have a response membership value equal to 1 (yes). The set strain is fuzzy since the transition from membership to non-membership is gradual rather than abrupt. A fraction between O and 1 would then represent the grade of membership of each of the responses in the set strain.

There are variety of methods by which MWL may be measured but physiological and performance measures appear to be the most useful for the purpose of MWL assessment.

Let the universe of discourse $X = \{x_1, x_2, ..., x_n\}$ be a vector comprised of a concatenation of physiological and performance measures or vector of responses. The mental workload may be represented by fuzzy set wit membership function $h : X \rightarrow [0,1]$. Chignell and Hancock (1986) defined two fuzzy subsets lobeled "excessive MWL" and "insufficient MWL" respectively. We then have two corresponding membership functions i.e. $h_E : X \rightarrow [0,1]$ and $h_I : X \rightarrow [0,1]$. The membership functions h_E for instance, associates with each measure (response) $x_i \in X$ a number $h_E(x_i)$ in the interval $[0,1]$ which represents the grade of membership of x_i in the fuzzy subset "excessive MWL". The closer this number is to 1, the stronger the indication that the individual is under excessive mental workload.

When considering each measure (response) x_i seperately, the interpretation of number $h(x_i)$ is quite simple. However, the interpretation (meaning) of the fuzzy set MWL as a whole is more complicated. There arise a few questions. What is the threshold value t for which $h(x_i) \geqslant t$ means that the operator is under strain or "excessive MWL"? Is the threshold common or different for each response? The operator'sio MWL is "high" when - "at least one $h(x_i)$ is greater than threshold" or "most (few) of $x \in X$ are above threshold"? Which of two fuzzy sets $h_1(x)$ and $h_2(x)$ means that the steady state is more destabilized? There are two of possible approaches of developing there problems i.e. aggregation of fuzzy set into one number and utilization of the whole fuzzy set representing MWL. One method of combining the measures $h(x_i)$ is to use a weighted average, thereby allowing for differences in the relative importance of each x_i. It is suggested (Mital and Karwowski, 1986) that expert oppinion be used to derive these weightings.

As mentioned before, there is no uniform pattern of a strain or MWL limits, therefore there is no unique (common) interpretation of fuzzy set representing MWL for all individual as well. It has been accepted that all responses (measures) $x \in X$ can not be equally important and importance of $x \in X$ may differ for particular individual also. Nowakowski (1986) proposed an approach based on concepts of fuzzy measure and fuzzy integral.

Consider a set of n information elements $X = \{x_1, x_2, \ldots, x_n\}$. If a set function $g : 2^X \rightarrow [0,1]$ has the following properties, g is called a fuzzy measure. (Sugeno, 1974).

$$g(\phi) = 0 \text{ and } g(X) = 1 \qquad (1)$$

$$A, B, \in 2^X \quad \text{and } A \subset B \rightarrow g(A) \leqslant g(B) \qquad (2)$$

Now, the following inequality will hold for any two elements x_i and x_j, $x_i \neq x_j$.

$$g(\{x_i\}) \vee g(\{x_j\}) \leqslant g(\{x_i, x_j\}) \leqslant 1 \qquad (3)$$

where "$\vee$" means "max" operator. In other words, $g(\{x_i, x_j\})$ can take any value as long as it satisfies (3). For instance possibility measure and probability measure are special cases of fuzzy measures. So, $g(\{x_i\})$ represents the grade of importance of measure x_i (symptom), while $g(A)$, $A \subseteq X$ is interpreted as the grade of importance of response subset (syndrom), i.e. importance x_i or $A \subseteq X$ as an infprmation input for MWL assessment.

Let. $S = \{s_1, s_2, \ldots, s_m\}$ be a set of MWL, $X = \{x_1, x_2, \ldots, x_n\}$ be vector of measures (responses) and $C = \{C_1, C_2, \ldots, C_p\}$ be a set of classes of MWL. We assume that these sets are finite ones and that any MWL is characterized by membership function $h_s : X \rightarrow [0,1]$. Each class C_k is defined by (profile of the class); i.e. by a vector g^k such that its i-th element g^k_i gives the grade of importance of element X_i for the class C_k. Nowakowski (1986) suggests using a fuzzy integral to determine the grade of agreement between requirements of k-th class of MWL and real picture of MWL i.e., profile $s \in S$, or to determine the grade of similarity of a pattern to a prototype. The grade of agreement (similarity) is defined by

$$g^k(s) = \bigvee_{i=1,2,\dots,n} \left[h_s(x_i) \wedge g(K_i) \right] , \tag{4}$$

,where $g(K_i) = g(\{x_1,x_2,\dots,x_n\})$, and "$\wedge$" - min operator. The equality (4) is valid iff h is rearranged decresaingly, i.e. $h(x_1) \geqslant h(x_2) \geqslant \dots \geqslant h(x_n)$. The algorithm for computation of fuzzy integral may be found in Sugeno's papers, (see also Nowakowski, 1986).

Using the above interpretation the following classification rule may be adopted (Wierzchoń, 1981):

- the MWL s belongs to the class of MWL, C_k iff the grade of agreement $g^k(s)$ is greater than grades of agreement for all remaining classes, i.e.

$$g^k(s) > g^l(s) \ , \ l \neq k, \ l = 1,2,\dots,p \tag{5}$$

Suppose that we accept a MWL s for a class C_k iff $g^k(s) > t_k$, where t_k is a threshold value for this class. The condition is satisfied if there exists subset of responses E with the grade of importance for class C_k not less than t_k $(g^k(E) \geqslant t_k)$ and simultaneously minimal intensity of responses from subset E is not less than t_k.

3. PERSONALIZED RECOGNITION OF MWL AND FUZZY MEASURE LEARNING IDENTIFICATION

The important problem consists of assigning profiles of the classes i.e., identification of fuzzy measures characterizing the classes (such as "excessive MWL" and "insufficient MWL" and so on). Up to this moment we have assumed that the fuzzy measures characterizing the classes of MWL are given. These fuzzy measures may be found in a number of ways. One of possible way is to use apriori experts' knowledge which requires identification algorithms (Wierzchoń, 1983). Another way is to use a learning procedure (Wierzchoń, 1981) utilizes in (Nowakowski, 1986). As input of learning procedure we use patterns of MWL which may be treated as the most representative for a given class. This procedure produces fuzzy density function for this class.

Formally, for each class C_k of MWL the set of responses (measures) may be divided into two subsets. The subsets X^k and $\bar{X}^k$ of X such that. $X^k \cup \bar{X}^k = X$, $X^k \cap \bar{X}^k = \emptyset$ and for each $x_j \in \bar{X}^k$ and for each

$x_i \in X^k$ there hold $g_i^k \gg g_j^k$ will be called relevant and irrelevant for class C_k (Here " $\gg$ " means "much bigger than"). For instance, if a given class is connected with an individual, then X^k and $\bar{X}^k$ represent subsets of responses then X^k and $\bar{X}^k$ represent subsets of responses relevant and irrelevant for this individual. Having a representative set of MWL for an individual we are able to find a set of responses relevant for this person or for personal class of MWL. Moreover, using another algorithm (Nowakowski, 1986), which allows for classification of objects when a number of class is not known in advance, one may divide the set of MWL into a few subclasses of MWL. It may be done for finding mutations of the same class of MWL.

Sugeno (1985) proposed totally different approach to identification of fuzzy measure. New algorithm based on the model reference adaptive approach is developed to identify the fuzzy measure with full degrees of freedom. Fuzzy measure Learning Identification Algorithm (FLIA) is shown to be much foster than, and superior to, the conventional identification schemes. Sugeno (1985) defined overlap coefficient and necessity coefficient and used them to characterize fuzzy measure. These coefficients well characterize "personal fuzzy measures" i.e. fuzzy measures characterizing personal classes of MWL.

Consider a fuzzy measure $g = 2^X \longrightarrow [0,1]$. The coupling coefficient μ_{AB} is defined as follows.

$$\mu_{AB} = \frac{g(A \cup B) - [g(A) + g(B)]}{g(A) \wedge g(B)} \tag{6}$$

where $A, B \in 2^X$ and $A \cap B = \phi$ and $A, B \neq \phi$.

The coupling coefficient μ_{AB} indicates the degree of additivity that the fuzzy measure holds between subsets A and B.

There overlap coefficient $m_{ij} \in [-1, 1)$ is expressed as follows.

$$m_{ij} = \begin{cases} \mu_{ij} & i \neq j \text{ and } \mu_{ij} \leqslant 0 \\ \dfrac{\mu_{ij}}{\mu_{ij}+1} & i \neq j \text{ and } \mu_{ij} \geqslant 0 \\ 0 & i = j \end{cases} \tag{7}$$

where $\mu_{ij} = \mu_{\{x_i\}\{x_j\}}$ and $x_i, x_j \in X$.

The overlap coefficient m_{ij}, with its range of $[-1, 1)$, is a symmetrical scale of the degree of additivity that g holds between elements x_i and x_j e.g. $m_{ij} = 0$ indicates that g is additive between x_i and x_j. The overlap coefficients m_{ij} form coefficient matrix $M = (m_{ij})$.

The degree of overlap n_j of each element x_j is defined as follows:

$$n_j = (\sum_{i=1}^{n} m_{ij}^3) / (n-1) \qquad j = 1,2,\dots,n \tag{8}$$

The coefficient n_j shows the average overlap between x_j and other elements. A negative n_j shows that x_j is redundant information element, while positive n_j indicates x_j's independence.

The neccesity coefficient $k_j \in [-1, 0]$ of the element x_j is defined as follows.

$$k_j = n'_j \, [1 - g(\{x_j\})] \tag{9}$$

where

$$n'_j = \begin{cases} n_j & n_j \leqslant -0.4 \\ 0 & n_j > -0.4 \end{cases} \tag{10}$$

If $k_j = 0$ then x_j is absolutely necessary element. If k_j is clase to -1 then x_j is redundant information.

The fuzzy measure $g(\cdot)$, the overlap coefficient matrix M, and the necessity coefficient k_j characterize the personal MWL well; they can generally be interpreted in the following manner:

- $g(\{x_i\})$, $x_i \in X$ indicates the importance of response x_i as an information input for MWL assessment "importance" implies total evaluation x_i's sensitivity and accuracy for a given class of MWL.
- The overlap coefficient matrix M shows the interaction between each information element (response). A small m_{ij} implies that the level of importance does not increase even when x_i and x_j are put together, i.e. they possess large overlap. Conversly, if m_{ij} is close to +1 the importance is amplified when x_i and x_j are put together.
- The necessity coefficient k_i reflects the importance of an information element x_i and its overlap with other elements $k_i = -1$ indicates a low importance and large overlap, hence implies that x_i is not necessary. On the other hand, $k_i = 0$ indicates the opposite, and thus, implies that x_i cannot be dropped.

4. CONCLUDING REMARKS

The derivation of valid MWL assessment is the major barrier to the implementation of knowledge - based adaptive mechanism in human-computer system. We advocate the use of method oriented towards personalized recognition of MWL. It is obvious, the presented approach requires calibration based on experimental works.

5. REFERENCES

Chignell, M.H., and Hancock, D.A., 1986. Comparison of Mental Workload and Available Capacity in Complex Person-Machine Systems. In: W. Karwowski and A. Mital (Editors), Applications of Fuzzy Set Theory in Human Factors, Elsevier, Amsterdam / Oxford / New York / Tokyo, pp. 271-288.

Greenstein, J.S., et.al., 1986. An Empirical Comparison of Model-Based and Explicit Communication for Dynamic Human-Computer Task Allocation. Int. J. Man-Machine Studies, 24, pp. 355-363.

Mital, A., and Karwowski, W., 1986. Towards the Development of Human Work Performance Standards in Futuristic Man-Machine Systems: A Fuzzy Modelling Approach, Fuzzy Sets and Systems, 19 (in press).

Nowakowski, M., 1986. Classification and Recognition of Fuzzy Strain, In: W. Karwowski (Editor), Trends in Ergonomics / Human Factors III, Elsevier, Amsterdam / Oxford / New York / Tokyo, pp. 127-133.

Sugeno, M., 1974, Theory of Fuzzy Integrals and its Applications, Ph.D. Thesis, Tokyo Institute of Technology (unpubl.).

Sugeno, M., 1985, A Model of Human Evaluation Process Using Fuzzy Measure, Int. J. Man-Machine Studies, 22, pp. 19-38.

Wierzchoń, S.T., 1981, Classification and Feature Selection Using Fuzzy Measure, Paper presented at th 12-th Meeting of the EURO Working Group on Fuzzy Sets, Hamburg (unpubl.).

Wierzchoń, S.T., 1983, An Algorithm for Identification of Fuzzy Measure, Fuzzy Sets and Systems 9, pp. 69-73.

Social, Ergonomic and Stress Aspects of Work with Computers,
edited by G. Salvendy, S.L. Sauter and J.J. Hurrell, Jr.
Elsevier Science Publishers B.V., Amsterdam, 1987 — Printed in The Netherlands

VDT WORK - AN OCCUPATIONAL HEALTH HAZARD?

B. G. KNAVE
National Board of Occupational Safety and Health Research Department
S-171 84 SOLNA (Sweden)

ABSTRACT

Visual display terminals (VDTs) have become a major element in the modern work environment as an interface between man and computer. The discussion as to whether work at VDTs can affect human health has been centered on different types of effects such as eye damage or discomforts, neck and shoulder discomfort, adverse reproductive outcomes, skin disorders and different stress reactions. In the present paper a short review is given of these health effects under discussion as well as some alleged radiation hazards from the VDTs. Finally, some aspects are given with respect to future VDT work and health.

1 HEALTH EFFECTS UNDER DISCUSSION

1.1 Eye discomfort and musculo-skeletal postural discomfort

Several studies report different types of eye discomfort and musculo-skeletal complaints from the neck, upper part of the back and shoulders. Also in objective measurements (electromyography) an increased strain of the neck muscles has been reported. The mechanisms behind the eye discomfort are not fully understood, but we know that the image quality of VDT screens is inferior to that of black text on white paper. The postural strain might be associated with change of work plane when working (i.e. looking) at VDT screens placed 10-20 degrees above the horizontal table plane. The line of sight and the head must then, consequently, be lifted up correspondingly. At the same time the work may be more repetitively monotonous and sedentary, which will not improve work conditions.

1.2 Skin

In Scandinavian reports some common skin diseases (e g acne, rosacea, eczema) have been associated with VDT work. The electrostatic charge of the operator (and a low room humidity) has been suggested as a possible contributing causal factor. The charge will increase facial deposition of air contaminants in the operator which, in turn, will cause a skin reaction especially among sensitive individuals. Also work stress has been mentioned in this connection. Several large scale epidemiological studies are in progress now, the results of which will be ready for presentation in the next couple of years.

1.3 Pregnancy outcome

Lately much attention has been paid to the possibility of adverse pregnancy outcome being linked to VDT work. Recent and on-going epidemiological studies have not given any support to these allegations, however. No study has demonstrated any significant increase in spontaneous abortion or congenital defects among VDT operators compared to respective controls. Neither there have been any /non-significant/ trends in the results of these studies.

Results of experimental studies, in chicken, mice and rats exposed to electromagnetic fields similar to those from CRT-based VDTs, have been too contradictory to allow drawing any far reaching conclusions. Further studies, epidemiological as well as experimental, are in progress however, in different countries.

The discussion on pregnancy and VDT work has resulted in a widespread concern and worry among VDT operators. Due to this concern - but not to any defined risk factor in VDT work - the availability of alternative work during pregnancy has been motivated. It should be stressed that good working conditions - in terms of absence of stress factors and good ergonomics - while of importance for all workers, may be especially important for a pregnant woman.

1.4 Work stress

Working conditions which might lead to stress reactions and stress related diseases are common in many VDT work routines. And increased frequencies of stress related symptoms have been found among VDT operators. It is possible that stress reactions play a role in the development of eye discomfort and postural discomfort as well as in the worsening of the above mentioned skin diseases among operators.

2 ALLEGED RADIATION RISKS FROM VDTs

A number of careful scientific studies have been focussed on the measurement of electromagnetic radiation or fields due to the VDTs, while some limited attention has also been given acoustic radiation.

2.1 Soft X-ray radiation

Soft X-ray radiation is produced within the CRT. The glass material of the tube, however, effectively prevents any leakage of X-ray radiation outside of the tube during operation. Thus, X-ray radiation from VDTs is not detectable with normal measuring instruments.

2.2 Ultraviolet radiation in the near region (UVA)

UVA radiation can be detected from certain VDTs. The levels are, however, insignificant compared to present occupational standards, and also insignificant compared to emission from other sources (e g sunlight through the windows). In one investigation, VDT operators were found to be exposed to lower levels than those not working with VDTs, attributable to the fact that the former often draw the window curtains.

2.3 Visible radiation

Visible radiation can be measured and is necessary in order to perform the intended function of the CRT - to provide a visual display. Luminance levels recorded are far below current exposure limits, thus precluding (according to present knowledge) the possibility of pathological injury due to excessive exposure.

2.4 Infrared (IR) radiation

IR radiation is emitted from all warm bodies, and since all surfaces of the VDT are at room temperature or slightly above, IR radiation can be detected, although at levels far below any limits of concern for potential hazards.

2.5 Radiofrequency radiation and Extremely Low Frequency Fields

In the radiofrequency and the extremely low frequency regions, electric and magnetic fields have been detected. The dominant sources are the power supply and the horizontal and vertical sweep arrangements (at frequencies of some 15-35 kHz and 50-80 Hz, respectively). These fields do not appear to represent any risk factor when evaluated by comparison with current standards, guidelines and recommendations for occupational exposure. Compared to fields in many industrial or household situations, the fields around VDTs do not correspond to high exposure situations. Some ongoing animal studies on the possibility of effects of these fields do, however, require further attention to exposure to low frequency magnetic fields in various situations including VDT work.

2.6 Electrostatic fields

Electrostatic fields at VDT workplaces have been suggested as a possible cause of skin disorders. A relationship between electrostatic fields and skin disorders must, however, still be regarded as hypothetical. Research conducted hitherto has indicated that the electrostatic charge of the operator might be a relevant factor. Some possible mechanisms have been discussed, and investigations are presently under way to attempt to clarify whether any relationship exists between different factors in VDT work and skin disorders.

2.7 Ultrasonic radiation

Airborne ultrasonic (acoustical) radiation is produced in VDTs as a result of mechanical vibra- tions generated in the core of the flyback transformer (responsible for the horizontal sweep of some 15-35 kHz). The sound pressure levels found are considerably below existing exposure levels. Some sensitive individuals may detect this sound or a subharmonic as an annoying factor.

2.8 Conclusive remarks

Effects which have in the debate been suggested as caused by exposure to electromagnetic radiation or fields include adverse pregnancy outcome, skin disorders or cataracts. Comparison of the occurrence of cataracts or of adverse pregnancy outcome among VDT operators to those of controls have failed to show an excess occurrence due to VDT work. In some countries, a number of VDT operators have experienced skin disorders. The relationship of these to VDT work is not known.

In conclusion, and based on current biomedical knowledge, there are no health hazards associated with radiation or fields from VDTs. Thus, there is no scientific basis to justify shielding, radiation monitoring or dedicated eye examinations to search for ocular pathology in VDT operators. However, since a large number of people are involved in VDT work, it is important that further knowledge is attained on certain areas where our knowledge must be regarded as incomplete. These areas include:

(1) Further investigations into the possibility that skin disorders may be related to VDT work, and if so, the factor(s) involved.

(2) The possibility of interactions between low frequency magnetic fields and biological systems. Considerations should be given magnetic fields in various situations, and not be restricted to VDT work situations.

3 ALTERNATIVE DISPLAYS

Nearly all the screens used hitherto have been CRT-based. LCD (=liquid crystal displays), plasma and electroluminescence displays are alternatives for the future. They have the advantage of low weight (easy to carry around) and very little if any electromagnetic fields. Their disadvantages are poor visual ergonomics (low contrasts and small viewing angles) and long change-over times. Current surveys indicate that it will take the new displays another five years to reach an acceptable stage of development.

4 FUTURE ASPECTS

Today the VDT density in Swedish company offices is 1 VDT on 4 employees (a somewhat lower density in the public sector, though), a ratio which is believed to increase to 1:1 in the next 5-year-period. In spite of this increase the physical VDT work environment will certainly improve during this time. For instance, the screen image quality will be better, and the postural work position will be less fixed to the screen, i e the operator will be more mobile at his/her work station with the new, light "alternative" screens. Probably the electrostatic charge of the VDTs and the operators will be eliminated, diminishing the risk of developing any skin disorders. Hopefully also the psychological work environment will improve, i e in the work organization emphasis will be put also on the well-being of the employee, in addition to job efficiency. This will of course mitigate known adverse stress health effects, but also be of beneficial importance for pregnancy outcomes and skin disorders among operators. In spite of the wide-spread interest in human-computer-interaction today the role of soft-ware design for operator health has hardly at all been studied. In the years to come issues of this kind ought to be investigated.

CONTENTS

Social, Ergonomic and Stress Aspects of Work with Computers,
edited by G. Salvendy, S.L. Sauter and J.J. Hurrell, Jr.
Elsevier Science Publishers B.V., Amsterdam, 1987 — Printed in The Netherlands

THE BENEFIT/COST OF VDT HIGHLIGHTING

D.W. MARTIN, D.R. MCDONALD, and C.R. PATTON
Department of Psychology, Box 3452, New Mexico State University, Las Cruces, New Mexico 88003 (USA)

ABSTRACT

A simple visual search task was used to investigate the benefits and costs of highlighting cues. Subjects were required to find a target in a linear array of five single digit numbers. In some conditions targets were highlighted with high or low validity. A comparision of valid and invalid trials to neutral trials allowed the benefits and costs of attending to the highlight to be assessed for the high validity conditions. The benefits and costs of attempting to ignore the highlight could be assessed in the low validity condition. Highlighting cues were: color, blinking, boxing, reverse video, and arrow tagging. Response times and errors were measured. General findings were: The benefit/cost paradigm provided a good way of evaluating highlighting options. The cost of ignoring cues was small. Cues had markedly different effects on discriminability of targets. Color provided some benefit without cost.

INTRODUCTION

An advantage of using computer generated displays is that large arrays of information can be displayed to the user. However, when the user's task is to search the display for a particular item of information, search times can be long for large arrays. For this reason, it is often desirable to draw the user's attention to a particular spatial location. For example, a menu of items might be presented from which the user is required to make a choice. If a system designer knows that a particular menu item has a high probability of being chosen, a choice might be made to use a highlighting cue (Morse, 1979). Highlighting is typically used to indicate: priority of information, unusual information, information change, organization of information, or the nature of alarms or warnings (Parish, Gates, and Munger, 1981). The use of highlighting is especially important when the user's major task is to monitor the changing status of a dynamic display in order to locate and respond to potential targets.

Many highlighting options are available to the system designer: reverse video; color differences; brightness differences; character size, case, or font differences; underlining; blinking; bordering with lines, graphics, or sýmbols; tagging with cursors, arrows, or symbols; and positon displacement (Hemingway, Kubala, and Chastain, 1979; Knapp, Moses, and Gellman, 1982; Stewert, 1976). While a few of these options such as color (Carter, 1982; Christ, 1975; Tullis, 1981) and blinking (Smith and Goodwin, 1971) have received some study, the guidelines for choosing among these options are sketchy at best (Engel and Granda, 1975). Results from research on the potential benefits of highlighting cues could be of value to the system designer in choosing which cues to use and under what circumstances to use them.

A second concern for the designer is the potential cost to the user's performance when the cue is either wrong and the user does not know it until examining the item or irrelevant to the task and the user knows it. For example, in the latter case a designer might highlight a menu item in order to help a novice user make an appropriate response. However, an experienced user who wished to choose another item would have to ignore the cue incurring a cost due to the distracting nature of a highly salient cue.

A methodology from cognitive psychology, cost-benefit analysis, has the potential for providing useful information about various highlighting cues (Posner and Snyder, 1975; Jonides and Mack, 1984). Cost-benefit analysis has typically been used to study the effects of preparation on responding. A cue is given to subjects prior to presenting a stimulus which requires a response. This cue contains information of potential value in making a correct response. When the information is correct, the cue is valid, and it leads the subject to respond more quickly and/or more accurately, thus providing a benefit compared to an uncued, or neutral, condition. When the information is incorrect, the cue is invalid, and the subject may incur a performance cost. By manipulating the proportion of valid to invalid trials, subjects can be induced to pay attention to or ignore the cue. In the research reported here, highlighting provides the cue. The highlighting cue occurs simultaneously with the stimulus array rather than preceding it, but in other respects the methodology is comparable to classic cost-benefit analysis.

METHOD

Presentation and timing of stimuli was accomplished using Pascal programs implemented on either a Terak model 8510 microcomputer or an IBM-PC. Each trial began with the presentation of a fixation + in the center of the VDT. After 500 ms, a horizontal array of five single-digit numbers appeared. The middle number replaced the +; the others were 2 and 4 degrees to the left and right of center. The subjects' basic task was to search the array for the single target digit, a number 1-4. Upon finding the target, subjects indicated its value by pressing one of the four response keys upon which the fingers of their right hands rested; index finger for 1, etc. After each response, feedback indicating response time and correctness was displayed. After one second a new trial began.

Three types of trials were presented. On neutral trials no highlighting cue occurred. On valid trials a cue occurred at the position containing the target. On invalid trials a non-target position was highlighted. All subjects received the three types of trials. Neutral trials appeared in blocks of 20. Valid and invalid trials were mixed in blocks of 100. The proportion of valid to invalid trials was a between-subjects variable set at either 80/20 or 20/80. In the former condition, subjects should have been induced to use the cue because it predicted the target location on 80% of the trials. Indeed, a comparison of the data reported here with a pilot experiment having only 100% valid trials for one set of subjects, and neutral trials for a second set, shows comparable benefits, indicating that subjects did consistently use the cue in the 80/20 condition. In the 20/80 condition, the cue is only randomly predictive of target location. Thus, subjects should have attempted to ignore the cue.

After being instructed, subjects were given a block of 20 neutral practice trials followed by a block of 100 highlighted practice trials. Within each block, targets occurred equally often at each position. Experimental trials consisted of five replications of this sequence. Thus, the basic experimental design for each type of highlight and each validity proportion consisted of the within-subject factorial combination of three levels of trial type (valid, neutral, invalid), five blocks, and five positions.

Five separate experiments were conducted to assess the effects of five different highlighting cues. The arrow cue consisted of an arrow of approximately the same size as the digit characters placed above one of the digits. The blinking cue consisted of an on-off cycle at 5 Hz with equal on off times. Boxing consisted of a box placed around the digit with the edges in close proximity to the character. Color consisted of a highlighted red digit that contrasted with the light green color of the other digits. Reverse video consisted of a boxed area around the digit within which the digit appeared as a black number surrounded by a light green background. These highlighting cues were chosen to represent some of the more commonly used cues on existing computer systems.

RESULTS AND DISCUSSION

The dependent measures of response time (RT) and errors were subjected to analyses of variance. Separate analyses were conducted on each experiment representing different highlighting cues and validity proportions. The two panels of Figure 1 illustrate a complete RT

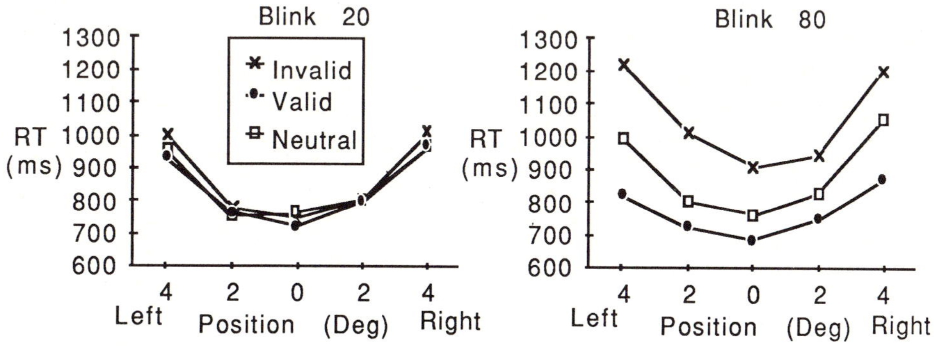

Fig. 1. RT data for blinking highlight cue condition as a function of trial type and position.

graph for the blinking cue. The benefits and costs of highlighting can be seen for each proportion at each position. The position effect shown here is typical of that found in each of the highlighting conditions. For all conditions, the position effect on RT was large and statistically significant at the p <.01 level. In all but two cases there was not a significant effect of position on errors. The two cases in which it was marginally significant were due to a slightly higher number of errors at peripheral positions. These position effects were not unexpected in that eye movements were probably required at the outside locations. Indeed, a supplementary experiment in which subjects were required to read a single digit number at these locations following tachistoscopic presentation, indicated that at the outside positions approximately 25% errors were made without eye movements compared to approximately 5% and 0% errors at the inside and central locations. Apparently subjects start checking for targets at the central position and then work their way toward the periphery. When a target is validly highlighted, the position effect is somewhat attenuated. In a few cases there was also some indication of a left-to-right reading strategy (see Figure 3).

Because the position effects are all similar to the one shown and the few marginally significant Position X Trial Type interactions are of the same nature as that shown, histograms will be used to represent the overall results. Figure 2 illustrates the mean RTs for each trial

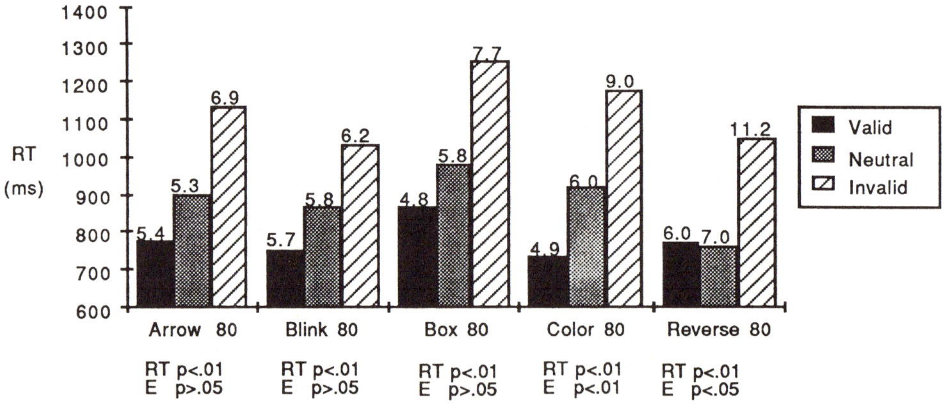

Fig. 2. 80/20 condition RTs by highlight and trial type. Mean errors are shown above the bars. Significance levels of trial type main effects for RT and errors (E) are shown below the graph.

type and each highlighting cue for the 80/20 valid/invalid proportion subjects. Mean percent errors are given above the bars. The significance levels for the main effects of trial type for both RT and errors (E) are shown below each condition. All conditions are based on data from 12 subjects except blink with 13 subjects and color with 14.

The data support the notion that the benefit/cost of highlighting cues can be assessed using the procedure employed here. The effects are large and orderly. There are also differences in the size of the benefits obtained. Color provides a benefit of 186 ms compared to arrow, box, and blink in the 111-125 ms range, and reverse video at -10 ms. In general, the error effects are nonsignificant, and where they are significant, they do not tradeoff with speed. Where they do vary, the size of the benefit/cost may be underestimated. The lack of benefit for reverse video may seem surprising but can be understood by examining results of the 20/80 condition and will be discussed shortly.

Overall, the lack of large benefit/cost effects in the 20/80 condition shown in Figure 3 indicates that subjects were following the expected strategy of attempting to ignore the highlighting cue. To evaluate whether the cue was distracting the subjects and interfering with their target search, the three trial types need to be examined. Distraction would be evident when there is considerable cost and little or no benefit. While the cost could be explained either by distraction or by a strategy of using the cue, lack of a benefit supports only distraction. The largest effect of this type can be seen for blinking. It is, perhaps, surprising that this effect is so small given the finding in previous research that abrupt onsets in visual displays tend to capture attention (Yantis and Jonides, 1984). An overall conclusion is that for the cues studied here, the cost of attempting to ignore irrelevant highlighting cues is small or nonexistent.

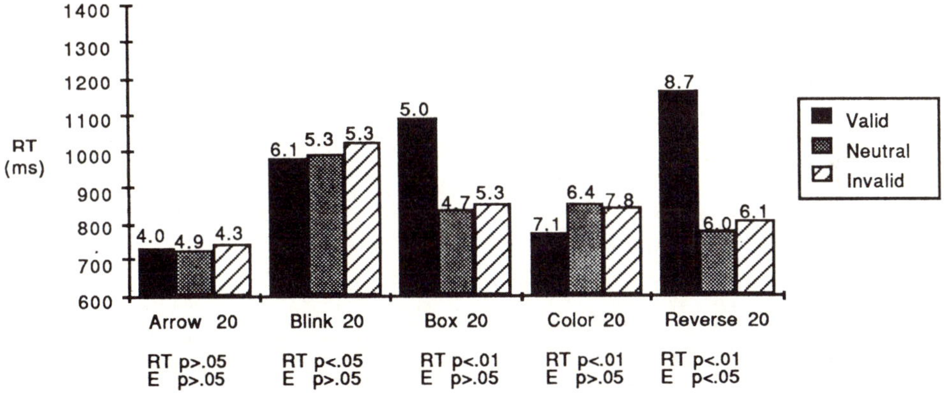

Fig. 3. 20/80 condition RTs by highlight and trial type. Mean errors are shown above the bars. Significance levels of trial type main effects for RT and errors (E) are shown below the graph.

The most striking effect found in the 20/80 data shown in Figure 3 is the large negative benefit for reverse video and boxing. This effect means that for valid trials, when the cue fell on the target location, subjects' responses were slowed. It is likely that the effect is due to the cue interfering with the subjects' ability to read the target number, that is, to perceptual rather than attentional causes. It is important for the system designer to realize that some of the highlighting techniques available affect not only the users' direction of attention, but also the users' ability to extract information from that location. Reverse video, which may cause reduced discriminability because brightness "splash" reduces the effective stroke width of the characters, seems to be particularly prone to perceptual degradation. Yet Engle and Granada (1975) in their Guidelines for Man/Display Interfaces recommend: "Provide maximum contrast of a highlighted item with a non-highlighted item. This seems best done with text by reversing the image of the item specified." It is undoubtedly perceptual degradation that counteracted any attentional benefit in the reverse video 80/20 condition. In this case, use of highlighting produced only cost. Putting a box around the item also apparently causes some perceptual degradation. For boxing, the problem may be due to lateral masking or contrast effects.

A particularly interesting finding in the 20/80 condition is the relatively small (80 ms) but statistically significant benefit of color with no associated cost. This finding cannot be due to a strategic factor of search order or there would be a cost to non-highlighted items on invalid trials. If this finding is due to attentional factors, it would seem to be analogous to Posner and Snyder's (1975) concept of pathway activation, although it would be difficult to apply that concept directly in this context. This finding would certainly be of both practical and theoretical interest if due to attentional factors. One far less interesting explanation for the finding is that the digits displayed in the highlighted red color may have simply been easier to read due to perceptual factors such as contrast or brightness. No attempt had been made to equate these factors for the red and green colors used. To assess whether perceptual factors could account for the finding, two supplementary experiments were conducted.

SUPPLEMENTARY EXPERIMENTS

In the first supplementary experiment a neutral, no-highlight condition was used throughout. A single group of 10 subjects first received a block of 20 trails in the light green color used in prior experiments and then, a block of 20 trials with all items in the red color formerly used only for the highlighting cue. This sequence was repeated one more time for a total of 40 practice trials. The 200 experimental trials consisted of repeating the sequence five more times. The RT means are shown in the left panel of Figure 4. An analysis of variance indicated that color was a significant factor on RT, $F(1,9) = 6.7$, $p < .01$. There was no significant effect of color on the error scores, $F(1,9) = 1.5$, $p > .05$.

The superiority of the neutral red over the neutral green indicates that it might have caused some of the original 20/80 color highlight benefit. However, the size of that effect was 80 ms compared to the 31 ms found here. Hence, there is still a possibility that the original finding was not due solely to perceptual factors.

In a second supplementary experiment the original 20/80 color highlight experiment was repeated except that the colors were reversed. That is, light green was used to highlight against a background of red numbers. If perceptual factors account for the original finding, then invalid and neutral trials should be faster than valid trials. If attentional factors account for the finding, then some benefit may be evident. The results are shown in the right panel of Figure 4.

The RT means for the valid, neutral, and invalid conditions are respectively: 848, 895, and 894 ms. An analysis of variance on the RT data indicates that trial type was a significant main effect, F (2,22)= 12.6, $p < .01$. The effect of trial type on errors was nonsignificant, F (2,22)= 0.82, $p > .05$. The finding that benefit without cost is obtained when the highlighting color is green on a red background as well as red on a green background clearly supports an attentional explanation. Apparently when subjects attempt to ignore the color highlighting cue, they still obtain some benefit from it and this benefit does not come at a cost to invalid trials and

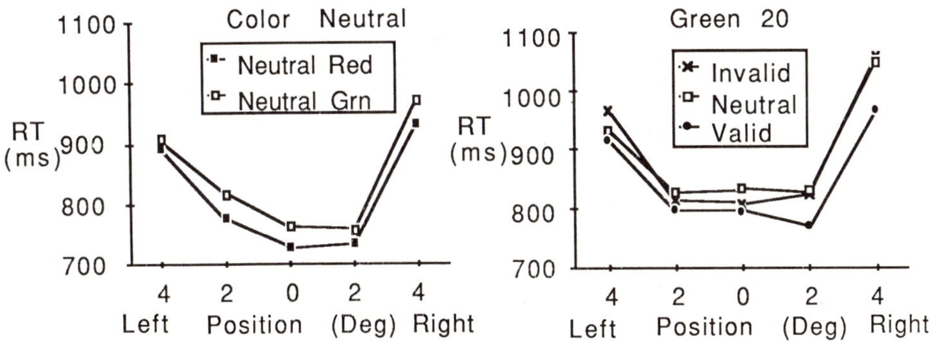

Fig. 4. RT data for supplementary experiments as a function of trial type and position.

thus, cannot be attributed to strategic factors.

At this point we can only speculate as to the processes that might lead to this pattern of results for color highlighting. Jonides (1983) has proposed a two-process model that might help explain our findings. It provides for two alternative modes of attention allocation. Eriksen and Yeh (1985) have recently found support for the model in a visual search task somewhat similar to the one used here. However, this model was proposed to account for data collected in experiments having a pre-cue, tachistoscopic presentation, and smaller angular displacements. It may not be directly applicable to our paradigm. The model suggests that attentional resources can be uniformly distributed over display locations in which case all are simultaneously processed in parallel. Alternatively, all resources can be concentrated in a circumscribed area leading to greater perceptual detail and faster processing rates. The system is flexible and it can switch back and forth between modes. Within the context of our experiment, it is possible to conceive of a system that simultaneously operates in both of these modes. While the subject is attempting to process all of the locations in a parallel fashion, the color highlighting cue might allow a certain amount of automatic (resource free) focusing of attention on the highlighted location. We are currently investigating the boundaries of the color highlighting effect to determine whether such a model can account for the findings.

CONCLUSIONS

1. A modified cost-benefit procedure can be used to assess the effectiveness of various highlighting cues.
2. Highlighting with a red color provides a large benefit compared to the other cues tested.
3. The cues tested here were not particularly distracting to the user when task irrelevant.
4. Highlighting factors that affect perceptual discriminablity should be considered when choosing highlighting cues, especially the potentially adverse effects of reverse video and boxing.
5. Color may provide a highlighting benefit without cost even if the user does not consciously try to use it.

REFERENCES

Carter, R.C., 1982. Visual search with color. J. of Exp. Psych.: Human Perception and Performance, 5: 127-136.

Christ, R.E., 1975. Review and analysis of color coding research for visual displays. Human Factors, 17: 542-570.

Engel, S.E., and Granda, R.E., 1975. Guidelines for Man/Display Interfaces, (TR 00.2720). IBM, Poughkeepsie Laboratory, Poughkeepsie, New York.

Eriksen, C.W., and Yeh, Y., 1985. Allocation of attention in the visual field. J. of Exp. Psych.: Human Perception and Performance, 11: 583-597.

Hemingway, P.W., Kubala, A.L., and Chastain, G.D., 1979. Study of Symbology for Automated Graphic Displays. US Army Research Institute Tech. Rep. 79-A18, Alexandria, Virginia.

Jonides, J., 1983. Further toward a model of the mind's eye's movement. Bulletin of the Psychonomic Soc., 21: 247-250.

Jonides, J., and Mack, R., 1984. On the cost and benefit of cost and benefit. Psychological Bulletin, 96: 29-44.

Knapp, B.G., Moses, F.L., and Gellman, L.H., 1982. Information highlighting on complex displays. In: A. Badre and B. Schneiderman (Editors), Directions in Human/Computer Interaction. Ablex, Norwood, New Jersey, pp. 195-215.

Morse, A., 1979. Some pricipIes for the effective display of data. Computer Graphics, 13: 94-101.

Parish, R.N., Gates, J.L., and Munger, S.J., 1981. Design Guidelines and Criteria for User/Operator Transactions with Battlefield Automated Systems, Volume IV: Provisional Guidlines and Criteria for the Design of User/Operator Transactions (Contract No. MDA 903-80-C-0094). US Army Research Institute for the Behavioral and Social Sciences, Alexandria, Virginia.

Posner, M.I., and Snyder, C.R.R., 1975. Facilitation and inhibition in the processing of signals. In: P.M.A. Rabitt and S. Dornic (Editors), Attention and Performance V. Academic Press, New York, pp. 669-682.

Smith, S.L., and Goodwin, N.C., 1971. Blink coding for information display. Human Factors, 13: 283-290.

Stewert, T.F.M., 1976. Displays and the software interface. Applied Ergonomics, 7: 137-146.

Tullis, T.S., 1981. An evaluation of alphanumeric, graphic, and color information displays. Human Factors, 23: 541-550.

Yantis, S., and Jonides, J., 1984. Abrupt visual onsets and selective attention: Evidence from visual search. J. of Exp. Psych.: Human Perception and Performance, 10: 601-621.

Social, Ergonomic and Stress Aspects of Work with Computers,
edited by G. Salvendy, S.L. Sauter and J.J. Hurrell, Jr.
Elsevier Science Publishers B.V., Amsterdam, 1987 — Printed in The Netherlands

ICONS IN EVERYDAY LIFE

WILLIAM T. WOOD and SUSAN K. WOOD
The Quality Group, 64 Windsor Street, Arlington MA 02174 (U.S.A.)

ABSTRACT

In this paper, we define the term "icon" as any symbol, image or pictograph used to represent a concept, idea or physical object. In this context, we can see that icons have been used for generations in familiar situations. The question is: How can we, as designers of graphic user interfaces, learn from this development and profit from the generations of experience with these everyday icons? This paper looks at three examples of iconic systems. First, we examine the store front signs which, in medieval times, served as a universal language. In our second example, road and highway signs, we examine the concepts of meta-icons, iconic sentences, and present ten criteria for evaluating an icon. Substantial human factor research has been done in this area which the computer designer is well advised to study. As a final example, we introduce some basic concepts from the Chinese language.

1 INTRODUCTION

"Icon", as a term associated with computer systems, was first suggested by Alan Kay and used by David Smith in "Pygmalion", his graphic programming language [Smith 1977]. For the purposes of this paper we would like to take a broader view of icons to include any graphic symbol or image that represents anything in the computer. In this sense, an hourglass symbol used to indicate that the CPU is occupied would be considered an icon for the concept "please wait".

Our working definition for an icon will be:

a symbol, glyph or pictograph used to represent a concept, idea, or object.

With this definition, we see that icons have been used for centuries in everyday situations. Road signs, mathematics, chemistry, map symbols, wash and wear instructions, and advertising logos are all icons that we deal with every day (see Figure 1).

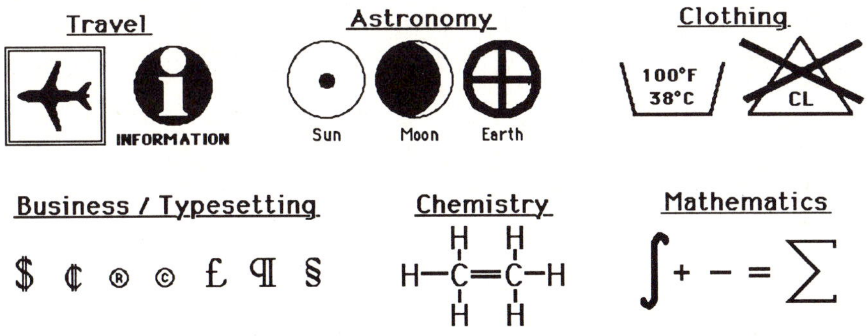

Fig. 1. Examples of "icons", or symbols, that we deal with every day.

This paper focuses on a small subset of the icons we face in our lives. In particular, we look at the pre-neon store front, which, while not the first recorded use of symbols, probably is the most direct cognitive ancestor of the computer-based icon. Our next example comes from the world of highway signs. The life-saving benefits of good road signs has led to a considerable store of human factors research into the effective design and use of symbols. This is research that we, as designers of computer icons and graphic interfaces should not ignore. Our final example comes from one pictographic language, Chinese. Chinese characters are one of the most highly developed symbol languages. A knowledge of some basic characteristics of the Chinese language is virtually essential to any researcher interested in symbols.

The icons that exist in everyday life have centuries of development behind them. The question is: How can we, as designers of graphic user interfaces, learn from this development and profit from the generations of experience with these everyday icons?

2 LESSONS FOR THE COMPUTER SYSTEM DESIGNER

The realization that icons have a centuries old history can be more than a cocktail party fact. As designers of graphic interfaces, we can use this knowledge to create recognizable, effective, and versatile icon systems.

2.1 Lessons from Medieval Signs

The hanging store sign provided a universal language in centuries past. Foreign travellers, illiterates, and locals could all understand a merchant's symbol. A peasant entering a strange town, faced with a row of shops all of which looked about the same, could accomplish the task at hand by recognizing the hanging shoe to mean "shoe repair", or the hanging anvil to mean "blacksmith".

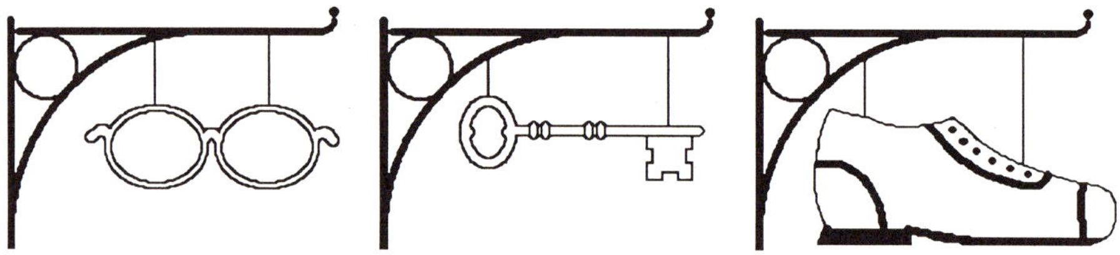

Fig. 2. In the days before neon was invented, store keepers would use three dimensional symbols to indicate their profession.

A graphic user interface should be similarly effective across nationalities and languages. In addition, computer "illiterates" and "foreigners" unfamiliar with a specific program or operating system can use icon-based systems without the training needed to learn complex command sequences.

2.2 Lessons from Road Signs

2.2.1 Meta-icons: Shape and Color May Convey Meaning.

Road signs convey information through the symbols on the sign, the color of the sign, and the shape of the sign. The color and shape represent classes of signs (e.g. yellow diamond = warning; blue rectangle = information). These classes can be considered meta-icons. (See Figure 3.)

For example, the yellow diamond meta-icon informs the motorist that there is danger ahead. The specific reason for the warning is held in the pictograph shown in the diamond. Even if the motorist cannot recognize the pictograph, the meta-icon (shape and color) conveys information, and will probably evoke the desired response of caution.

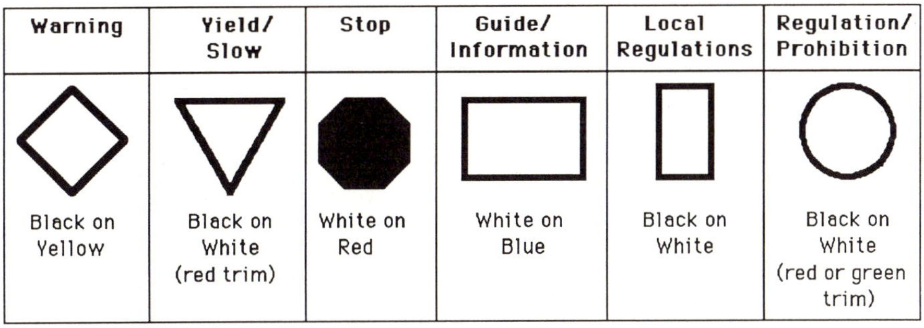

Fig. 3 The shapes and color for selected meta-icons used on U.S. roadways.

Color is one area that distinguishes road signs from computer icons. Road signs make generous use of colors to convey meaning. Bit-mapped computer displays have historically been in black and white. When color becomes more prevalent, computer icons could all be multicolored. The experience with road signs indicates that each icon should use only one or two colors, using the colors to distinguish *between* the icons.

2.2.2 Iconic Sentences: Combining Icons Conveys Complex Meanings. Another feature of road signs is the ability to combine symbol elements into complicated sentences. Figure 4 shows two examples where simple elements are combined into a complex concept. Research has shown that a motorist's cognitive process involves an actual parsing of these signs into their basic elements [Dewar 1976], rather than simple recognition of the sign as a single (albeit complex) image.

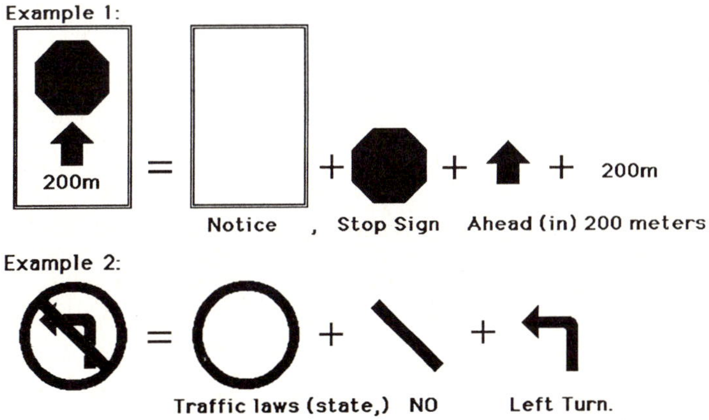

Fig. 4. Some examples of iconic sentences.

The important lesson here is that a good choice of basic icon elements in the design of a graphic interface can allow combinations that go beyond the simple semantics of these images.

2.2.3 Ten Criteria for Icon Evaluation. In 1962, the International Committee for Breaking the Language Barrier was founded to promote the use of symbols, including road signs, as an international language. This committee proposed ten useful criteria for the evaluation of symbols [Kato 1972]. We have paraphrased Mr. Kato's own description of these criteria below, as viewed from the perspective of human factors in computer icons.

(1) *Is it easy to associate the symbol with the message*? One class of icon is self-interpreting. These icons will contain *representational* symbols which present a fairly realistic picture of objects (such as a truck or a pedestrian). Representational symbols utilize a visual resemblance between symbol and meaning. *Abstract* symbols, which are often representational symbols reduced to their essential graphic features, also provide visual symbol-meaning association (when properly designed). Another class of symbols are *arbitrary* or invented and their meaning must be learned. Mathematics symbols are one example of arbitrary symbols. The "Do Not Enter" sign in Figure 5 shows one method that can be used to help people associate meaning with an abstract symbol: combining the arbitrary icon with words. Arbitrary symbols should not be considered as poor designs even though they require user training. These symbols are chosen to be as unique or compact as possible. Along highways, arbitrary symbols are used where misinterpretation would be disastrous: as in the red octagon STOP symbol.

(2) *If the symbol is to be used in a variety of cultures and situations, is it equally appropriate to them all?* Consider the familiar Red Cross symbol for medical facilities, which may be inappropriate for Moslem and Jewish populations.

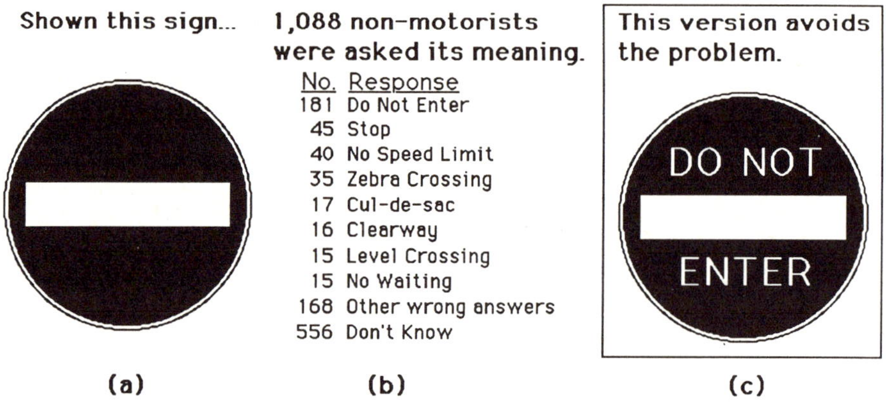

Fig. 5 . a) The abstract symbol used to represent the common concept "DO NOT ENTER". The symbol is very distinct from other road signs, however, the symbol cannot be self-interpreted, as the results in (b) show. The modification shown in (c) provides an explanation for motorists unfamiliar with the sign. (Data from [Mackie 1972].)

(3) *Will the symbol be appropriate in the future?* The world is rapidly changing and we must beware of developing icons that rely on a metaphor that may disappear. Many of the symbols developed for flowcharting (such as the punchcard symbol for input) are not relevant for the new generation of programmers and computer users.

(4) *Is the symbol pleasing and noncontroversial?* For example, the use of exclusively male images whenever an icon deals with people should be avoided.

(5) *Is the symbol in accordance with existing international standard symbols?* In other words, don't make new symbols if one already exists. Using the international "INFORMATION" symbol

(shown in Figure 1) as the computer icon for "HELP" is one possible application of this.

(6) *Can the symbol, or its elements, be applied systematically for a variety of interrelated concepts?* Good choice of symbol elements can leave a designer with a rich symbolic language. Simple and universal elements can be combined to form more complex symbols. Examples from Chinese language (Figures 7 and 8) illustrate simple elements which can be combined to express hundreds of related concepts.

(7) *Is the symbol easy to reproduce in a variety of environments and situations?* Symbols should transfer easily to different systems with different screen resolutions and graphic capabilities. Even on a single system, symbols should be capable of being enlarged or reduced without loosing their crispness or detail. When using colors as meta-icons, we must be particularly attentive to their effectiveness on various systems.

(8) *Is the symbol distinguishable from other symbols?* If two symbols have different semantics we want them to be syntactically different. An interesting illustration is provided by the First Aid and Crossroad signs shown in Figure 6. The basic symbol element, the cross, is a very ambiguous symbol with a variety of religious and secular interpretations. Yet the two signs shown are easy to differentiate because of the context provided by the color and shape of the signs.

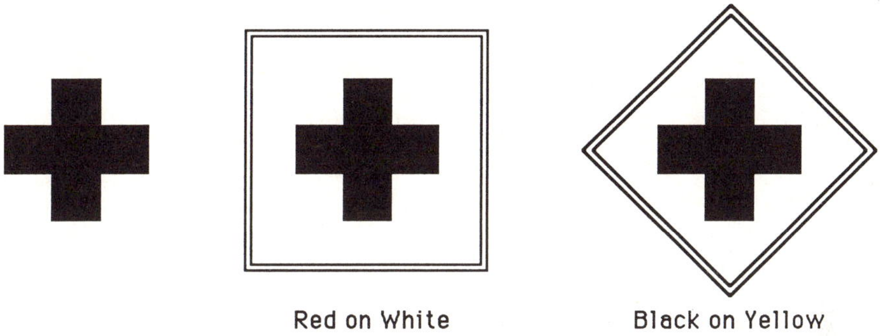

Fig. 6. The cross on a sign has many possible meanings, such as "First Aid" and "Crossroads ahead". The semantic difference is conveyed by the meta-icons of sign shape and colorr.

The remainder of the rules relate to the physical nature of road signs, and do not apply to the computer world. However, they are included for the sake of completeness. (9) *Can the symbol be perceived from many different distances and angles, and in different conditions?*, and (10) *Does the symbol stand up to vandalism?*

2.3 Lessons from Chinese Pictographs

The process of transmitting meaning through stylized pictures has been developed over millennia in such languages as Mandarin Chinese. The earliest Chinese characters were pictographic and bore a clear visual resemblance to the object or idea being expressed. Figure 7 presents Shang dynasty (1500 B.C.) characters which are pictographic and indicative of their meaning. These characters have changed only slightly in 3500 years.

In modern Chinese, the simplest character, a single stroke means "one" (Figure 7). The most complex character, with 64 strokes, also has some relation to its meaning: "verbose".

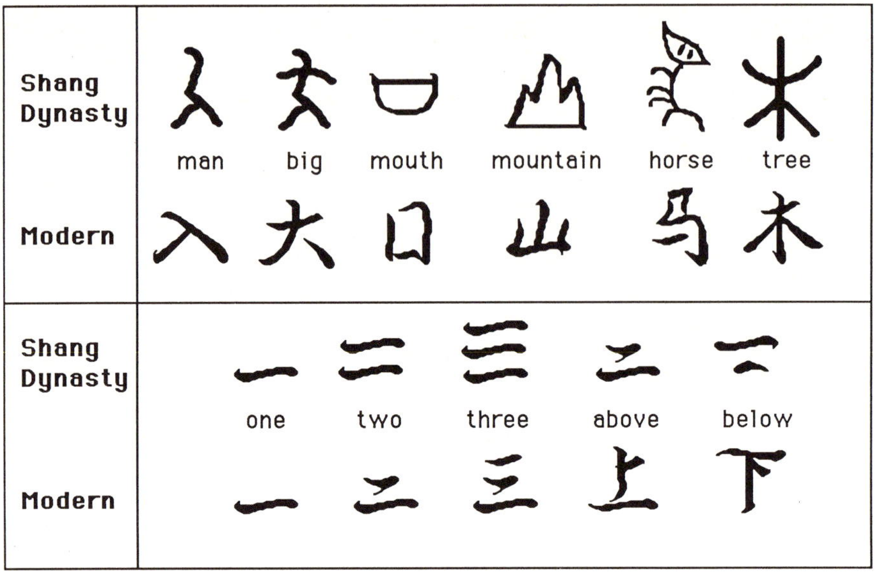

Fig. 7. Shang dynasty and modern representations of pictographic Chinese characters.

<u>2.3.1 Icons Can Represent Nouns and Verbs</u> The Chinese language includes characters which represent complex, abstract concepts. In some cases, the character may convey the image of the verb. One example of this is the character for "crying", which might be construed as large eyes and even a tear above the character for "human" [Huang 1983]. However, most verbs and abstract concepts bear little or no resemblance, visually, to their meaning. Their meaning is learned, semantic, rather than pictographic.

Computer systems also utilize icons to represent complex functions. The hourglass, watch, and coffee cup have been used to indicate "CPU is busy processing". However, we should be cautious about proliferating new icons which will be as difficult to learn as new words.

<u>2.3.2 Combining Characters Creates New Meanings</u>. By combining familiar characters, new meanings are created in Chinese. The character for tree, repeated three times, becomes a new character meaning "forest" [Kato 1972]. Combining the character for sun with that for moon , creates a new character which means "bright". (See Figure 8.) Similarly, "yellow" joined with "sickness" takes on the new meaning "jaundice" [DeFrancis 1984]. The character for "crisis" is the union of "danger" and "opportunity" [Huang 1983].

Fig. 8. Combining familiar characters creates new characters with new meanings.

In computer systems, familiar images may also be combined to convey new meanings. Figure 9 shows that a number of Macintosh icons utilize a standard shape to convey the meaning "application", often combined with small images which effectively convey "writing", "painting", or "drawn objects". Through this combination, the more complex meaning "writing application" is created by effectively combining a meta-icon with a detail symbol.

2.3.3 Pictographs Evolve to Become Semantic-Phonetic Words. Chinese characters have evolved from pictographs to semantic-phonetic words. Character meaning must be learned, and the image is less reminiscent of the idea or object . In artifacts from the Shang dynasty, 66% of the characters deciphered had a pictographic or indicative meaning. This contrasts markedly with the 18th century census where only 3% of the characters maintained a visual resemblance to their meaning. (Simultaneously, the number of characters catalogued, increased from 977 to 48,641. Thus, through the 3500 year evolution, the absolute number of pictographic icons remained fairly stable.) [DeFrancis 1984]

One sees the same evolution in computer system icons. Figure 9 shows a collection of increasingly arbitrary icons from the Macintosh. The icons on the right do not convey meaning through shape, or visual resemblance to their meaning. The user must learn that the LVT icon means "outlining application".

3 STANDARDS VS. UNIQUE DESIGNS

A major force that opposes any effort to standardize computer icons is the desire of many developers to be unique. Every designer wants his application to stand out, and quite frankly it is more fun to design a special set of icons without trying to imitate others. Often a company wants to reflect its corporate logo in its icons. There has also been an effort by some creative designers to copyright striking and innovative glyphs. It is the user who suffers most from the lack of conformity on individual computers and between computer systems.

An illustrative example is found in the Macintosh environment. As Figure 9 shows, the original designers proposed a framework of meta-icons to represent generic concepts such as "document" and "application". Many application developers have designed effective icons within this framework, as shown in the figure. However, other developers have created increasingly arbitrary icon designs, several of which are also shown. It is not even clear whether some of these icons represent documents or applications. Even an experienced Macintosh user can get confused when trying to deal with large numbers of arbitrary icons.

The world of computers does not have the legislated standardization imposed on the highway departments of the world. When a user misinterprets an icon a file may be lost, but when a motorist misinterprets a sign, lives are at risk. The risks involved have dictated the use of uniform standards in the world of highway signs: the more dangerous the situation, the more widely standardized the sign representations. For example, there is much more diversity in airport direction signs than in STOP signs.

4 CONCLUSIONS

We have identified a number of ways in which the lessons from generations of experience with familiar icons may be applied in the design of computer icons. Effective computer icons enable the naive user to successfully accomplish tasks, just as the illiterate traveler could in medieval times. We have demonstrated, using examples of road signs, that shape and color as meta-icons can convey meaning very effectively. Standardization of meta-icons can make applications on one system easier to recognize, or make "foreign" systems seem familiar to a new user. Complex meanings may be created by combining icons, as evidenced in road signs and Chinese characters. Standardization of meta-icons need not eliminate individuality, rather, it allows each unique icon to have more meaning to a wider population of users.

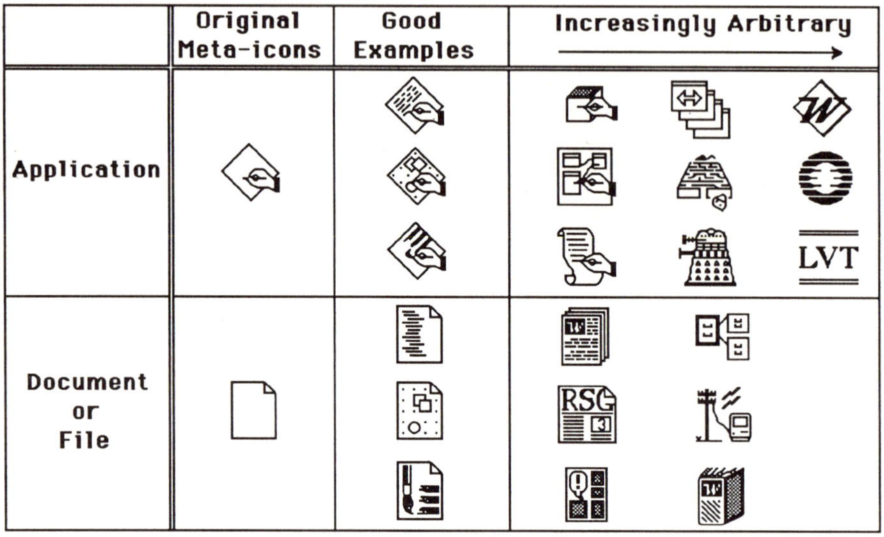

Fig. 9. Macintosh interface uses meta-icons for application and files.

5 REFERENCES

American Institute of Graphic Arts, 1974. *Symbol Signs.* (DOT-OS-40192) National Technical Information Service: Springfield, VA.

DeFrancis, John, 1984. *The Chinese Language: Fact and Fantasy*, University of Hawaii Press, Honolulu.

Dewar, Robert E., 1976. "The Slash Obscures the Symbol on Prohibitive Traffic Signs", *Human Factors*, Vol. 18, No. 3, pp. 253-258.

Dreyfus, Henry, 1972. *Symbol Sourcebook: An Authoritative Guide to International Graphic Symbols.*, McGraw-Hill Book Company, New York.

Huang, Chungliang Al, 1983. *Quantum Soup: A Philosophical Entertainment*, E.P. Dutton, Inc., New York.

Kato, Soichi, 1972. "A New Universal Language for the New Human Environment", *International Conference on Highway Sign Symbology*, International Road Federation and U.S. Department of Transportation, Washington, D.C., pp. 13-19.

Mackie, A.M. and Higgs, Marvis H., 1972. "Motorist's Understanding of the Meaning of Traffic Signs", *International Conference on Highway Sign Symbology*, International Road Federation and U.S. Department of Transportation, Washington, D.C., pp. 61-78.

Smith, D.C., Pygmalion, 1977. *A computer Program to Model and Simulate Creative Thoughtt,* Birkhauser Verlag, Basel and Stuttgart.

Social, Ergonomic and Stress Aspects of Work with Computers,
edited by G. Salvendy, S.L. Sauter and J.J. Hurrell, Jr.
Elsevier Science Publishers B.V., Amsterdam, 1987 — Printed in The Netherlands

SMALL WINDOWS ON A LARGE SYSTEM

DWIGHT P. MILLER

Sandia National Laboratories
Albuquerque, NM, U.S.A.

Abstract

The Particle Beam Fusion Accelerator II (PBFA-II) recently completed at Sandia National Laboratories is the most powerful pulsed-power accelerator in the free world. The facility comprises over 25 major subsystems that are controlled by a small operations crew through a computerized control/monitor system (C/MS), which operates or monitors over 1900 points in the various subsystems. The operators' primary interface with the C/MS is a console containing seven 19-inch color monitors with touch panels and other dedicated hardware. This paper discusses some unique problems associated with this human-computer-system interaction and describes several ergonomic design solutions. Application areas include color graphics displays, menu selection, soft-key coding, command structure, and approaches taken toward manual, semi-automatic, and automatic control modes.

Introduction

Completed and initially fired in December 1985, the Particle Beam Fusion Accelerator II (PBFA-II) is designed to deliver over 100 trillion watts on target, making it the only driver in DOE's Inertial Confinement Fusion research program theoretically capable of achieving ignition in the laboratory. The $48-million facility has 36 identical modules arranged like spokes of a wheel around a focal target section. When fired, high-energy pulses propagate inward to a diode which converts the energy into lithium ions and focuses the resultant beam on a target. Simultaneity of convergence is critical to acquiring the power density required to ignite a target. The timing is done on nanosecond time scales while power densities are measured in tens of terawatts per cm^2. Although current research is directed toward producing lithium ion beams, target experiments are scheduled to begin in the late 1980's-early 1990's.

The PBFA-II facility currently operates at about one experiment (shot) per week. This rapid pace demands close coordination among the scientists planning the shots, the engineers manipulating the hardware and collecting the data, and the technicians who prepare and operate the machine. Although shot preparation involves scores of people and takes three or four days, the actual charging and firing requires only three or four technicians and takes only about an hour. It is these technicians for whom the user interface of the Control/Monitor System (C/MS) was designed.

Efficient user-system interaction is important since one human-induced system error could delay the shot schedule (at best) or lead to serious equipment damage or personal injury (at worst). Unlike many papers in this field which describe the ergonomic design process, this article presents the results of a multi-year design effort with the intent that others may benefit from utilizing techniques appropriate to their own applications.

Control Room

Due to the intense electromagnetic radiation produced during firing, the C/MS electronic equipment resides in a windowless, RF-isolated control room. The equipment was arranged according to ergonomic criteria using an operational concept and considering frequency of use, functional relatedness, access for maintenance, lighting requirements, noise reduction, and traffic patterns. Safety considerations dictate that the highbay, where the accelerator resides, be evacuated during final preparation and firing such that any human-system interaction must be performed remotely via hardware in the control room. The primary human-computer-system interface is the operations console which accommodates two operators who directly control all non-automatic aspects of charging and firing the accelerator. Their supervisor coordinates their actions with verbal direction from an elevated, seated position directly behind them. This location affords good visibility for the shot coordinator and the "hands-off" mode allows him to supervise other activities and tend to procedural checklists. Although occasionally VIPs will be allowed to spectate during a shot, the general rule requires all nonessential personnel observe control-room activities via closed-circuit television.

Operations console

The console, designed from available commercial hardware, consists of four full-height, 19-inch bays attached in a straight line. Wedge-shaped turrets were utilized to position the installed hardware optimally for both reach and visibility considerations. Seven of the eight large portals are occupied by Conrac 19-inch color graphics monitors (Model 7211) with integrated Carroll infrared touch panels. These monitors are the windows to which the title of this paper refers. The C/MS operates through a Hewlett Packard 1000, series-900 host computer with two series-600 satellite computers driving the color graphics. The remainder of the console space is taken up by dedicated hardware including communications equipment, surveillance camera controls and monitors, access-control keyswitches, and emergency manual-override controls.

Graphics Displays and Controls

Operational control of the accelerator can currently be performed using the touch-interactive displays in two modes--remote-manual, and semi-automatic. Remote-manual operation involves the direct manipulation of individual subsystem components or control points by means of a menu-driven hierarchy of display pages. The three levels represent choice of subsystem, groups within subsystems, and individual points within groups. In the semi-automatic mode, the C/MS controls entire operations, ensures conformity to operational interlocks, and performs critical timing functions. This mode allows the operators parallel access to displays representing four phases of operation (Preparation, Charge, Fire, Abort), an alarm summary, and access control information. Although an automatic mode is planned for future implementation, its display designs are discussed below along with those currently being used. First, however, techniques generic to displays in all three modes will be described.

Menu grammar and phrasing

The command syntax follows the object-function selection sequence and allows only valid object-function pairing to occur. When quantitative values are required, a SET key is pressed prior to object selection and a keypad appears on the screen for input. When the operator is satisfied with the displayed value, he presses EXECUTE to send the command. These keys, as well as other standard functions (CANCEL), and display selection keys (PREVIOUS, NEXT) reside within a dedicated strip at the bottom of every screen. Due to the large number of system elements and the potentially severe consequences associated with erroneous commands, a typeahead option was not employed.

The names of menu items reflect the adopted terminology of the operations crew. Redundancies in menus, such as "system" are avoided and the wording is kept concise in order to reduce display clutter and visual search time. Abbreviated words are avoided and whenever possible, related items are grouped within a menu.

Soft-key techniques

Since location is such a powerful memory cue for finding a desired item, soft keys are displayed whenever space allows. However, both shape coding and fill coding are used to differentiate the opportunities. While fill coding identifies the availability of key, shape coding identifies key type. Figure 1 demonstrates the coding used for invalid, valid, and latched keys. In addition to an instantaneous change of fill, the touch panel emits an audio tone as feedback whenever a valid key is properly hit. A lower-pitched, longer tone signals an invalid touch.

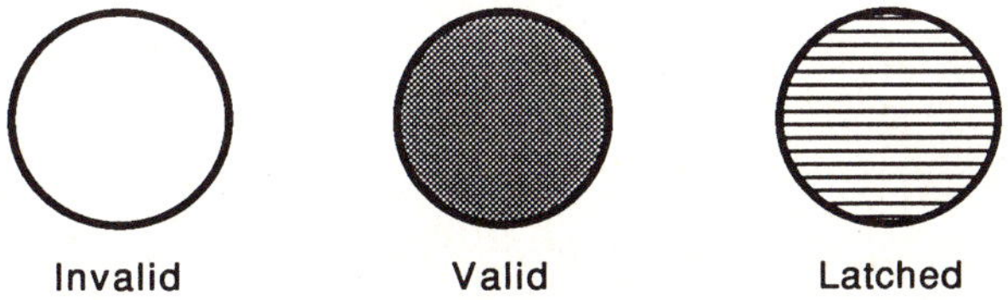

Figure 1. Soft-key fill coding.

Three geometric shapes are used to differentiate key types (Figure 2). Circles designate direct-action keys. Hexagons identify selection keys which are used to choose objects (latch) prior to issuing a command. Squares are used to call up other displays. Protected keys, requiring two hits, are displayed within a rectangle which disappears after the first hit. The assignment of shapes was intended to follow population stereotypes, if any exist.

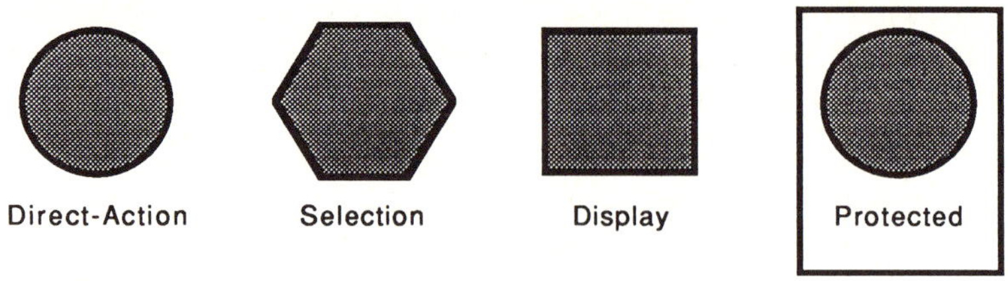

Figure 2. Soft-key shape coding.

Operators found that long menus with vertically stacked soft keys led to inadvertent hits below the desired key due to trailing fingers. Increased spacing would have helped, but at the expense of menu choices. Instead, horizontal staggering was used to vertically misalign successive menu choices. This technique had the added benefit of making menu choices easier to locate within the column.

Color coding and highlighting

Despite the technological capability of displaying hundreds of colors, only a few carefully-chosen hues are actually used. Only four colors are used to imply meaning, and three of these abide the traditional ergonomic color-coding scheme--*green* for satisfactory, *yellow* for caution, and *red* for alarm. Since three priorities of alarm are defined, *orange* was selected to indicate the intermediate level between red and yellow. *White* and *cyan* are used to indicate status but are neutral in terms of meaning. Reversed video, flashing, and character size are also employed to differentiate or highlight text.

Alarm handling

Alarms and faults were assigned one of three priorities based on the severity of the consequences or urgency of response. Priority-one alarms are characterized as potentially catastrophic and are coded in red. This category is displayed in a dedicated location on all displays during operations and requires acknowledgment and resolution before operations may proceed. These tasks are performed on a specialized interactive display which lists all the alarms in chronological order, as well as their priority and status. Priority-two alarms are considered critical and are coded in orange. Cautionary alarms and advisements are defined as priority-three and do not interfere with the progress of charging and firing the accelerator. All alarms are recorded using a line printer for post-shot analysis.

Computer-automated tasks

Algorithms have been written for computer control of selected operational tasks. In the semi-automatic mode the operator commands the C/MS to perform an operation and the computer coordinates the constituent tasks. For instance, if an operator wants the water section filled, the computer will open the appropriate valves, start the pumps, and monitor the water level until it reaches the target height. In the future, the automatic mode will coordinate several of these operations simultaneously.

Regardless of mode, the computer performs other tasks that require accurate timing and split-second decisions. Once the accelerator's power supplies are charged, the strategy is to fire the machine as quickly as possible to minimize equipment wear and avoid prefires. If a charge imbalance or other priority-one fault is encountered during this stage, the computer decides in milliseconds to abort the shot and safely shunt the charge to a dummy load, thus avoiding a wasted shot or damaged equipment.

Remote-manual displays

The remote-manual displays are essentially a hierarchical set of menus which, in addition to selections and control functions, include status information concerning the subsystems or control points. This information is displayed as color-coded bars beneath the subsystem name (Figure 3). The three positions are assigned to current subsystem control mode, recorder status, and fault status from left to right. When all conditions are correct for the particular kind of shot, all the mode flags will be green, and no fault flags will be present.

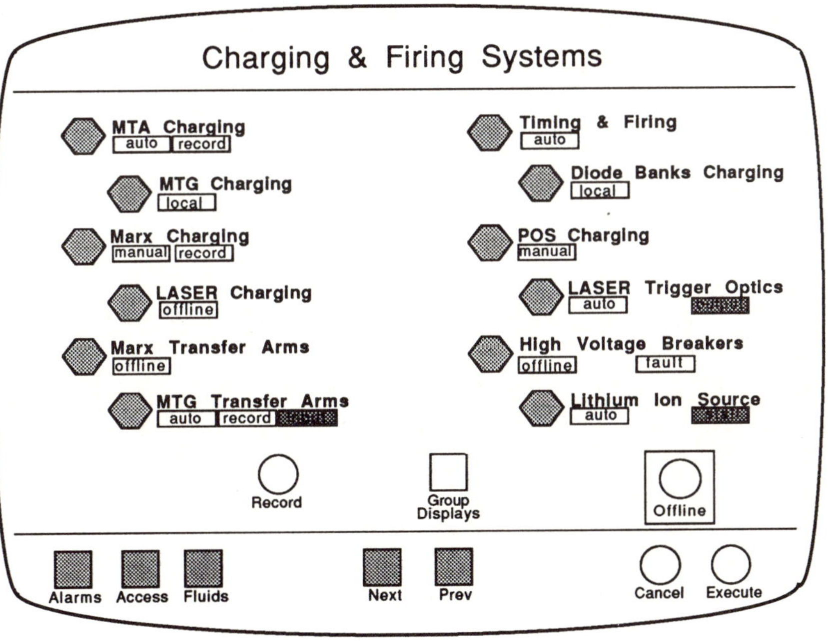

Figure 3. Remote-manual system menu.

Semi-automatic displays

The semi-automatic displays are oriented more toward supervisory control and consequently display the progress of semi-automatic operations within each subsystem. Unlike the remote-manual displays, these have no hierarchical depth, and are displayed concurrently. Each of the four screens shows six subsystems (Figure 4). Due to the modular approach taken in designing each subsystem display, different combinations can be used to create customized screens for the various kinds of shots. Although each module is different, all follow a similar format. Just below the subsystem name is the status bar, followed by analog and digital quantitative graphics in the center. Valid soft keys are displayed either at the bottom or right-hand side, depending on whether the bar graphs are electrical properties (horizontal), or fluid quantities (vertical). In keeping with the greenboard approach, the graphical fill colors are either blue, indicating an operation still in progress, or green, indicating completion. If a process has not yet begun, the graphic remains unfilled. Color-coded alarms pop up when appropriate in a dedicated strip just below the status bar.

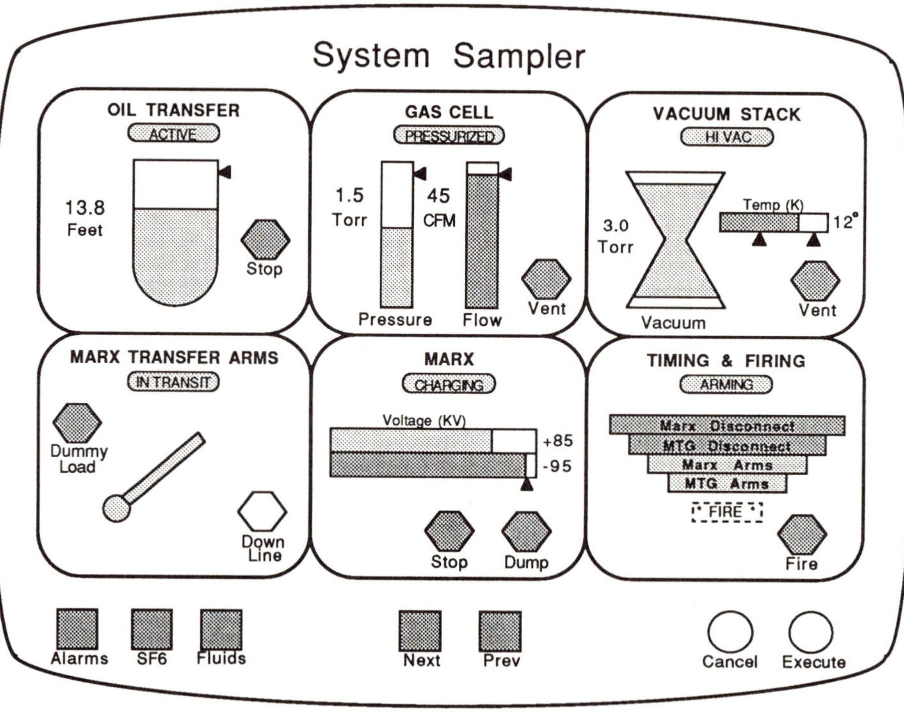

Figure 4. Mixed sample of semi-automatic modules.

Automatic displays

Since the C/MS performs most operations in the automatic mode, the operators essentially monitor the progress of major preparation activities. Using themes from the manual and semi-automatic displays, the automatic mode shows less detail and condenses many subsystems into a single display. Except for specialty displays dealing with access control and alarms, only two displays are necessary to operate the accelerator in the automatic mode. One shows the relative state of completion of all major preparation tasks. The other is used during charging and firing (Figure 5). In addition to the overall status display and the control soft keys on the right-hand side, status indicators for the systems critical to charging and firing occupy the central portion of the display. As each system reaches its ready state, its associated rectangle turns from blue to green in the upper triangular block. When this block turns entirely green, the machine fires, and the white FIRED rectangle appears in the gap. As quick-look diagnostics are completed, the lower triangular block fills with green. If a prefire or misfire occurs, the anomaly is indicated with a pop-up red flag adjacent to the FIRED rectangle.

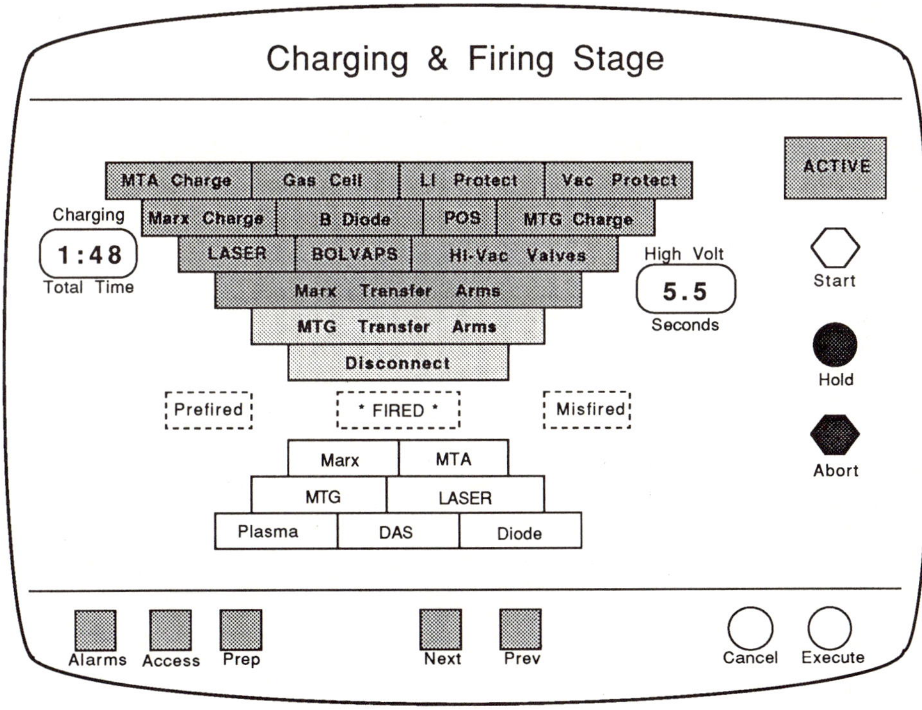

Figure 5. Automatic charging and firing display.

Two elapsed-time displays are positioned on either side of the screen. The one on the left displays elapsed time since the first power supplies in the charging sequence received charging current. The right-hand display indicates how long the entire system has been at full charge prior to firing. This display is updated only at half-second intervals so as not to overload the operators with a rapid cycling of illegible digits. A well coordinated shot would minimize both of these elapsed times.

In addition to relaying all the necessary information to the operators, the charging and firing display was designed to look like the hour glass-shaped center section of the accelerator. Note that the mimic diagram of the vacuum stack surrounding the center section is also shaped like an hour glass (Figure 4). The accumulation of green rectangles above the gap emulates the build-up of enormous potential energy awaiting release. And finally, the activation of the FIRED box represents the great white flash which occurs at T_{zero}.

In conclusion, the PBFA-II research program has enjoyed many benefits as a result of early human factors involvement. Its control room is more habitable, and its human-computer-system interface is easier to understand and operate. The technicians who know the subsystem hardware can relate to the graphical shapes, recognizable terminology, and consistent coding incorporated into the computer-driven interface. These concepts increase the transparency of the computer in the human-computer-system interface and should result in fewer operator errors and an enhancement of overall operational efficiency. Future human factors efforts will involve improving experimental strategies and developing means of displaying complex shot performance data in more comprehensible formats.

Bibliography

Galitz, W.O., 1985. Handbook of Screen Format Design. QED Information Sciences, Wellesley.

Miller, D.P., 1985. Human Factors Guidelines for PBFA-II CRT Displays. Sandia National Laboratories Internal Memorandum.

Miller, D.P., 1981. The Depth/Breadth Tradeoff in Hierarchical Computer Menus, Proceedings of the Human Factors Society 25th Annual Meeting, Rochester, NY. Human Factors Society, Santa Monica.

Shneiderman, B., 1987. Designing the User Interface: Strategies for Effective Human-Computer Interaction. Addison-Wesley, Reading.

Tufte, E.R., 1983. The Visual Display of Quantitative Information. Graphics Press, Cheshire.

VanDevender, P., 1985. Ion-Beam Focusing: A Step Toward Fusion, Sandia Technology, Vol. 9, No. 4, Sandia National Laboratories, Albuquerque.

Social, Ergonomic and Stress Aspects of Work with Computers,
edited by G. Salvendy, S.L. Sauter and J.J. Hurrell, Jr.
Elsevier Science Publishers B.V., Amsterdam, 1987 — Printed in The Netherlands

IS BIGGER BETTER?
THE EFFECTS OF DISPLAY SIZE ON PROGRAM READING

J. F. REISEL and B. SHNEIDERMAN
Department of Computer Science, University of Maryland, College Park, Maryland 20742.

ABSTRACT
An experiment was conducted in which four window sizes (10, 22, 60, and 120 lines) were compared for their effectiveness for program reading. The largest window, in which the entire program could be seen at one time, was found significantly better, in both time to complete the reading task and the number of correct answers in the reading task. Subjects also preferred the larger windows to the smaller windows.

1. Introduction

The traditional computer screen is small. Measuring less than twelve inches diagonally, it holds no more than 24 lines of 80 characters. This modest size has been proposed as one of the reasons why reading from computer screens has been less satisfactory than reading from paper. However, new technology has permitted the development of larger screens capable of holding many windows equal in size to the traditional screen or one window much larger than the traditional size. Examples of such screens include the Apollo or Sun workstation and the IBM 3290 Plasma Display. If programmers had bigger screens or multiple windows, would they be able to do their tasks, such as comprehension and debugging, faster or more accurately? Would they feel more satisfied? How much of the difference between screen and paper is due to the limited number of lines available on the screen?

Previous research on computer terminal screens has focused on two main areas: formatted displays and text reading. Formatted displays are defined as displays in which the display is selectively searched for information rather than read continuously. Formatted displays differ from programs in several ways. First, designers of formatted displays can lay them out to enhance readability, an option not necessarily available to the programmer, whose programs must adhere to a specified syntax. A related difference is that formatted displays require little interpretation while programs must be studied carefully to be understood. That means that the users of formatted displays do not have to do as much searching for information as do programmers. Finally, formatted displays are often meant to be used by a heterogeneous audience, while programs are generally read only by programmers, an audience with more skills and knowledge.

The second area of relevant research has been in the reading of natural-language text on a computer display. Again, text differs from programs in several ways. First, text is free-format, in that every line is filled to capacity; sentences follow one after the other. Most programs are not written that way; each statement starts on a separate line of the screen; and, any space at the end of a line after a statement is left blank. Second, text is usually read serially; most readers start at the beginning and continue straight through to the end. It is fairly easy to understand text by reading that way. However, most programs are read in a series of jumps. The reader reads the data declarations, then skips to the main program, then jumps around from procedure to procedure as needed. Third, text is composed mainly of alphabetic characters, while programs have many special symbols and operators with special meanings that must be understood.

Haas and Hayes [1985] found that text reordering could be done significantly faster on a 50-line screen than on a 24-line screen, and that reading comprehension scores were higher on a 50-line screen than on a 24-line screen. They found no difference between screen sizes when the task was proofreading. However, their 50-line screen had more differences with the 24-line screen than just the size. Duchnicky and Kolers [1983], studying 1-line, 2-line, 3-line, 4-line, and 20-line screens, found that text on the 4-line and 20-line screens was read faster than text on the 1-line and 2-line screens. No other comparisons were found significant. Based on these two studies, we can roughly group these screen sizes as 1-2, 3, 4-20-24, and 50, in increasing order of suitability for reading text.

Finally, a small body of experimental results has evolved that actually examines the specific issue of reading programs when more lines are available. Weldon, Norman, and Shneiderman [1986] found no difference in program comprehension between 24-line and 48-line displays when the 48-line display was split between 2 24-line screens. It should be pointed out that this experiment was not designed to study the issue of larger display sizes directly, and their lack of results may be due to the fact that their experiment did not focus on the relevant factors.

The first study to look specifically at the issue of program reading on larger screens was performed by Don Preuss and Phil Shafer at the University of Maryland in 1984. They used the IBM 3290 Plasma Display in three configurations: one window of 22 lines, a second window of 60 lines, and a third window of 120 lines. They tested 21 novice computer users (7 per cell) on program comprehension by seeing how many questions could be correctly answered in a fixed time. They used two programs, both of which required the user to scroll the program regardless of the window size.

Their results found no significant difference between the window sizes. They attributed their finding to two facts: the familiarity of the 22-line window to the novice users and the clutter on the screen caused by the larger window sizes. However, there were other factors that may explain their result. First, their experiment fixed the total time for each subject and counted the number of correct answers to a set of questions. Their questions contained very difficult questions intermixed with very easy questions. There was no way for a subject who got stuck on one difficult question to go on to an easier one. These two conditions made it quite likely that their subjects would get stuck early in the experiment, with no way to advance to the next question. Since the scores for the experiment were extremely low, it is reasonable to assume that that is what happened in many cases. A related concern is that the users were novice computer science students. It is quite possible that the programs being studied and the questions being asked were too difficult for the novice users to answer. A third factor is that there were only seven subjects per cell.

We learned a great deal about designing such experiments from Preuss and Shafer's work, and decided to look again at the question of reading programs on larger screens.

2. Materials

The IBM 3290 Plasma Display terminal has a display screen which measures 11.5 inches by 14.5 inches, and is capable of holding 62 lines of 160 characters. Two font sizes are available, which change the number of lines and the number of characters per line that can be seen. Also, provision is made for dividing the screen into up to four independent windows, the sizes of which may also vary.

For this experiment, four screen configurations were developed. The first consisted of two windows of 62 lines by 80 characters. The first sixty lines being viewed were displayed in the left window, and the next sixty lines were displayed in the right window. This configuration displayed 120 lines of program text at one time. The second consisted of one 62-line by 80-character window centered in the screen, which displayed 60 lines at one time. The third was a 24-line by 80-character window centered both horizontally and vertically. This configuration, the traditional size of computer screens, displayed 22 lines at one time. The fourth was a 12-line by 80-character window, centered both horizontally and vertically, which displayed 10 lines at one time. (In each display, the lines not used for text are used for title and command lines.)

We used the same font size for each display with a 6 by 12 dot matrix for each character. That is the only way we could implement the 120-line display, and we wanted to keep the font constant so that any effects would be due to the number of lines visible. In the cases of the 22-line and 10-line displays, this meant that there was an large amount of blank space surrounding the text; and, we believe that increasing the font size in the smaller displays might improve performance of people using them.

3. Preliminary Study

Our preliminary study looked at the same three displays examined by Preuss and Shafer. These were the 120-line, the 60-line, and the 22-line displays. Responding to the observation that the questions of Preuss and Shafer were of varying levels of difficulty and measured different levels of expertise, we divided the task of program comprehension into two separate levels, following the syntactic/semantic model of Shneiderman [1980]. On the syntactic level, we devised "program reading" questions, which required the subjects to answer questions merely by examining the code itself. An example of such a question would be, "In which procedure is the variable 'Distance' declared?" These questions are syntactic in that the subject did not need to be able to understand the program in order to answer them. On the semantic level, we devised "program comprehension" questions, which required the subjects to perform a trace of the program to answer the questions. An example of this type of question would be, "Given this input, what would the output of the program be?" These questions are semantic in that the subject needed to know not only what the individual statements meant, but also how the program as a whole worked.

The subjects for this study were twelve junior or senior undergraduate computer science students who were taking a senior-level course in data structures. All subjects had at least four semesters of computer science classes. Each subject answered a set of reading questions and a set of comprehension questions about a program. The program was a 118-line implementation of Dijkstra's Algorithm for finding the shortest path between all pairs of points on a graph. The program was short enough to fit entirely in the 120-line window. Most of the functions of the IBM XEdit editor were disabled, allowing the subjects to use only the page-up and page-down commands. This was done primarily to prohibit subjects from minimizing the negative effects of a smaller window by using the Locate command to find a string.

There was no significant difference found in any of the dependent variables among the screen sizes. In the case of the correctness measure for the reading task, we speculated that that was due to the nature of the task. The questions were such that everybody did well, regardless of the window condition. In the cases of the other three dependent variables, we speculated that that was due to the combination of the lack of repeated measures and the well-established phenomenon of individual variability, that is, low statistical power. The differences in individual ability so expand the range of performance that it becomes impossible to isolate the effects of the independent variables; and the absence of repeated measures prevents controlling variability by comparing the same subject's performance under different conditions. equally to all conditions. We found ranges of 3.5:1 and 4:1 in the time measures, which confirms the earlier study of Curtis [1981].

4. Main Experiment

Because of the difficulty in the preliminary study caused by programmer variability, we decided to use a repeated-measures design for the second experiment. When each subject did both the reading questions and the comprehension questions, the time per subject exceeded one hour, with the reading questions taking fifteen minutes and the comprehension questions taking over forty-five minutes. By going to a repeated-measures design, we would be increasing the time per subject to over three hours. We decided to focus on reading questions. Since the reading questions would take only about fifteen minutes for a subject to answer, we used this opportunity to add a fourth display, the 10-line display.

With our new design being a four-level, repeated measures experiment, we needed to have four programs. We also wanted to test the effects of being able to see the entire program on the screen at one time, so we decided to have all the programs be less than 120 lines long. These programs were:

(1) a 118-line implementation of Dijkstra's Algorithm for finding the shortest path between all pairs of nodes on a graph (the same program used in the preliminary study);
(2) a 114-line simulation of a vision system looking at planar surfaces at random angles;
(3) a 118-line implementation of Conway's Game of Life;
(4) a 116-line implementation of an electoral system using a transferable-votes scheme.

Not all reading questions are alike. Different skills needed to answer reading questions were identified, and it was decided to use four types of questions. These question types required the subject to locate:

(1) All occurrences of a specified feature, which could be located anywhere in the program. A typical question of this type would be "How many Writeln statements are in this program?"
(2) All occurrences of a feature, which could only appear at specified points in the program. A typical question of this type would be, "Which procedure uses local variables that are not of type integer?"
(3) A specific feature, that could appear only once in the program, but could occur anywhere in the program. A typical question of this type would be "What expression is assigned to the variable Distance?"
(4) A specific feature, the identity of which could be determined by studying two separate sections of the program. A typical question of this type would be "What variables are declared in the main procedure but never used in it?" In order to answer this question, the subject had to study the data declarations of the main procedure and the body of the main procedure, separated by the other procedures and functions.

In this experiment, each subject would answer one question of each type for each program, each program appearing on a different screen configuration. The actual questions used are in the appendix; in each case, Question 1 is of type 1, Question 2 is of type 2, Question 3 is of type 3, and Question 4 is of type 4. Also, each subject answered the questions in the same order.

In summary, the independent variables were:

(1) Screen Configuration/Window Size (four levels)
    (a) 10-line configuration
    (b) 22-line configuration
    (c) 60-line configuration
    (d) 120-line configuration
(2) Question Type (four levels)
(3) Program (four levels)

We measured the following dependent variables:

(1) Correctness of answers
(2) Time to answer each individual question
(3) Total time to answer each set of four questions.

The subjects were 16 students in an advanced computer science course on human factors in computer and information systems. Each subject was given an instruction sheet which included a table of the Program Function (PF) keys used in the experiment. He/she was then told to concentrate on the specific questions being asked and not to try to understand the whole program. The terminal was set to the appropriate display, and the first question, written on a separate 3 x 5 card, was given to the subject. The subject read the question, and when she/he had understood it, pressed a PF key which caused the program to appear on the screen and the clock to begin timing. The subject would look through the program, scrolling if necessary, and followed the instructions to answer the question. Both the subjects' answers and the time involved were recorded on the computer. After each set of four questions had been answered, the subject recorded his/her subjective impressions about the configuration just used. Finally, after all four configurations had been used, the subject filled out a background survey.

The answers were graded such that questions of types 1 and 2 were worth two points, and questions of types 3 and 4 were worth three points. Partial credit was given for certain answers.

5. Results

When we examined the correctness scores (16 per display size) for all four programs, we found:

Display Size	Mean Correctness Score	Standard Deviation
10	6.00	1.76
22	6.94	2.46
60	7.44	2.20
120	7.75	2.19

Although the average correctness score is increasing, the difference is not statistically significant, as shown by a one-way ANOVA. Apparently all of the programs were not equally difficult. The transferable votes program produced statistically significantly lower scores than the other programs in all display sizes. Examining only the scores (12 per display size) on the programs of similar difficulty, we found:

Display Size	Mean Correctness Score	Standard Deviation
10	6.08	1.71
22	7.75	1.79
60	8.50	1.19
120	8.58	1.75

A one-way analysis of variance revealed statistically significant differences ($p < .01$) for the factor of display size, and an HSD test revealed that both the 120-line and the 60-line displays were superior to the 10-line display. None of the other comparisons were significant by the HSD test.

The time results were analyzed by individual question type, and we received the following results (in seconds):

		Display Size				
		10	22	60	120	
	1	120.4	170.9	130.6	107.6	(Mean)
		44.4	122.3	71.7	48.4	(Std Dev)
	2	110.2	105.6	92.1	95.2	(Mean)
		33.7	69.9	61.1	66.6	(Std Dev)
Question Type	3	92.9	90.4	61.6	70.0	(Mean)
		60.4	78.7	25.1	49.1	(Std Dev)
	4	226.9	184.1	188.0	120.8	(Mean)
		111.9	147.5	146.3	53.1	(Std Dev)
Total Time		553.1	551.0	472.3	393.6	(Mean)
		175.3	323.8	240.4	157.1	(Std Dev)

A two-way ANOVA found significant main effects for question type, ($p < .01$) and display size ($p < .05$.) There were no significant interaction effects.

A count was also kept of the number of page turn commands each subject issued, according to display and question type:

		Display Size				
		10	22	60	120	
	1	17.50	13.19	2.44	0.06	(Mean)
		10.40	11.18	2.12	0.24	(Std Dev)
	2	21.38	10.38	1.44	0.25	(Mean)
		7.45	5.07	0.61	0.97	(Std Dev)
Question Type	3	21.25	9.69	1.44	0.00	(Mean)
		9.38	7.87	0.86	0.00	(Std Dev)
	4	53.81	38.50	6.88	0.00	(Mean)
		33.70	29.90	9.95	0.00	(Std Dev)

The dramatic differences in the number of page turns is apparent. Most of the difference is due to the amount of scrolling needed to see the entire program. With a 120-line display, no scrolling is necessary for these short programs. With a 10-line display, repeated scrolling is necessary to see the program even once. With type 4 questions, which require the study of two separate parts of the program, repeated scrolling is necessary with the 10-line display just to find the separate parts.

6. Subjective Reactions

Nineteen people filled out subjective evaluations of the screen configurations. These included the sixteen subjects plus three people whose data could not be used in the analysis due to external factors. These evaluations consisted of little more than free-response subjective answers to the question "What do you think of the display," asked after each trial. There was a general preference for seeing as much code as possible. A sizeable minority, however, were not convinced that seeing 120 lines of code would be beneficial in all circumstances; they cited the dangers of clutter and of clumsiness. Nevertheless, almost all the subjects expressed a preference for configurations bigger than the traditional 24-line terminal. This was unexpected, because twelve of the respondents had reported using the traditional terminals almost exclusively and that it was their favorite size.

The observations of the experimenter tend to bear that out. When the subjects were using the twelve-line configuration, the subjects became generally frustrated with having to scroll through the program in so many small steps. This was especially noticeable when the subjects were answering type 4 questions, and they would have to go from one end of the program to the middle. In contrast, users of the 120-line display merely looked from location to location in a regular pattern, signifying that they were able to recall the location of specific points of the program on the screen. It should also be pointed out that the subjects were generally surprised to see the 120-line display fill up the entire terminal, particularly after they had seen two or three of the other displays.

7. Discussion

The correctness result supports the hypothesis that the bigger windows are better. We did not find a significant result when all four programs were included because the transferable votes program produced correctness scores that were significantly lower than those for the other three programs. When we focused on the three programs that produced similar scores, we found a statistically significant result.

The total time, with no question breakdowns, did not produce a significant difference. Again, we speculate that this is due to the range of times the various questions produce. The two-way ANOVA revealed a statistically significant difference among questions with type 4 questions taking significantly longer to answer than the others. Whether that is due to the larger screen, the absence of scrolling with these particular programs, or some other factor is an issue for further experiments.

The number of page-turns should decrease as the window size increases; subjects needed fewer page-turns to see an entire program if the window size was bigger. One might expect that with the 120-line display no subject would have to turn pages with these programs, which have fewer than 120 lines. That is what happened; the five page-turns recorded are instances in which the subject hit the wrong key accidentally. It was speculated that the page turns would slightly decrease through question types 1-3 as the subject became more familiar with the program and then dramatically increase with question type 4, which was harder and required more page turns. This pattern appeared with the 22-line and 60-line displays, but did not appear with the 10-line display.

This may indicate that users may have trouble locating specific points in a program, as opposed to counting features which may occur anywhere, with a small display. Or, it may be due to the increased use of top and bottom commands with the ten-line display, which were counted in the page-turns as the maximum number of pages possible. The use of those commands also led to an occasional overshoot of the target and meant even more page turns. The top and bottom commands were most often used with the 10-line display.

A common pattern of looking at programs on the 10-line display was to slowly page through the program once to learn where everything was, and then take advantage of that knowledge to use the top and bottom commands later. A subject who moved from the end of the program to the second page using page-turns would need ten keystrokes, whereas turning to the top and then paging once forward required only two keystrokes.

8. Suggestions for Further Experiments

As mentioned earlier, the programs studied all had less than 120 lines, which meant that they could fit in the 120-line display without any page turning. The features of the editor were disabled, allowing the users only to turn from page to page. The font size was kept constant. The results of this experiment may not hold in different conditions.

The following are candidates for follow-up experiments:

(1) Repeating this experiment with longer programs, so that even on the 120-line configuration, page turning is necessary.

(2) Repeating this experiment, but enable the editor commands. Any "search" or "locate" command would be especially useful for finding tokens or expressions.

(3) Repeating this experiment, but increasing the emphasis on program comprehension, rather than program reading. Replications with other tasks such as composition, debugging, and maintenance would also be worthwhile.

(4) Allowing subjects to gain familiarity with larger screen sizes over a one week period before conducting the experiment.

In addition to the development of larger windows, there are other uses for the larger screens that should be studied. Some of them are given below.

Multiple windows. Instead of making one large window, use the screen to open up multiple windows. Each window can allow vision of different parts of one file or of different files. Issues to be looked at include that of synchronized scrolling vs. independent scrolling vs. fusion [Weldon, Norman, and Shneiderman, 1986] and that of tiled windows vs. overlapping windows [Bly and Rosenberg, 1986].

Novel display strategies. One of the uses of larger screens might be to provide additional information about a file without detracting from the ability to see the contents of the file. Such strategies include the hierarchical browser and embedded selection. [Shneiderman et al, 1986]

9. Conclusion

This study has demonstrated that even in a one hour period student programmers can become familiar with larger screen displays so that their performance in program reading tasks improves. Small displays, such as 10 lines, seem to inhibit performance because substantially increased page turning is required. These results support the current efforts to provide larger screen sizes with more information visible, but there may be a danger of overwhelming the user with too large a screen.

Acknowledgements: This work supported in part by IBM Federal Systems Division, Bethesda, MD. We appreciate the constructive comments of Kent Norman and Marvin Zelkowitz and the technical assistance of Carl Smith.

References

Bly, Sara A. and Rosenberg, Jarrett K., "A Comparison of Tiled and Overlapping Windows," *CHI '86 Proceedings,* (April 1986) pp 101-105

Duchnicky, R.L. and Kolers, P.A., "Readability of Text Scrolled on Visual Display Terminals as a Function of Window Size," *Human Factors 25,* (1983) pp 683-692

Haas, Christina and Hayes, John R., "Reading on the Computer: A Comparison of Standard and Advanced Computer Display and Hard Copy," CDC Technical Report No. 7, Carnegie-Mellon University (February 1985)

Haas, Christina and Hayes, John R., "Effects of Text Display Variables on Reading Tasks: Computer Display and Hard Copy," CDC Technical Report No. 3, Carnegie-Mellon University (March 1985)

Haas, Christina and Hayes, John R., "What Did I Just Say? Reading Problems in Writing with the Machine," *Research in the Teaching of English Vol. 20, No. 1,* (February 1986) pp 22-35

Mills, Carol B. and Weldon, Linda J., "Reading Text from Computer Screens," Manuscript submitted for publication, (1986)

Shneiderman, Ben, *Software Psychology,* Little, Brown and Co., Boston, MA (1980), ch. 3,10

Shneiderman, Ben; Shafer, Philip; Simon, Roland and Weldon, Linda, "Display Strategies for Program Browsing: Concepts and Experiment," *IEEE Software Vol 3. No.4,* (May 1986)

Weldon, Linda J; Norman, Kent L. and Shneiderman, Ben, "An Empirical Test of Display Strategies for a Multiple Screen Programming Environment" (1986)

Social, Ergonomic and Stress Aspects of Work with Computers,
edited by G. Salvendy, S.L. Sauter and J.J. Hurrell, Jr.
Elsevier Science Publishers B.V., Amsterdam, 1987 — Printed in The Netherlands

THE MODELLING OF INFORMATION PROCESSING ON CHINESE VDT*

S.-L. HWANG and C.-C. Ho
Department of Industrial Engineering, National Tsing Hua University, Hsinchu, Taiwan, Republic of China

ABSTRACT

A queueing theory was employed to model Chinese information processing on an automated control system. Information rate and the number of information categories were manipulated as variables in the model because these factors would affect the performance of the operator in processing information. By simulatior method, the limit of human mental workload was quantified using ρ value. This model provided system designer a reasonable mental workload level and can be used to evaluate the safety and efficiency of the system.

1. INTRODUCTION

The Chinese video display terminal (VDT) is at an early stage of development. Since the population of Chinese VDT end users can not be neglected, the information display strategies on Chinese VDT should be standardized to provide a better human-computer communication interface.

A study concerning Chinese character information display has been done by Hwang et. al. (1987). In addition to considering the optimal character information display, the present study emphasizes on dynamic information display in an automated system. After the implementation of computers in automated systems, the tasks of human operators have been changed. Instead of operating complicated machines, the operators supervise the system by communicating with computers. The human operator would recognize the critical information displayed by computer, make decisions, and give commands to the computers. Usually, the operators process the information from each subsystem according to some priorities. This is a kind of multitask control situation where each task can be considered as the specific information from main system or subsystems. As the system getting larger, more and more information need to be displayed, which may cause operator's mental overload. Therefore, the purpose of the present study is to discover the limit of operator's mental workload in order to assure

---

*This research was supported by the National Science Concil of the Republic of China under contract Number NSC76-0415-E007-05.

the safety and efficiency of the system. Because queueing theory deals with multitask where the human allocating attention among a variety of task or information (Rouse, 1980), the present study would apply queueing theory to quantify mental workload in information processing.

2. INFORMATION PROCESSING IN AUTOMATED CONTROL SYSTEMS

Figure 1 shows the stages of information processes in an automated control

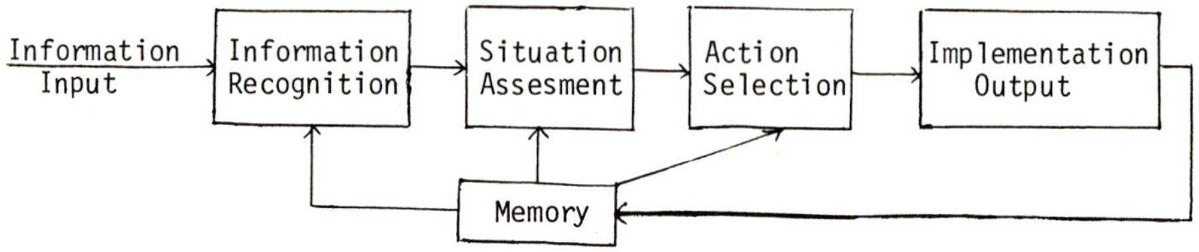

Fig. 1. Information processing of controllers (Schmidt, 1978)

system. The operator recognizes the information displayed by computers, and then evaluates the current system conditions before taking an action. From recognizing information to selecting an action, the operator usually compares the information with knowledge base in the memory. The final stage of the processes is the implementation of output, i.e. giving commands to the computer.

During processing the information, the performance of operators might be affected by some factors such as information rate and the number of information categories. Increasing the rate of information could decrease operator's idle time and thus improves performance. But if the rate of information is too high, the performance would decrease. In other words, there is an upper bound for the increment of information. As to the relationship between performance and the number of information categories, the large number of information categories would increase the complexity of decision making and thus decrease performance. Although one needs information to make decisions, there exists a critical value for information rate/category. The performance is improved by the increment of information rate/category below the critical value while decreased if the information rate/category is above the critical value. In short, the critical value is the maximum information rate/category the operator is able to handle, and is the limit of human mental workload.

3. THE APPLICATION OF QUEUEING MODEL

There are three main parameters in queueing theory: (1) interarrival time distribution, (2) service time distribution, and (3) service discipline.

Queueing system are usually classified as follows (Rouse, 1980):

$$(X/Y/Z):(U/V/S)$$

where X = interarrival time distribution

Y = service time distribution

Z = number of parallel servers

U = service discipline

V = maximum number of tasks allowed in the system

and S = size of task population

One of the important problem in queueing model is performance measures. The present study uses three measures:

W = servide time

W_q = waiting time

and ρ = server occupancy or utilization

When applying queueing theory to an automated control system, the "information" in the system can be viewed as "event" in queueing model and the "information processing time" is equal to "service time" in queueing model.

Some researchers have applied queueing theory to evaluate pilot performance in flight control system (Carbonell, 1966; Carbonell, et. al., 1968; Senders and Posner, 1976; Walden and Rouse, 1978). Walden and Rouse (1978) employed queueing system described as $(M/E_k/1)$:(PRP/N/N) to model pilot decision making in a multitask flight control system. Schmidt (1978) analyzed the workload of air traffic controllers by queueing model. Schmidt used ρ as an index of workload to help in scheduling the appropriate number of controllers.

In the present study, the different category of information in an automated system can be viewed as different type of task. Under this multitask situation, the arrival rate and the service time of each category of information might be different. It is assumed that the information arrival time distribution and service time distribution are Erlang distributions. The model is described as $(E_k/E_k/1)$:(NPRP/∞/∞). Figure 2 indicates the combination of information processing stages and queueing model. In this figure, W refers to response time which is the summation of waiting time (i.e. W_q) and servie time (i.e. $1/\mu$). Waiting time will be increased by the increment of information arrival time and service time (or information processing time) will be increased by information rate and the number of information categories as discussed before. Therefore, the response time is affected by information arrival time, information rate, and the number of information categories. Among these three factors, information arrival time is a fixed parameter while the information rate and the number of information categories are variables in this study.

4. SIMULATION PROGRAM

The model of the present study is solved by simulation since an analytical solution is not available and the service rate (i.e. $1/\mu$) is unknown. A SLAM simulation language can be used where the input parameters are

N : number of information category

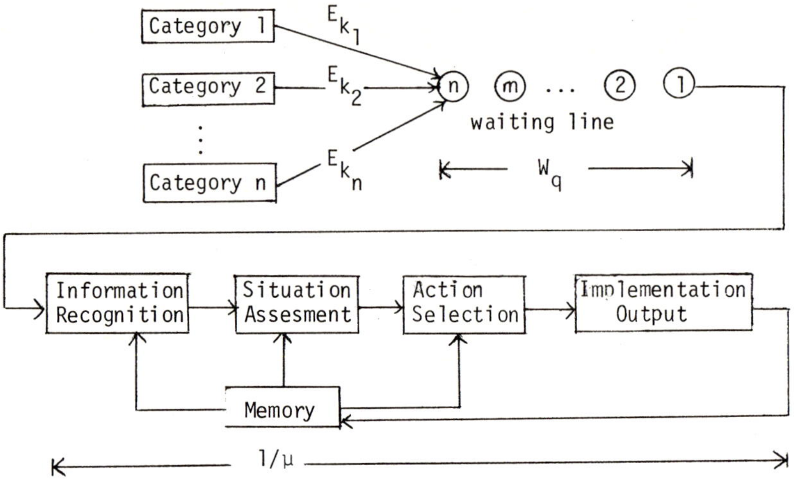

Fig. 2. A queueing model in information processing

S_i : information rate

E_{k_i} : probability distribution of information arrival rate

λ_i : information arrival rate

p_i : priority

$i = 1,2,\ldots,N$

and the output parameters are

W_q = waiting time

W = response time

ρ = utilization

5. CONCLUSION

This model can be applied to evaluate the safety and efficiency of the current system and to help system designer in calculating reasonable mental workload.

For the future research, an experimental study is recommended to verify this model.

6. REFERENCES

Carbone 11, J.R. 1966, A Queueing Model of Many - Instrument Visual Sampling, IEEE Transactions on Human Factors in Electronics HFE-4(No.4), 157-164.

Carbonell, J.R., Ward, J.L. and Senders, J.W. 1968. A Queueing Model of Visual Sampling: Experimental Validation, IEEE Transactions on Man-Machine Systems, MMS-9(No.3), 82-87.

Hwang, S.-L., Wang, M.-Y., and Ho, C.-C., 1987. The Study of Chinese Character Information on Video Display Terminals, Annual International Industrial

Ergonomics and Safety Conference '87, June 9-12, U. of Miami Conference Center, Miami, Florida, U.S.A.
Schmidt, D.K. 1978, A Queueing Analysis of the Air Traffic Controller's Workload, IEEE Transaction of Systems, Man and Cybernetics, SMC-8(No.6), 492-493.
Senders, J.W., and Posner, M.J.M., 1976. A Queueing Model of Monitoring and Supervisory Behavior. In Monitoring Behavior and Supervisory Control, (T.B. Sheridan and G. Johannson, Eds.) New York: Plenum Press.
Rouse, W.B., 1980. Systems Engineering Models of Human-Machine Interaction, Elsevier North Holland, Inc.
Walden, R.S. and Rouse, W.B., 1978. A Queueing Model of Pilot Decision Making in a Multi-Task Flight Management Situation, IEEE Transactions on Systems, Man, and Cybernetics, SMC-8(No.12), 867-875.

Social, Ergonomic and Stress Aspects of Work with Computers,
edited by G. Salvendy, S.L. Sauter and J.J. Hurrell, Jr.
Elsevier Science Publishers B.V., Amsterdam, 1987 — Printed in The Netherlands

MAKING CUSTOMER-INITIATED VARIETY FEASIBLE FOR CIM : CUSTOMER-ORIENTED PRODUCT DESIGN SOFTWARE

ROBERT A. KARASEK
Industrial and Systems Engineering Department, University of Southern California, Los Angeles, CA, USA

ABSTRACT

A system of software tools is proposed which could facilitate design-for-variety for customers of moderate levels of sophistication ("smart customers"). The Subcomponent Manipulator operates on "virtual objects," which are idealized components for conceptual design that are similtaneously optimimized with respect to the structure of customer's functionality needs and the company's producability requirements . The software tools used to create the virtual objects suppliment existing knowledge of product producability (embodied in Group Technology Coding Systems , for example), with structured information about customer needs. These tools are the Functional Taxonomy Tester, Function Data Coding System, and the Virtual Object Compiler. A Real Object Compiler tests the design based on virtual objects for final producability and cost criteria.

1. INTRODUCTION

The first assumption upon which this paper rests is that the primary justification of the use of complex computer assisted production technologies, such as CIM (computer integrated manufacturing), is their capability of yielding high levels of *variety* of output at relatively low cost. The importance of this capability is reflected in the common references to *flexilble* manufacturing based on CIM. However, at present most of the economic justifications for use of CIM technologies are based on the expectation that reduction of direct labor costs or inventory holding costs within existing, non-flexible, product lines will more than offset the costs of the large CIM capital investments . These advantages of CIM technology are probably far too limited to justify its implementation in most settings (of course, fixed ,"Detroit" automation is cheaper for large lot production). What CIM technologies can do more effectively than previous generations of production technology is to generate variations in the product's manufacture that are of value to a specific customer . Without being able to take advantage of CIM's reprogrammabiity in terms of gaining increased "rewards" for variety - paid by the customer to the firm - the primary justification for the utilization of CIM is lost. This paper discusses product design strategies, and the related software aides, that could facilitate the realization of customer-driven product variety from CIM production technologies.

There are several major hurdles which can hinder the ability of manufacturing firms oriented toward conventional large lot manufacturing from reaping variety-based "rewards" from the customer . Such firms often suffer from limitations imposed by their rigid, hierarchial structures. Such firms can have difficulties in dealing with the huge information volume associated with many customized product variations. Such organizations typically deal with only three or four pieces of information per order: price, quantity, d elivery date, and product number. In customer-initiated product variety, there are numerous aspects of the product or its subcomponents that might have to change. This requirement implies a much broader "bandwidth" of communication between customer and shop floor production facilities. This in turn, probably requires major transformation of the organization's communications style and organizational structure itself, in to decentralized forms {note: these forms are not congruent with the often displayed "maps" of new hierarchial data structures that will be introduced alongside CIM machinery }. Many organizations acquiring CIM have done little thinking about these types of organizational change, and could easily be imagined to resist them. However, such new organizational forms represent the necessary platform for use of the more detailed, design process tools we discuss below (Karasek, l983) .

2. DESIGN-FOR-VARIETY: INCORPORATING "FUNCTIONALITY" REQUIREMENTS

The second hurdle for CIM technology acquiring firms which do have variety-producing goals, is that their product lines are primarily filled with products that cannot easily or cheaply accept customer-initiated product variations. One problem may be the "tight-interconnectedness" of the product's subcomponents. Changing one minor component to satisfy a customer can require costly changes in all the other components - even though the customer has no interest in those secondary changes (and could not pay for them). The goal of eliminating interdependendencies between subcomponents introduces the need for a new "design-for variety" philosophy of product design. Modular design is an approach which minimizes unnecessary connections between components, but raises the immediate question of where module boundaries should be drawn (we address this question below in definition of virtual objects).

A second new element of any design-for-variety approach is the need to engage the customer in the design process (creating the new role of the"customer/designer"). This is a communications process requirement for design activity. Wthout this integration the process of developing a set of final design specifications ("where the customer finally knows what he wants") would be prohibitively lengthy and expensive. The crucial challenge is to facilitate the customer's review of different prototype designs, with a rapid convergence to solutions which are also feasible in terms of producability and cost. Ideal design components would be "modules" that are easy for the customer/designer to learn about, and thus effectively manipulate. Customers think about product use and function. For the customer's mental maps to be congruent with design elements, these ideal design components or modules should reflect the structure of *functionality* of the product - not just convenient manufacturing decsions. We will call this type of subcomponent-like module a "virtual object" in the discussion below because it is an idealized subcomponent: typical of but not necessarily physically the same as any of the company's existing subcomponents. Finally, the customer/designer must be given quick feedback on his initial design ideas in terms of cost, performance and production feasibility.

Of course, customers can come in many levels of sophistication , and each level could easily require a different set of design aides(Karasek and Chignell, l986). At a low level the customers may be a sales representative or knowledgable consumer; at a high level the customer may be a teamof experienced engineers in another part of the same company . Our design aides in this paper are oriented to a customer who has (or is willing to acquire) an intermediate level of sophistication in terms of "producability " understanding . Such a customer , like the VLSI chip customer below, may be willing to invest some significant "learning " t ime in order to master use of a set of design aides that will potentially allow him to gain the advantages of a flexible but inexpensive custom design.

3. FUNCTIONAL DESIGN SOFTWARE TOOLS TO FACILITATE CUSTOMER -VARIETY

Two types of software tools might facilitate the use of the easily manipulatable and conceptualizable modules that were referred to above.. The most important tools, of course, are those that allow the new "smart" customer to use the system: i.e. to design for his needs. These are the run-time tools: the Subcomponent (virtual object) Manipulator which allow the customer to play with alternative configurations of the components of the product , and the Real-Object Compiler which makes the final tests for producability of a selected combination of virtual objects (these are further descussed below). The most difficult sotware aides to develop, however, will probably be those which assist systemmatic "computation" of an optimal set of "virtual objects" from the information on subcomponent producability on the one hand, and the structure of customer needs for change and maintence on the other.

We foresee four different software aides to facilitate the difficult process of virtual object definition. The difficulty is alluded to by the fact that such attempts have been called an "art" not a science in the Prodevent publications(Bosrup and Sandell, l976)which outline such manufacturing design proceedures in a non-software oriented format . The first software tool already exists but will have to modified for our use: Group Technology Classification Systems for categorizing of "producability" information. The second and third software aides attempt to develop something analygous to the group technology classification system, but in the area of the customer's functional needs (one tool to assist refinement of the logic of the functional requirements taxonomy [the Functional Taxonomy Tester] and another to facilitate coding of function data to product subcomponents{Function Data Coding Aide}). The most difficult software aide to develop will be the the Virtual Object Compiler, which defines the optimum boundaries "virtual objects"on the basisof both producability and functionality information.

3.l Functional Taxonomy Tester and Functional Data Coding Aide

The goal of the Functional Taxonomy Tester is to reflect the structure of customer's needs for product changes in a rigorous, logically structured manner. Initially, data on functional requirementswill have to be input from market surveys of typical customers. Let us take for an example a pump for fluids, which might have many subcomponent variations depending on the customer's needs. Suppose that our functional data review shows that customers rarely have any explicit requirements to pose about the nature of the pump's ball bearings - other than that they work efficiently , without lubrication, leakage or need for replacement. Since the customer is not likely to specify variations in the bearing category, the function taxonomy can be quite sparse in this region. The Function Taxonomy Tester will help structure the interrelated functions of the pump - at different logical levels. It could determine the extent to which the pump function data fits into a tree-like hierarchial structure for example. Tests for the regularity of the taxonomy structure can be made by implementing the presumed logical relations between a set of functions in a logic programming language such as Prolog (see Karasek and Chignell (l987) for an initial pump example). This logic program structuring of the functional data (incompletely shown as "function templates " in Diagram l) will be necessary for the later operation of the Virtual Object Compiler. This compiler will use information about subcomponent functionality, organized on the basis of this taxonomy , as an input to the determination of optimum virtual object boundaries.

Another related software tool will be a Function Data Coding Aide which will do for the function data that which group technology software tools do for producability data. It will facilitate the orderly and efficient input of coded data (usually by engineers) on the functional properties of each subcomponent.

Information about the producability of manufactured components can often be organized on the basis of the machine processes, or other processes, that are necessary for their production. These Group Technology Coding Systems are already in existence for many production situations and should serve as a basis for input of the producability information to the Virtual Object Compiler. However, the manner in which the information is aggregated may differ for our use of Group Technology Coding Systems. Also, higher level data about producability may also be needed. Information will be aggregated on the basis of production operations that need to be undertaken in a tightly interconnected manner in order to feasibly produce a particular subcomponent. Typical aggregating rules might involve grouping together : (a) all sets of components that contain precision machined metal components and electrial circuitry that must be tested together to determine performance acceptability ; or (b) all sets of components that should be produced together to reduce raw material waste.

3.2 The Virtual Object Compiler

The software operations that are to be performed by the Vrtual Object Compiler are represented by visual analogies in Diagram I. The basic rules for final virtual object boundary determination are that the highest order of both producability and functionality boundary constraints that that can be *similtaneously* delineated , should be chosen as virtual object boundaries. In the case of conflicts (which would often arrise) the next lowest order of boundary in terms of either producability or functionality must be chosen, increasing the number of virtual objects. The customer will be probably most easily able to understand product design when manipulationing a small number of virtual objects rather than a large number (with the upper bound equal to the number of original subcomponents -overwhelming to the customer and defeating the value of the sustem altogether). Thus, the goal is to have the smallest number of "virtual objects" that incorporate the highest orders of non-conflicting producability and functionality constraints at the same time. Another constraint is that the selected virtual objects, when combined, should give the greatest range of functional capability and production possibility.

In Diagram I we can see that on the basis of functionality (left hand side) two major subdivisions of the pump seem to be adequate, while three divisions appear necessaryfrom the standpoint of producability (right hand side). In terms of functionality there are few reasons to subdivide the shaft from the impeller: the function may be to move a fluid of a certain viscosity with a certain volume using the rotating parts of the pump as a unit. However, in terms of producability it is clear why there could be several good reasons to separate them : the shaft is probably easily machined from bar stock (or purchased) while the impeller may represent a casting with much special machining. The resolution of this conflict would probably involve selection of three vitual objects, so as to adequately reflect customer choices that would affect pump costs through shaft and impeller production decisions.

The flexibility of the virtual objects is , of course, related to the number of subdivisions that are made. The fewer subdivisions that need be employed the fewer virtual objects that will be created, but the greater the rigidity they will have with respect to either functionality or producability. Thus the goal of each of those input classification systems is to capture the most variance within the fewest categories (a task complicated by the fact that several logical levels must probably be defined for both producability and functionality).

As can be seen, development of software that could accomplish the goals of the Virtual Object

Diagram 1 **VIRTUAL OBJECT COMPILATION - PUMP EXAMPLE**

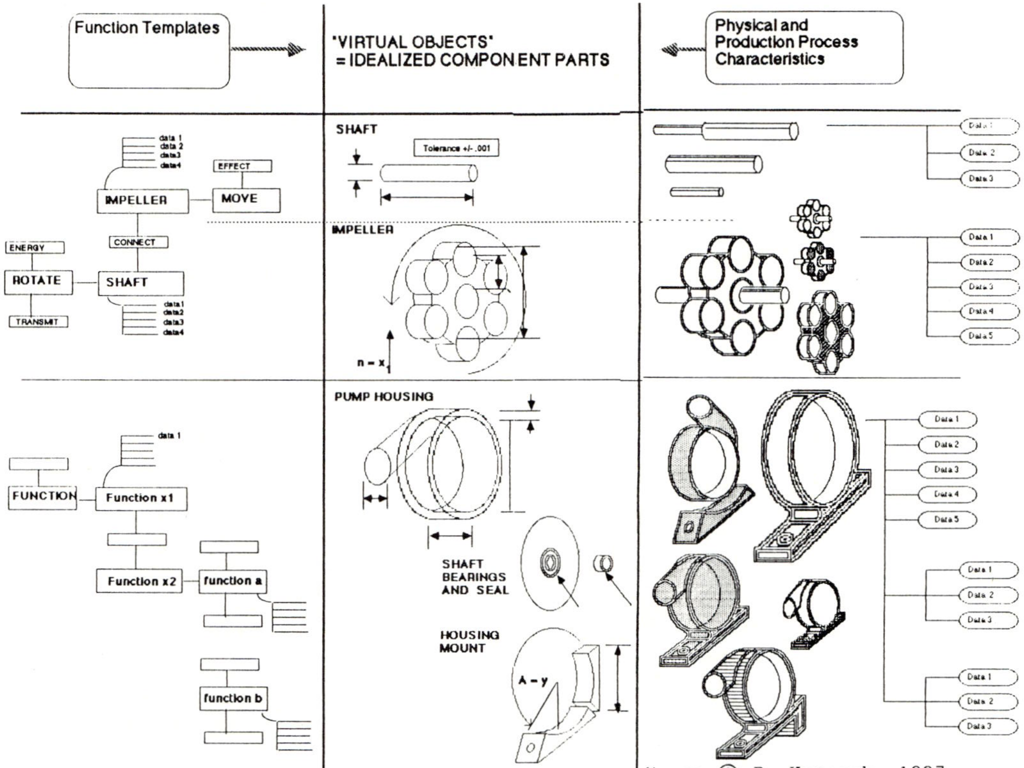

diagram © R. Karasek, 1987

Compiler would be a major challenge, beyond the capabilities of existing known design tools (however, Towne [l986]describes a useful starting point: his networkof modularly -designed, functionally-operative and interconnected subcomponents in VDT-based systems analysis software). Fortunately, the "runtime' software aides, discussed below, already exist within the range of existing computer-based design software.

3.3 The Subcomponent Manipulator

The first software tool needed at "runtime" by the customer or designer is the Subcomponent Manipulator. This is software which gives the designer a "kit of parts" to manipulate; with limited feedback on both the production difficulties involved (cost, design posibility) as well as iformation on functionality (performance effectiveness). An important characteristic of this tool is the level of sophistication of the customer or designer (as discussed in section 2). While software that makes custom product design possible for a customer relatively unschooled in product producability is new, several examples are now in existence. One recent example is software being developed for use on Macintosh personal computers for use by Buick dealers. It allows the customer to specify on the computer whether their primary functional concern is trunk space or price, for example, and then to complete a set of customized Buick specifications that can be used as the product order (Farber, l987). A system that involves somewhat more sophistication in understanding of the product subcomponents is a program for designing customized truck freight compartments developed by the Stenhagens Company in Sweden(Olsen, l986). Here the typical user is a truck-rig dealer or a knowledgable truck user who specifies his needs for particular functional characteristics such as the overall size of the compartment, the number of side panels, the numberber of side doors required etc. The program computes the feasibility of a particular configuration, and after a "second pass" can compute costs for the customer and generate production shop drawings.

For the more expert designer, such as the truck-rig customer or even truly expert designers such as the engineers who utilize Digital Equipment Corporation's R1 System (McDermott, l982) to configure minicomputer components to custom requirements, the Subcomponent Manipulator can rely on a more detailed understanding of the types of components that are likely to be available, and their production difficulties. This could mean that the virtual objects originally selected could be more numerous and less abstract - since the designer will be more skilled in the manipulation of such objects. This in turn means that the process of developing virtual objects will be simpler, and that they will be able to reflect without error a wider range of both production limitations and functional cababilities. Of course, if taken to the extreme, such virtual objects would simply become the existing set of detailed product subcomponents - highly flexible and realistically producable, but giving no advantages over conventional design approaches requiring staffs of highly trained engineers.

At the other extreme, for unsophisticated customers, we see two possible directions that might be taken in development of custom design software. The firstdirection that such programs may take with niave customers is that the virtual objects become relatively few in number and restricted in flexibility. In this case the lack of any understanding about subcomponents means that the customer/designer can manipulate ony a small number of relatively well-behaved virtual objects. Defining such a set of virtual objects from amoung the existing array of subcomponents may be computationally difficult for the company, placing a substantial burden on the engineering staff. Or it may require that many interesting but irregular possibliities are simply omitted from the capabilities of the virtual objects. Of course, by selecting only a small number of "prepackaged" virtual objects, the design system is moving in the direction of offering the customer a simple catalogue of product variations, with optional equipment packages, for example. This extreme solution also takes use back to one of the most common existing conventional approaches to product design and selection.

A second approachwhich retains both customization and the ability to deal with "niave customers" is to utilize an education program to first train them to be "smart customers". In this case the customer must be interested enough in the benefits he expects to obtain from custom design devote time to training in the use of a *design language*. This direction is illustrated by Texas Instrument's VLSI chip design system (see for exampleTexas Instruments, l983). They have a circuit design language based on a system of predefined circit types and functions (analygous ot our virtual objects). This design language, once mastered, will insure that the customer/designer can state his needs in the form of producable VLSI chip specifications. (The Texas Instrument language is meant to be understood by trained electrical engineers, so still further investment would be required to make this system usable by niave customers . Of course, a custom VLSI chip is an extremely sophisticated product). Since many of the software tools for custom design would require some investment in time and energy on the consumer's part, such systems may be primarily of use to customers who expext to request a design more than once, or for whom the product will be very important. This may not be a major limitation, however, since this group could includes car and home buyers, as well as many industrial customers of highly specialized products.

3.4 The Real Object Compiler

The Real Object Compiler translates well-structured representations of possible designs into "real" solutions on the basis of a set of software simulatable physical production operations. Two diverse examples of software that performs an analygous founction which are already in existence are generic process planning software for metal machining; and VLSI circuit design and testing software in electronics. Once the customer or designer has identified a set of virtual objects which satisfy rough functional requirements and rough producability constraints, it is necessary to check the detailed feasibility of the virtual objects design by translating it into a real (albiet still simulated) physical design. This is equivalent to sending a "conceptual design" through a more detailed production analysis in conventional terminology . (Much of the required detailed physical information might be easily relinked in the case of virtual objects which had been developed from existing sets of subcomponents).

Since the virtual objects are abstractions, much of this detailed information in the virtual design stage existed in the form of default parameters. While the conceptual feasibility of a design at the virtual object level may be determined with many of these parameters set to default levels, at the Real Object Compiler stage more specific information will need to be input by the customer or designer. An example is a virtual object pump shaft : while its general properties are already known, its specific diameter, length and capacity must be specified at this stage. The specific values of each of these papameters is associated with specific information of production difficulty, cost and connectivity restrictions. For example, a shaft of diameter 1", a length of 5", and a capacity of 20 horsepower, will be associated with a set of materials and production tolerances compatabile with the rest of the design and overall pump performance specifications. These will, in turn, allow the cost of the shaft and other producability information to be determined: these are the output of the Real Object Compiler.

This feedback to the customer will often lead to design refinements - but at a detailed level. The basic selection of a combination of subcomponents (the virtual design) is not in question at this stage. After a sequence of such refinements, the abstract "virtual design" is finally translated into a real physical design. This design may be satisfactory to the customer, or its deficiencies may force him to return to a more conceptual virtual design phase - testing a new combination of virtual objects to obtain the performance and cost levels desired.

4. CONCLUSION

A set of structured intellectual proceedures have been outlined for defining and using a small set of "typical" subcomponents -on the basis of their production similarities and their functional similarities - that have the greatest potential for simple "combinability" into a large number of product variations in the hands of a relatively niave customer. We cannot say how completely any particular product will find itself amenable to "virtual object" defi nition in this manner. I n many cases the "software tools" described above might have to be human conceptual processes, because of the difficulty in fitting them into even the moderately rigid logical structures we discuss. Whatever specific computational approaches that might be employed by a Virtual Object Compiler or our other software tools, it is clear that such programs would require substantial machine "intelligence," as well as provision for the intervention of a highly trained engineer to resolve many conflicts.

Indeed, design of the optimum set of virtual objects, and their periodic revision, would become the primary design task of the design engineer in this approach to custom design. Another new role for the engineer would be development of the educational interfaces, for example: development of the optimum combination of training activity and Subcomponent Manipulator complexity. To borrow common terminology from the software design field (which has many analogies to this discussion), the engineers who engage in design-for-variety activties must both design user interfaces and "write modular code" for systems of "user friendly" sub components.

REFERENCES

Bosrup, L. and Sandell, P., l976. Coordinated Customer Directed Manufacturing (Samordnad Produktframtagning). Prodevent Project, IVF-76604. Sveriges Mekanforbund, Stockholm

Farber, D., l987. Software Ads [discussion of SoftAd Group software, Sausalito, CA]. Macworld, April,p113

Karasek, R., l983. Product Variety and Work Life Quality using Computer-Aided Technologies: Organizational Prerequisites for Effective Design. Mimeo: Department of Industrial and Systems Engineering, USC, Los Angeles

Karasek, R. and Chignell, M., l986. The Function/Structure Approach to Design for Computer-Integrated Manufacturing. Society of Manufacturing Engineers Technical Paper MS86-973. Ultratech Conference,September, l986, Long Beach, CA

Karasek, R. and Chignell, M., l987. Functional Design for Assembly: Tool Analysis and Development. Working Paper: Department of Industrial and Systems Engineering, USC, Los Angeles, CA

McDermott, J., l982. R1: A Rule-based Configuer of Computer Systems. Artificial Intelligence, 19:39-88

Olsen, K., l984. Stenhagens Flak[Advertising brochure]. Stenhagens, Drottninggatan 3, Uppsala, Sweden

Towne, D., l987(in press). The Generalized Maintenance Trainer: Evolution and Revolution. in W. Rouse(ed), Advances in Man-Machine Systems Research, Vol 3, JAI Press

Social, Ergonomic and Stress Aspects of Work with Computers,
edited by G. Salvendy, S.L. Sauter and J.J. Hurrell, Jr.
Elsevier Science Publishers B.V., Amsterdam, 1987 — Printed in The Netherlands

A NOTATION FOR THE INPUT AND STORAGE OF MATHEMATICS: —AN APPLICATION OF SGML—

JAMES R. ROESSER
Science Typographers, 15 Industrial Blvd., Medford, N.Y. 11763, USA.

ABSTRACT
Microcomputers, word processors, and laser printers provide the basis for a greatly improved communications system within the scientific community. To realize the full potential of this technology it will be necessary to establish a notation for the electronic storage of scientific (frequently mathematical) information. The purpose of this paper is to develop such a notation based on the following principles:

1. the notation must be usable by both technical and nontechnical people,
2. the notation should require only a standard ASCII keyboard,
3. codes (tags or prefixes) must be inserted at the places where changes occur,
4. the notation must be consistent with SGML.

1 INTRODUCTION

Microcomputers, word processors, and laser printers provide the basis for a greatly improved communications system within the scientific community. The full potential of this technology has not been realized for two reasons. First, the device compatibility problem continues to exist. Second, there is no standard notation for electronic communication.

These problems may be overcome with a generic notation which is device, system, and application independent. The Association of American Publishers has defined such a notation for electronic manuscripts consisting of text and special symbols.[1] It is an application of the Standard Generalized Markup Language (SGML). A chapter within the AAP document deals with the markup of mathematical formulas. This paper began as a working paper within the committee which was responsible for that chapter.

Let us begin by considering who uses mathematical notation. Although mathematicians, scientists, and other technical people will be the major users, they will not be the only users. Librarians, editors, printers, and secretaries, among others, will also write, read, or edit mathematical files. Thus the first principle in developing a notation is that it must be usable by both technical and nontechnical people.

Next consider the large and rapidly growing number of devices used for computer input and output. An efficient communications system must be able to utilize all of

[1]"The Association of American Publishers Standard for Electronic Manuscript Preparation and Exchange", 1987, Association of American Publishers,Inc., Washington D.C.

these devices. This leads to the second principle: The notation should require only a standard ASCII keyboard for input. In addition, files should be stored using only standard ASCII characters. Output from such files will depend on specific driver programs. Non-ASCII characters in output should be called for by special input codes composed of ASCII characters.

A third principle is that codes (tags or prefixes) must be inserted at the places where changes occur. For example, where a paragraph ends and another begins, a tag is required. This idea is simple in that the author knows whenever a perturbation occurs, exactly at that point a code is required. We may think as we go through a file: now I am starting a paragraph so I must insert a code; now I begin displayed math so I must insert a code. In general we begin passages with a code. In some cases a code is required to begin and end a passage. Those codes which require both a beginning and an end are clearly indicated.

A final principle is that the notation should be consistent with SGML. It is helpful to think of a three tier system. The highest level is SGML which establishes basic rules for communication, the most important of which is the separation of form and content. The second level is the AAP level. Communications in AAP notation may readily be transmitted between sites. The AAP notation may also be used for input and output. Many organizations may choose, however to use a different notation internally (the third level). If this notation conforms to the principles above then it will be reasonably easy to translate to the AAP level.

Section 2 will outline the reasoning behind the development and a handout will provide the specific details of the notation for mathematics.

2 THEORY

The input and storage of mathematical information require the solution of two problems: (1) how to indicate the positioning of characters (symbols) on a two dimensional space and (2) how to uniquely identify specific characters. The solution should be reasonably complete, logically consistent, and easy to understand.

2.1 Problem of Positioning

The problem of positioning has two general solutions. The first we will call the "literal" and the second the "geometric". The goal of the literal solution is to input mathematics as it would be spoken. That is, we would input "sum from i = 1 to infinity" for $\sum_{i=1}^{\infty}$. The advantage is simplicity in both input and reading of the data file.

There are disadvantages. The most important is that this solution violates the first principle. Even if a nontechnical user understood the meaning of $\sum_{i=1}^{\infty}$, what would happen with $\bigvee_{i=1}^{\infty}$? Or with $\mathbb{W}_{i=1}^{\infty}$? Or with $\bigcup_{i=1}^{\infty}$? Clearly this is not a solution which may be used by nontechnical people.

A second disadvantage is that no common language exists for "speaking" mathematics. One might say for $\frac{a}{b}$: (1) a over b; (2) a divided by b; or, (3) the fraction a over b. Similarly, $\sum_{i=1}^{\infty} x_i$ might be verbalized as "the sum over x". As the mathematics become more complex the number of possibilities increase. Thus if we were to accept the literal approach we would first be required to define a verbal language. It should be noted that it was the difficulty of developing and using such a language that led to mathematical notation.

Finally the literal approach is inconsistent with principle three in the introduction. Note that in the input form "a over b" does not have a tag until after the change (perturbation) has occurred. Indeed we do not know that a fraction is beginning until the numerator is completed.

The second possible solution is the geometric. Here mathematical entities are grouped according to their geometric properties. For example a fraction has the properties that a segment (the numerator) is centered above a line (the fraction bar) and a second segment (the denominator) below the line. The input requires three pieces of information: (1) that the entity (fraction) is beginning; (2) where the numerator ends; and (3) where the denominator (as well as the fraction) ends. At each of these places a change occurs and, therefore, a code is inserted.

The geometric solution has the advantage that nontechnical users can be easily trained to recognize the small number of geometric forms required. Thus it is usable by everyone involved in the communications system. The solution has the further advantage that it inserts tags at the points of change.

There is a second type of geometric solution where the symbols are moved into position on a CRT screen. Because there an infinity of possible paths the entropy is high. Therefore files created in this way are incompatible with SGML which is a much lower entropy system.

This paper recommends the geometric solution. Mathematical symbols are then positioned in a small number of geometric forms (for this discussion, called *frames*). An example of a frame is the fraction. This frame consists of a numerator, a denominator, and the fraction bar. The numerator and denominator are parts of the frame called *chains*. The boxes in the following diagram are chains and the overall construction is a fraction frame.

$$\frac{\boxed{\textit{Numerator}}}{\boxed{\textit{Denominator}}}$$

The positioning problem is then reduced to the identification of a complete set of frames and deciding on the required tags (codes). Table 1 gives a possible set of frames with recommended tags.

Positioning tags convey different amounts of information and therefore must be of variable length. Consider the three tags required for a fraction. The first simply tells that a fraction is beginning. The second tells that a "separation" (change from numerator to denominator) has occurred. The third tag tells that the fraction frame has terminated (again two pieces of information).

Consider a further way in which the amount of information within a frame tag may be increased. There is a frame for a matched set of vertical bars:

$$\left|\frac{A+1}{B}\right|.$$

TABLE 1. Basic mathematical frames

Frame	*Example*	*Code*
Array	$\begin{matrix} a & b \\ c & d+1 \end{matrix}$	<AR>
Box	$\boxed{x \geqslant y}$	<BX>
Fences	$\left\langle \begin{matrix} 0 \\ 1 \end{matrix} \right\rangle$	<FE>
Formula		<FO>
Fraction	$\frac{1}{a+b}$	<FR>
Limits	$\sum_{n=1}^{\infty}$	<LI>
Overbar	$\overline{xy}$	<OV>
Radical	$\sqrt{xy}$	<RD>
Underbar	$\underline{xy}$	<UN>

But there are a number of variations on this frame. Consider:

$$\left|\frac{A+1}{B}\right|, \quad \left\|\frac{A+1}{B}\right\|, \quad \left|\!\left|\!\left|\frac{A+1}{B}\right|\!\right|\!\right|, \quad ¦\frac{A+1}{B}¦.$$

$$\left|\frac{A+1}{B}\right., \quad \text{and} \quad \left.\frac{A+1}{B}\right|$$

Thus a frame tag may carry three types of information: (1) the particular frame involved; (2) whether the tag represents the beginning or end of the frame; and (3) possible variations (attributes) on the basic frame.

Because the tags carry a variable amount of information they must have beginning and ending delimiters. We will use the SGML forms: **<...>** to begin a frame, and **</...>** to end it. The frame involved will be identified by two letters (i.e., **FR** for fraction). This two-letter code is the generic identifier (GI) for the frame. Variations on a frame will be indicated by following the GI with a space and a second two letter code called an attribute.

For example consider the fraction $\frac{a}{b}$. It would be input as: **<FR>a<>b </FR>**. (The tag which ends the numerator and begins the denominator we may call a separator. It may take this form because the SGML definition for the fraction frame specifies that a denominator must follow a numerator and hence the short form is unambiguous.)

2.2 Problem of Identification Mathematical files may contain 15 or more separate type fonts (italic, bold, Greek, script, etc.) and over 400 unique symbols. As with the positioning problem, there are two approaches. First the mnemonic where one keys a word (or part of a word) which corresponds with the name of the symbol. For example, alpha (from the Greek font) might be keyed as **α** (the ampersand indicating that a mnemonic follows). This procedure has a certain simplicity. Frequently the author is able to key exactly what he is thinking (at this point in his writing he is thinking "alpha" and he keys `α`). It has the further virtue that readers of the file can often understand the special symbols without reference to a look-up table.

There are several disadvantages. First, there are so many special symbols that it is impossible to uniquely specify or remember a mnemonic for each one. It is simple to remember plus, minus, and sum. But what does one call $\precnsim$? Thus the author of a complex manuscript must still refer to a list of codes. The use of this mnemonic list requires some knowledge of the symbol name and is therefore useless to the nontechnical user. Second, more keystrokes are required for mnemonics. This is a necessary feature of mnemonics because they attempt to relate to the name of the symbol or font.

The second possible solution (the use of precedence codes) would key alpha as (for example) `&ga;`. The ampersand, within the formalism of SGML, heralds an entity reference, but from the user's point of view it may be taken to indicate that a special symbol or font follows. (Note that a special symbol really means a change to a font containing special symbols. Thus it is logical to consider the problems of special symbols and font changes together.) The `g` identifies the new font as Greek and the `a` indicates the particular Greek character. Note that this solution is, in fact, a modified mnemonics in that the letter indicating the character usually (but not always) is the first letter of the character's name.

When more than one font is applied to a character we may "stack" . That is a bold alpha would be either `&BGa;` or `&GBa;`. Further, a bold, slanted alpha could be `&BIGa;`, `&IBGa;`, `&GBIa;`, etc.

There are several advantages to the use of precedence codes. The most important is that it is logically satisfying. An ampersand always indicates a change of font. The next character always indicates the new font and the third character always indicates the specific character. This consistency simplifies both keying and parsing. The use of precedence codes along with a well-designed character layout allows nontechnical users to key complex mathematics. At the same time its logic should appeal to the technical user.

The main disadvantage to the use of precedence codes is that the user must always have the list of codes available. On balance it seems that mnemonics is more suited for material with relatively little (or simple) mathematics. For heavy mathematics precedence codes have the advantage. Thus the AAP solution will recommend precedence codes but allow a defined set of mnemonics as a user option.

In addition to this overall philosophy for identification there are several specific problems.

Generic identification. In some cases symbols may be identified generically. For example, a vector might well be identified as a vector rather than as boldface. This has the advantage that the file containing this identification could in one application set all vectors as boldface and in a second application set all vectors with arrows above. The problem is that there are a very large number of possible generic prefixes. Most individuals dealing with technical information (beginning with the author) are trained to identify symbols by font. Thus to completely change to a totally generic system of identification would be confusing.

It will be better to identify most special symbols by their type font but allow for a small number of mathematical entities which can be identified generically. These exceptions represent symbols which do not have a standard type representation.

Cute codes. Some coding systems allow the use of "cute codes" such as a hyphen followed by a greater-than sign (->) for a right arrow. There are several hundred math symbols, of which only a very small number can be keyed in this way. Thus cute codes cannot be made into a complete and consistent set, and should not be used.

Locks. Special symbols may occur one at a time or in groups. In mathematics they are more likely to appear alone. It is therefore convenient to have change font codes apply only to one character. When a string of characters will be in the new font then a "lock" will be required. These locks are defined in the AAP standard.

Access codes. It will reduce confusion if prefixes used for identification have a unique form. One may then immediately recognize an identifier (access code). In general there will be three kinds of codes within a file. The first are the SGML tags for entering and leaving mathematical passages. The second are tags for positioning (markers in what we will call mathematical *frames*); these also may have the form of SGML tags. The third are access codes—identifiers for characters that are not on the keyboard. The form for an access code is an ampersand to indicate a symbol or font change, followed by a single character indicating the new font, followed by a single character specifying the particular character within the font, followed by a semicolon to indicate the end of the special access.

The complete mathematical notation developed from these considerations may be found in the mathematics chapter of the AAP document.[1] I would like to acknowledge the contributions to this effort by the members of the joint AAP and GCA committee on the markup of mathematical formulas.

CONTENTS

Social, Ergonomic and Stress Aspects of Work with Computers,
edited by G. Salvendy, S.L. Sauter and J.J. Hurrell, Jr.
Elsevier Science Publishers B.V., Amsterdam, 1987 — Printed in The Netherlands

COUNTERINTUITIVE CRITERIA FOR VISUAL DISPLAYS

HARRY L. SNYDER
Displays and Controls Laboratory, Department of Industrial Engineering and Operations Research, Virginia Polytechnic Institute and State University, Blacksburg, VA 24061 USA

ABSTRACT

Existing handbooks, texts, and design criteria for visual displays do not generally address the variables necessary to assure compatibility between appropriate software interface design and user capabilities. Dependence on visual science and human factors engineering data sources can be misleading and can lead to less-than-optimal results. This paper discusses visual display research results which are imp;ortant to human-computer compatibility, yet are counterintuitive.

1 INTRODUCTION

As professionals concerned with human-computer interaction, we spend considerable time looking at visual displays. We have learned to judge quite critically the differences between good quality displays and poor ones, and we have even generalized that critical judgement to other forms of visual presentations such as slides at meetings, photographs, and perhaps works of visual art. In our work, we follow the standard textbooks and military or commercial standards to assure that we use characters of adequate size and contrast, avoid the wrong colors for color coding, and suggest only those displays that do not flicker, swim, or have noticeable distortions. In a word, we use and recommend only <u>good</u> displays in accordance with our scientific principles.

These scientific principles are supported by hundreds of research studies which show the effects of many variables on user performance in a number of fundamental perceptual tasks and in more complicated system simulations. The results of these tests are conveyed consistently in tables of published guidelines or standards.

When we deviate from symbolic displays and printed text to literal images such as photographs, we have a different, yet equally large, body of experimental data to follow. Among others, Eastman Kodak scientists have been researching the concepts of image quality for monochrome (e.g., "black and white") photographs for decades; those researchers have contributed to standards of image quality for color photographs for entertainment purposes as well. In fact, a similar body of research, although less accessible, exists in photographic image quality for reconnaissance purposes to assure the photointerpreter of the best and most usable images for his task.

In spite of this huge body of data, in recent years we have seen the development of systems to which the old guidelines do not apply; that is, many imaging systems and alphanumeric displays are electronic in nature and not just electronic in the form of the home television system or the personal computer CRT display. Rather, we now find that displays exist in color and monochrome for alphanumerics and symbolics and literal images, in forms ranging from the liquid crystal and LED wristwatch to the varied quality of PC images to the very high resolution and quality of some computer-aided-design (CAD) workstations.

The capabilities and limitations of these displays derive from their intrinsic hardware design. The quality of the displayed image is thus a product of both the hardware characteristics and the software which commands the image. The software, however, cannot transcend limitations in the hardware. As a result, particular hardware limitations may severely curtail the utility of the system in spite of well-designed human interface software. For these reasons, it is imperative that we ask some critical questions about the current and future capabilities of electronic display hardware. What are the specifications for a good (acceptable, excellent) electronic display? Are the specifications task dependent? What human factors engineering guidelines and rules exist for selection and evaluation of these displays? Do our personal or expert opinions provide good answers? How can visual science data be used to predict both task performance and user acceptance of some of the electronic displays that we will be dealing with in the AI, CAD/CAM, or desktop publishing environments?

A little thought to these questions clearly leads to the conclusion that existing handbooks, standards, and texts do not address these issues, nor do they offer any useful approaches to the problem. To the contrary, one must look to the primary research literature in the hope of finding the answers.

Questions such as these have spurred considerable research for a variety of applications ranging from the simplest desktop PC to the most flexible high-resolution image-processing color display. Researchers have attempted to apply models and concepts from visual science, photographic science, electrical engineering, and simple human factors sources to these problems. Many experiments have been conducted to evaluate the models or to produce basic task data for given system applications. Most of the research has been useful and directly applicable to the problems and theories of interest.

Perhaps of greater importance, however, is the fact that some of the research is especially interesting in that it has not produced the expected answers but, instead, answers that are totally unexpected and counterintuitive. These counterintuitive results have caused us to rethink some of the established concepts and to develop new and different guidelines.

This paper describes a few of these experimental issues and illustrates through them how our scientific methods once again are important, indeed critical, in generating data through controlled experimentation rather than intuition, to develop solutions, concepts, and models for better system designs. To illustrate, I will take examples from several dimensions of visual display design and evaluation: the spatial, chromatic, and temporal domains.

2 SPATIAL DOMAIN RESEARCH

In display and visual research, the spatial domain refers simply to things which are visually separate in (typically) two-dimensional space: vertical and horizontal. This type of encoding applies to nearly all visual displays and is the only form of information coding on most displays. Thus, we make use of concepts such as luminance contrast and visual acuity, spacing between lines of text, and the like--concepts which are not new.

We also evaluate displays, both objectively and subjectively, in terms of their image quality or perceived visual sharpness. In photographic prints, we select those that are free of grain or specks and rough or blurred edges, and which appear to have good "sharpness." In fact, there are ways to measure sharpness (Dainty and Shaw, 1974), which can be shown to correlate with both visual task performance and subjective image quality (Snyder, 1987).

In electronic displays, we also frequently look for sharpness in the image. Unfortunately, the sharpness to which we generally attend is in the wrong part of the image which leads to poorer task performance regardless of whether the display is a flat-panel solid-state dot-matrix display or the more common raster-scan CRT display. This is so for the following reason.

Most of our computer-addressed displays are based on the television form of a raster scan--one composed of a large number of sequentially scanned parallel lines, each containing a large number of addressable locations (picture elements or pixels). As each pixel on each line is made brighter or dimmer (or a different color), a pattern is developed. Conversely, a dot matrix display is composed of individually fixed dots (also pixels) in both the horizontal and vertical dimensions. Thus, while the television-type raster is continuous in the horizontal dimension and discrete in the vertical dimension, the dot matrix display is discrete in both dimensions. This distinction is important in assessing image quality.

By coordinating brightness or chrominance of pixels, both along the raster line and across the many raster lines on a raster display, or in the two spatial dimensions of the dot-matrix display, a two-dimensional pattern is developed. The pattern, which is of utmost importance to the user, is formed by the locations which are changed in brightness and color, not by the spatial positions of the raster lines or the matrixed dots themselves (although the two are obviously not independent). Thus, we must keep in mind that the individual locations of brightness and color must be designed appropriately and that the raster lines or matrixed dots and their visibility are only a means to this end. This issue is best illustrated by several studies.

F. T. Thompson (1957) performed an interesting, yet not well known, experiment. He asked 50 people, one at a time, to enter a studio in which there was a television set displaying a commercial program. A single chair in front of the set could be easily moved toward or away from the set. Each subject was asked to place the chair at the distance he felt most comfortable in viewing the television. Those distances were recorded.

Thompson then replicated the experiment with an additional 50 subjects and a slightly modified television set. In this portion, he added what has been called a "spot wobble" to the television to cause the scanning electron beam to rapidly move up and down a small amount as it moved across each raster line. That is, it filled in the dark space between raster lines, essentially by elongating the spot vertically or defocussing it in the vertical dimension. Those subjects, one at a time, placed their chairs approximately 40 percent closer to the television set than did the first group of subjects.

Obviously, it can be seen that the subjects sat closer to the television set when the raster lines were not visible than when they were visible. The situation is not unlike looking into a neighbor's yard through a chain link fence. Placing your eye at a distance such that the fence is in focus does nothing to permit you to see into the yard; however, should you move back far enough so that the fence structure is too small to see, then the yard image becomes clearer. The objective of moving back is to put the interfering image of the fence outside the size or spatial limits of sensitive vision.

To understand this issue better, it is useful to consider the concept of the visual contrast sensitivity threshold, which relates the amount of contrast required to see a sine-wave grating or pattern to the size of the pattern. Because the visual system generally works in angular units, we can best measure the size of the pattern in cycles of the light and dark bars of the pattern in degrees of angular subtense to the eye. A psychophysical experiment to measure this function will generally produce results like those shown in Figure 1. The upper and lower functions are estimates of the 90 percent normal vision population limits around the threshold function. Note that there is a range at which the eye is most sensitive (about 2 to 4 cycles per degree). At lower spatial frequencies (or larger patterns), the eye requires more contrast to see the pattern; similarly, for smaller bars in the pattern, or higher spatial frequency, the eye also requires more contrast. The spatial frequency at which the eye no longer can see the pattern, even with 100 percent contrast, is the limiting resolution of the eye, which is about 60 cycles per degree, or one cycle per minute of arc which is our conventional rule of thumb.

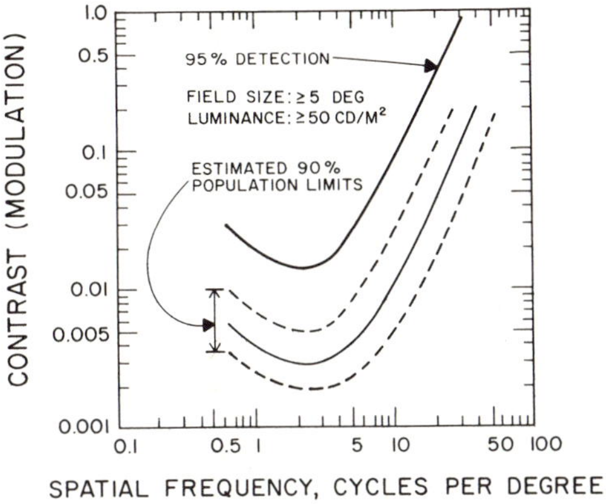

Fig. 1. Typical contrast threshold sensitivity function.

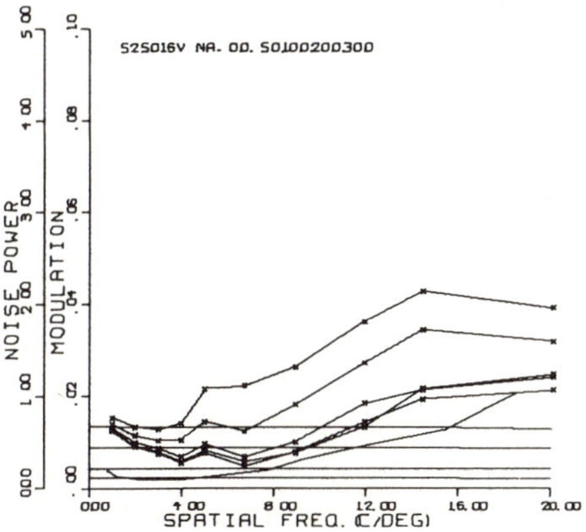

Fig. 2. Contrast threshold curves for grating perpendicular to raster.

To understand how the visibility of raster lines interferes with visual detection of a pattern, we measured contrast sensitivity thresholds for a number of raster-scan displays. Figures 2 and 3 illustrate only a small part of the results of this research (Keesee, 1976). The contrast threshold was determined for a standard 525-line television display with the bar pattern perpendicular to the display raster; that is, the bar pattern the subject was to detect was oriented vertically. The results (Figure 2) show the typical threshold curves, in this case for five video noise levels. When the bar pattern to be detected is rotated 90 degrees to the horizontal position, however, two things happen (Figure 3). First, the contrast required to see the bar pattern increases substantially. Second, the higher frequency bar patterns (smaller bars) are no longer detectable at any contrast--the visual interference of the raster lines prevents detection of the pattern. This interference, of course, is due to sum and difference spatial frequencies, or aliasing, of the display by the discrete sampling of the raster. Simply stated, the visible raster interferes with visual performance.

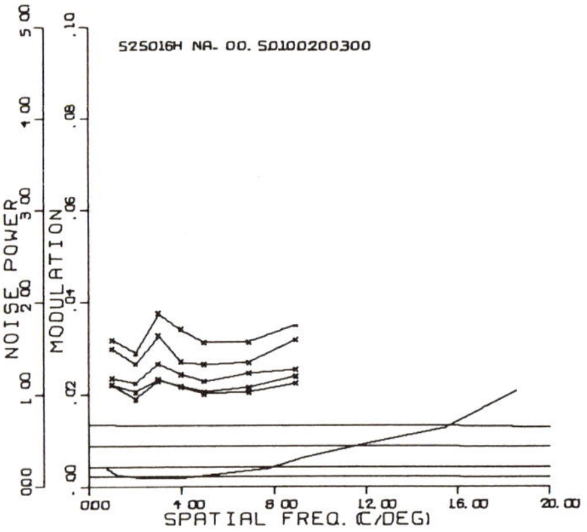

Fig. 3. Contrast threshold curves for grating parallel to raster.

Of course, Thompson was not a psychologist or human factors engineer and was therefore content in knowing what people liked on the display. While such information is interesting, we need to know how and if raster visibility affects performance. To determine this, we performed two studies on the effects of spot wobble and raster contrast on visual task performance (Beamon and Snyder, 1975; Snyder, Beamon, Gutman, and Dunsker, 1980). In both studies, as the raster modulation decreased, the display appeared less sharp or more fuzzy. Yet, task performance for otherwise noise-free displays increased significantly for the fuzziest image. That is, the least sharp appearing displays performed best!

What does this mean in terms of the now common dot-matrix display? We all know how annoying it can be to read a coarse, printed letter or report done by a draft quality dot-matrix printer. Is it merely aggravating to read such or is there a reduction in reading performance from such printed text? The answer is the latter, even though the dots in the dot-matrix printed page are quite sharp and visually clear as shown by some parallel research on dot matrix electronic displays.

Snyder and Maddox (1978; 1980) simulated a number of dot matrix display parameters on a high resolution graphics CRT and validated the results using a variety of dot-matrix displays. Thus, using the CRT they controlled the size and shape of individual dots, as well as the spacing between them, both vertically and horizontally. Several legibility and readability experiments were conducted.

As part of this program and noting that a one-dimensional discrete (raster) display interferes with visual performance, they determine that a similar two-dimensional effect could be obtained. One of the studies investigated the effects of the size of the individual dots in the dot matrix characters and the spacing between the dots on text reading performance. The results (Figure 4) clearly indicate that the greater the space is between the dots, the longer is the reading time, if the reading time is normalized to that with (typewritten) stroke character printed text. Again, the larger the space between the information bearing elements (the dots), the poorer the visual performance, even though the dots are visually sharp.

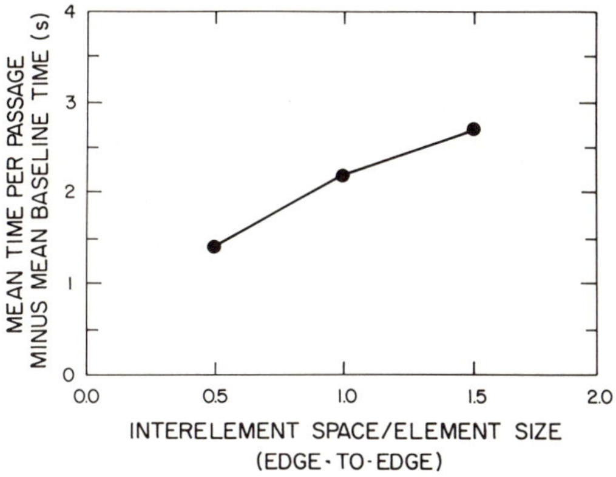

Fig. 4. Effect of spacing between dots on reading time.

These examples from the spatial domain research are only a couple of those that might be mentioned. However, they nicely illustrate a point. People traditionally evaluate electronic (e.g., television or PC) displays in terms of the perceived sharpness of the line (raster or dot) elements. In actual fact, those displays having the sharpest perceived image structure are indeed the poorest for visual task performance. Thus, the uninformed person still chooses the display with the sharpest raster lines or dots even though they interfere with visual information reception.

But progress is being made. The Human Factors Society draft ANSI standard for VDT workstations includes a requirement that the raster modulation of the VDT display not exceed 20 percent. In contrast ratio terms, that means that the ratio of the light part of the raster line to the dark space between raster lines is no greater than 3:2. Most "good quality" displays already meet this standard, and most of the very inexpensive and visually poor displays do not. A similar percent active area requirement is placed on dot matrix displays.

3 CHROMATIC DOMAIN RESEARCH

For many years, our literature has contained recommendations to avoid using red characters and, in particular, red and blue foreground/background combinations due to chromosteropsis. Also, this literature points to the need for luminance contrast in addition to any chromatic contrast to see objects against any background. In fact, these earliest recommendations, published in 1942, suggest that luminance contrast can be traded off against chrominance contrast to improve legibility of displayed symbols. Because there was no agreed upon metric of color contrast that adequately described this luminance/chrominance tradeoff, we conducted a five-year research program on the subject. The program was structured around a display format that required subjects to read strings of numerals of known chrominance and luminance presented against backgrounds of different known chrominance and luminance. Unique software and radiometric measurement concepts permitted control of all display parameters well under threshold differences.

Of initial interest was a comparison of achromatic, yellow-green, and red numerals against a variety of chromatic backgrounds. Contrary to the recommendation in the standard literature, the response speed was fastest to the red numerals, even when the red numerals were presented against a blue or magenta background.

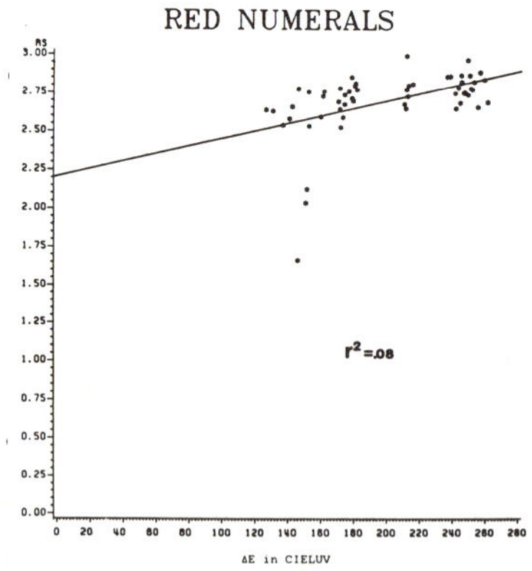

Fig. 5. Correlation between response speed (RS) and distance in CIE L*u*v* space for red numerals.

Presumably there should be predictive tools in the world of color science to explain these results. Turning to the standardized CIE color spaces, one notes first that the CIE (x,y) space is not psychophysically uniform as proven by many studies. But the 1976 CIE conversion to the dimensions of L*u*v* is supposed to be perceptually uniform to a reasonable degree. It turns out that the correlation between distance in L*u*v* and response speed (RS) is essentially zero as illustrated in Figure 5 (Lippert, Farley, Post, and Snyder, 1983). However, by using the axes of luminance (Y) and the chrominance (u' and v'), and by rescaling the axes one to the other, a very good prediction occurs (Figure 6).

Thus, the classic dimensions and predictions from the psychophysical literature do not permit us to predict legibility from combinations of chromatic and luminance contrast, but rescaled axes from empirical research do. space for red numerals.

This result has been replicated in two additional studies, including both positive and negative luminance contrast, and has been shown to predict experimental results from other laboratories as well (D. Post, personal communication, 1986). Because of the consistency of these results, this measure of color space for legibility has also been incorporated into the draft ANSI VDT standard.

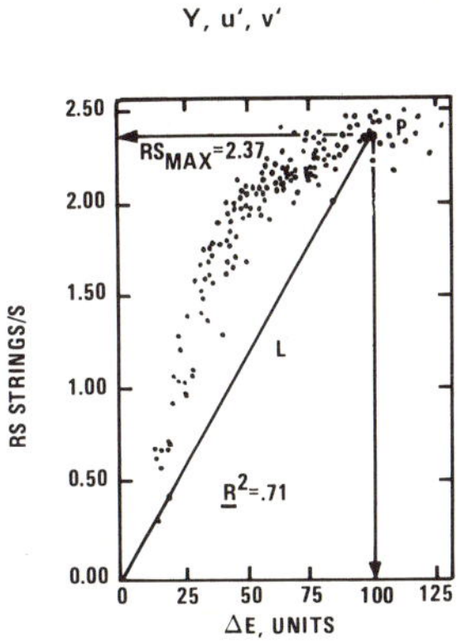

Fig. 6. Correlation between RS and distance in Yu'v' space.

4 TEMPORAL DOMAIN RESEARCH

The visual system has a temporal contrast sensitivity function similar to the spatial function. As illustrated in Figure 7, the visual system is most sensitive to frequencies near the middle of the range and less sensitive to those above and below this mid-point. Considerable data exist on the shape of this threshold curve, above which the perceptual response is flicker and below which the response is nonflicker or fusion. One can use these data to predict the likelihood of a display user to perceive flicker, or the designer can predict the presence of flicker from a knowledge of the refresh rate of the display, the mean luminance of the display, and the luminance temporal modulation over time (Snyder, 1980).

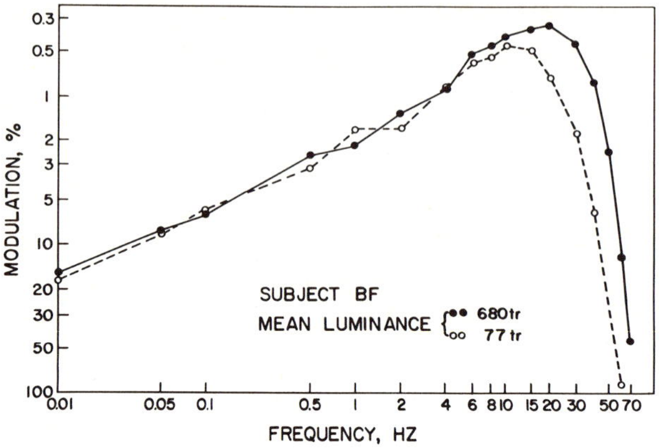

Fig. 7. The temporal contrast threshold function for different luminance levels.

Of scientific interest, although unlikely to be used for display design purposes, is the fact that the decreasing sensitivity of the visual system to lower frequency modulation continues at least to frequencies as low as 0.01 Hz (Almagor, Farley, and Snyder, 1979). Thus, the ability of the eye to integrate over longer intervals continues to decrease to intervals as long as 100 seconds. Figure 7 illustrates the sensitivity for one of several subjects over this wide temporal frequency range for a display subtending over 160 degrees at the eye.

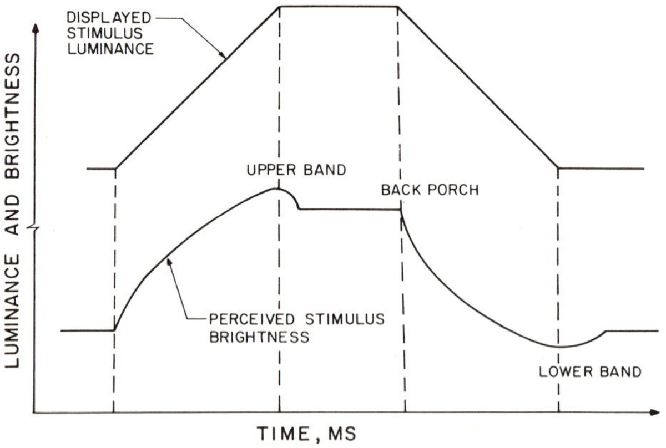

Fig. 8. Prediction of the temporal band.

Most interesting in this research is the issue of the internal control of the visual system in the temporal domain. Based on data of the form shown in

Figure 7, it was hypothesized that the integration time constant of the visual system would decrease with increases in luminance. Then, one predicts a temporal Mach Band phenomenon as shown in Figure 8. That is, as the luminance changes in a ramp function over time, the integrating time constant of the visual system locally changes, thus predicting an overshoot of sensitivity due to the lag of the control system. By measuring the duration of the overshoot using sensory cross-modality matching, we determined that the time constant of integration in the visual system is directly controlled by the luminance and luminance rate of change immediately preceding the measurement interval. (Shorter time constants exist for higher luminance levels.) Further, the critical flicker frequency (or modulation at threshold for any frequency) can be predicted accurately from this measured time constant.

Now, let us look at an application of these data to the work environment, particularly to the preferred luminance level of an individual's visual display. If we ask people to set the variable illuminance on their desk to the level they wish, we can then correlate the preferred illuminance with the length of their integration time constant, anticipating that persons with shorter time constants prefer more illumination. Such a relationship indeed turns out to be the case--at least for our sample of six subjects. Similarly, persons with short time constants will generally want brighter displays or at least displays with more contrast. Because the time constants of normal vision subjects vary by more than two to one, it is reasonable that room illuminance levels and display luminance controls have at least this range of variation.

Lastly, in the time domain, one additional observation may be of interest. Our data show clearly that visual integration time increases as the luminance level of adaptation decreases. Therefore, dynamically noisy displays should be presented at lower luminance levels so that the visual system may temporally integrate out the dynamic noise over time; conversely, high resolution, non-noisy dynamic displays should be presented at higher luminance levels so that the eye can better resolve the sharp image and better track the motion in the display. Thus, CAD/CAM displays require greater contrast and higher character luminance levels than do text processing displays. Our current system designs do not follow this logic.

5 CONCLUSIONS

This paper presents a few examples of how standard classic visual performance or human factors engineering data may not directly apply to the design and evaluation of modern high-performance electronic displays. Yet, these parameters of display design can have severe limiting consequences on the utility of the display or the effectiveness of the user interface.

An understanding of basic visual science may help, but it can also sometimes lead to wrong answers. Examples of those wrong answers are given above. As a result, there remains a demonstrated necessity for good empirical controlled research, both in the laboratory and in the working environment. After all, is this not the premise upon which our science was originally based? Should that premise be forsaken because of the development of new forms of technology? It appears that the examples contained in this paper, as well as others, restate the need for not replacing intuition, however insightful, with solid empirical research in some critical areas.

6 REFERENCES

Almagor, M., Farley, W.W., and Snyder, H.L., 1979. Spatio-temporal integration in the visual system. Technical Report AMRL-TR-78-126, USAF, Wright-Patterson AFB, OH.

Beamon, W.S. and Snyder, H.L., 1975. An experimental evaluation of the spot wobble method of suppressing raster structure visibility. Technical Report AMRL-TR-75-63, USAF, Wright-Patterson AFB, OH.

Dainty, J.C. and Shaw, R., 1974. Image Science. Academic Press, New York.

Keesee, R.L., 1976. Prediction of modulation detectability thresholds for line scan displays. Technical Report AMRL-TR-76-38, Wright-Patterson AFB, OH.

Lippert, T.M., Farley, W.W., Post, D.L., and Snyder, H.L., 1983. Color contrast effects on visual performance. Digest, Society for Information Display: 170-171.

Snyder, H.L., 1980. Human visual performance and flat-panel display image quality. Technical Report 80-1, Virginia Polytechnic Institute and State University, Blacksburg, VA.

Snyder, H.L., 1987. Toward the determination of electronic display image quality. In W. Rouse (Ed.), Advances in Man-Machine Systems Research, Vol. 4, in press.

Snyder, H.L., Beamon, W.S., Gutmann, J.C., and Dunsker, E.D., 1980. An evaluation of the effect of spot wobble upon observer performance with raster scan displays. Technical Report AMRL-79 91, USAF, Wright-Patterson AFB, OH.

Snyder, H.L. and Maddox, M.E., 1978. Information transfer from computer-generated dot-matrix displays. Technical Report, Virginia Polytechnic Institute and State University, Blacksburg, VA.

Snyder, H.L. and Maddox, W.E., 1980. On the image quality of dot-matrix displays. Proceedings of the SID, 21:3-7.

Thompson, F.T., 1957. Television line structure suppression. Scientific Paper 8-1041-F14, Westinghouse Laboratories, Pittsburgh, PA.

Social, Ergonomic and Stress Aspects of Work with Computers,
edited by G. Salvendy, S.L. Sauter and J.J. Hurrell, Jr.
Elsevier Science Publishers B.V., Amsterdam, 1987 — Printed in The Netherlands

DEVELOPMENT OF A WORK - REST SCHEDULE FOR VDT WORK

HELMUT T. ZWAHLEN and CHARLES C. ADAMS, JR.
Department of Industrial and Systems Engineering
Ohio University, Athens, Ohio USA 45701-2979

ABSTRACT

Subjective discomfort scores measured at different times during the workday from two VDT operator studies involving six and eight subjects were used to develop a Work-Rest Schedule Model for VDT work. Discomfort score changes per minute for work, rest breaks and lunch break were calculated, using the combined data, for the four discomfort questions (all musculoskeletal) which produced the largest discomfort scores at the end of the workday. A BASIC computer program is included which provides the user with the number and time duration of rest breaks needed to keep a given percentage of the VDT operator population below a selected discomfort score threshold value at the end of a workday.

INTRODUCTION

Zwahlen and Adams (1987) presented a Work-Rest Schedule Model which was based on the subjective discomfort scores of six VDT operators. This study updates that model by using data from eight additional VDT operators and presents a BASIC computer program to calculate the number and time duration of the rest breaks.

DEVELOPMENT OF THE WORK-REST SCHEDULE MODEL

The subjective discomfort score data is presented in Zwahlen, et al. (1984a) and Zwahlen and Kothari (1986). The studies used six subjects (all females) and eight subjects (7 females, 1 male), respectively, all experienced typists with normal uncorrected vision. They were paid (incentive pay, high workload) and tested over two full workdays consisting of two 90 minute work sessions in the morning and two in the afternoon (45 minute lunch break). The two morning (and afternoon) work sessions were separated by a 15 minute rest break. At the beginning and end of each 90 minute work session a subjective discomfort questionnaire was presented on the VDT (one question at a time) and queried the subjects as to the amount of discomfort they were experiencing at that moment. Subjects responded by pressing the minus (-) key from column 00 to column 72 according to their discomfort level. The scale was divided into various discomfort levels using verbal indicators at different points (column 05 "no, not at all", column 20 "only very little", column 36 "somewhat", column 49 "quite a bit" and column 64 "yes, very much so"). A data entry task or a file maintenance task was alternately performed. Files consisted of names of chemical compounds, stock numbers with 10 characters (eg. Q1M20Z1P1), addresses, phone and bin numbers, prices and an eight digit location code. The studies used an original hardcopy for two of the 90 minute work sessions and a split screen presentation for the other two 90 minute sessions. Zwahlen and Kothari (1986) also used two screen polarities. The workstation (IBM chair, adjustable keyboard height) was well lighted with non-glare lighting (Armstrong Tascon lighting fixture) and the VDT (IBM 3101) was fitted with a glare filter.

Four musculoskeletal discomfort questions (Question 1-Discomfort from back, arms or other parts, Question 3-Discomfort, stiffness or soreness in neck, Question 4-Discomfort, stiffness or soreness in shoulders, Question 7-Discomfort, stiffness or soreness in back) produced the highest discomfort levels at the end of a workday for both studies (see Table 1). The score range for the four questions is 42 to 56 (36="somewhat", 49="quite a bit"). Table 1 also indicates that the musculoskeletal discomfort questions produce much higher discomfort scores at the end of the workday than the ocular discomfort questions which range from 08 to 36 (05="no, not at all", 20="only very little"). Zwahlen, et al. (1984b) illustrated that in general the subjective discomfort of VDT operators increases over a workday and that scheduled rest breaks away from the workstation help to reduce the discomfort level. Table 2 shows that the discomfort score changes per minute for work and the two rest breaks are fairly constant over the workday and among the four selected musculoskeletal discomfort questions. The W-R Schedule Model is based on a representative average of the discomfort questions which produce the largest scores at the end of the workday. The general methodology can be applied to any similar VDT work situation where subjective discomfort scores have been measured for a number of VDT operators before and after each work session over an entire workday using the procedures outlined by Zwahlen, et al. (1984a) and Zwahlen and Kothari (1986). Subjective discomfort scores are used since accurate, reliable, and easily acquired objective measures do not

TABLE 1. Summary of Subjective Discomfort Scores for Both Ocular and Musculoskeletal Questions at the Beginning and at the End of the Workday from Zwahlen, et al. (1984a)-Study 1 (S1, N=12) and Zwahlen and Kothari (1986)-Study 2 (S2, N=16)

NO.		BEGINNING MEAN	BEGINNING SDEV	END MEAN	END SDEV	NO.		BEGINNING MEAN	BEGINNING SDEV	END MEAN	END SDEV
1	S1	1.5	0.5	42.2	9.9	10	S1	1.5	0.5	18.5	13.3
	S2	5.0	7.2	54.3	15.1		S2	10.8	12.1	25.3	18.7
2	S1	8.2	8.3	34.3	18.0	11	S1	6.2	5.7	20.2	12.6
	S2	7.9	10.8	39.3	19.4		S2	6.3	10.2	8.7	11.8
3	S1	7.7	6.4	52.4	17.1	12	S1	7.4	10.2	20.6	17.0
	S2	3.6	4.0	50.8	17.1		S2	7.9	10.8	13.9	17.6
4	S1	9.5	10.8	44.9	13.8	13	S1	7.4	7.1	19.9	17.4
	S2	5.4	9.2	48.6	15.2		S2	7.5	11.7	14.3	17.5
5	S1	7.3	6.2	32.3	17.6	14	S1	8.2	9.1	24.7	20.8
	S2	3.1	3.9	29.8	21.2		S2	17.1	19.6	29.8	17.3
6	S1	2.4	2.5	22.4	15.9	15	S1	9.6	10.9	32.3	18.2
	S2	1.9	2.4	19.8	15.1		S2	10.9	13.6	36.5	17.5
7	S1	10.3	10.6	47.7	15.5	16	S1	17.4	13.9	23.9	25.1
	S2	8.0	10.8	56.4	16.5		S2	20.7	16.8	27.0	17.3
8	S1	3.8	2.6	22.9	16.3	17	S1	14.8	16.4	11.0	12.4
	S2	2.8	4.1	28.4	19.4		S2	9.3	11.4	20.5	20.2
9	S1	3.6	4.2	14.1	15.5	18	S1	5.3	6.0	23.3	16.3
	S2	2.4	4.3	3.0	3.7		S2	7.6	12.0	18.7	22.7

Note: Discomfort scale has a range from 00 to 72

No 1: Discomfort back, arms other parts
No 2: Weariness or general fatigue
No 3: Disçomfort, stiffness, soreness in neck
No 4: Disçomfort, stiffness, soreness in shoulders
No 5: Disçomfort, stiffness, soreness in arms
No 6: Disçomfort, stiffness, soreness in hands
No 7: Disçomfort, stiffness, soreness in back
No 8: Disçomfort, stiffness, soreness in legs, feet
No 9: Headache
No 10: Discomfort from eyes
No 11: Pain above eyes
No 12: Dryness in eyes
No 13: Irritation in eyelids
No 14: Sensitivity to light
No 15: Hard to see sharply
No 16: Blink frequently
No 17: Eyes watering from eyeballs
No 18: Dull, aching feeling

exist at this time.

The W-R Schedule Model assumes that the morning and afternoon periods are of equal length, that all rest breaks are of equal duration and symmetrically distributed with reference to the lunch break and that all work sessions are of equal length. Symmetric spacing and equal duration are assumed from an administrative and motivational point of view. The number of rest breaks during the workday will either be 0, 2, 4, 6, etc. The discomfort scores of the population are assumed to be normally distributed (supported by data) at the end of the workday to make it possible to calculate a W-R Schedule so that a given percentage of the population is below a selected discomfort score threshold value. The shortest break period is 9 minutes, since VDT operators should leave their workstation during breaks and move around to change their posture and visual conditions. The longest rest break is 17 minutes since the data indicated that longer rest breaks, like the lunch break, appear to be less efficient in reducing discomfort.

TABLE 2: Discomfort Score Changes per Minute for Four Musculoskeletal Questions for Each 90 Minute Work Session, for Each 15 Minute Rest Break and for the 45 Minute Lunch Break

QUESTION		SESSIONS						BREAKS							LUNCH					
		STUDY 1 (n=12)		STUDY 2 (n=16)		WEIGHTED OVERALL			STUDY 1 (n=12)		STUDY 2 (n=16)		WEIGHTED OVERALL		STUDY 1 (n=12)		STUDY 2 (n=16)		WEIGHTED OVERALL	
	#	MEAN	SD	MEAN	SD	MEAN	SD		MEAN	SD	MEAN	SD	MEAN	SD	MEAN	SD	MEAN	SD	MEAN	SD
1	1	.36	.10	.32	.18	.34	.15	1	-1.34	.83	-.87	.90	-1.07	.89	-.41	.42	-.70	.48	-.57	.48
	2	.26	.16	.29	.13	.28	.14	2	-.73	.73	-.64	.81	-.68	.77						
	3	.19	.16	.32	.18	.27	.18	OV	-1.04	.82	-.76	.86	-.88	.83						
	4	.24	.13	.19	.13	.21	.13													
	OV	.26	.15	.28	.16	.28	.15													
3	1	.33	.15	.33	.20	.33	.18	1	-.68	1.04	-.85	.79	-.78	.89	-.30	.41	-.69	.42	-.54	.45
	2	.16	.24	.26	.14	.22	.19	2	-.75	1.03	-.72	.96	-.73	.97						
	3	.17	.18	.34	.22	.27	.22	OV	-.72	1.01	-.79	.89	-.76	.93						
	4	.23	.20	.16	.12	.19	.16													
	OV	.22	.20	.27	.17	.25	.19													
4	1	.22	.15	.25	.21	.24	.18	1	-.72	.92	-.38	.97	-.52	.95	-.41	.29	-.63	.49	-.54	.43
	2	.26	.13	.20	.17	.22	.15	2	-.79	.94	-.69	.76	-.73	.82						
	3	.20	.13	.34	.15	.28	.16	OV	-.76	.91	-.54	.87	-.63	.89						
	4	.16	.17	.16	.12	.16	.13													
	OV	.21	.15	.24	.17	.23	.16													
7	1	.20	.17	.29	.17	.25	.17	1	-.24	.79	-.48	1.06	-.38	.95	-.32	.36	-.66	.51	-.51	.48
	2	.15	.21	.21	.19	.18	.20	2	-.77	.85	-.77	.67	-.77	.74						
	3	.17	.19	.35	.15	.27	.19	OV	-.51	.85	-.63	.87	-.58	.85						
	4	.23	.18	.21	.12	.21	.14													
	OV	.19	.19	.27	.16	.23	.18													
OV	1	.28	.14	.30	.19	.29	.17	1	-.75	.90	-.65	.93	-.69	.92	-.36	.37	-.67	.48	-.54	.46
	2	.21	.19	.24	.16	.23	.17	2	-.76	1.02	-.71	.80	-.73	.83						
	3	.18	.17	.34	.18	.27	.19	OV	-.76	.96	-.68	.87	-.71	.88						
	4	.22	.17	.18	.12	.19	.14													
	OV	.22	.17	.27	.16	.25	.17													

AVERAGE BEGINNING SCORE (A) = 7.3 - STUDY 1; 5.5 - STUDY 2; 6.3 - UPDATED W-R SCHEDULE MODEL

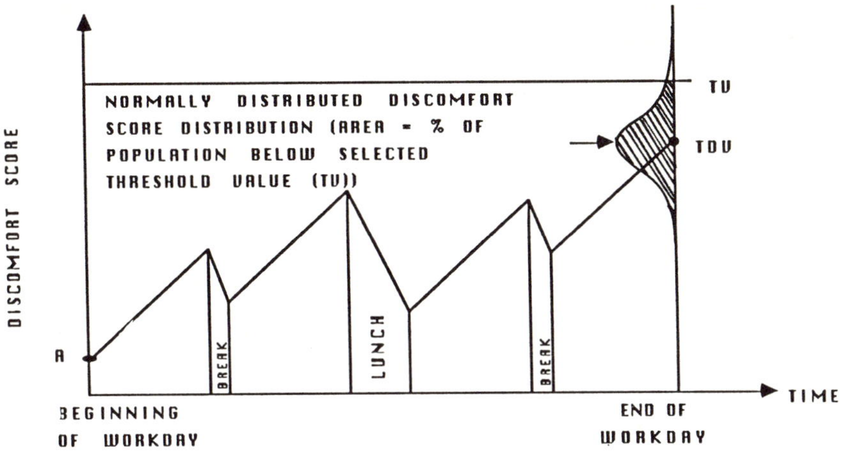

Figure 1. The Conceptual Work-Rest Schedule Model

The conceptual Work-Rest Schedule Model (Figure 1) illustrates how the average VDT operator population discomfort score increases over the workday. Zwahlen and Adams (1987) presented a flowchart for the W-R Schedule Model and it was used to develop a BASIC program as shown in Figure 2. To start, the program reads a standard normal table into an array. It then asks the user to enter the average initial discomfort value (A), the total number of minutes in a workday including rest breaks but excluding lunch (NW), the number of minutes for lunch (NL), the average discomfort score changes per minute during work, rest breaks and lunch (MUW, MUB and MUL, respectively), a discomfort score threshold value (TV), the percentage of the population to be below the threshold value (%BTV), a standard deviation estimate of the population discomfort score distribution (SD) and the maximum number of breaks to be considered (K). A target discomfort value (TDV) and a discomfort score which considers lunch (DSL) is calculated (see Figure 2) and the difference between TDV and DSL (DELTAS) is determined. If DELTAS is greater than or equal to zero, no rest breaks are needed but if DELTAS is negative then the total number of minutes break (TMB) is calculated. If TMB is less than or equal to 34 then the number of breaks required (J) is 2 and if TMB is less than or equal to 18 then the minutes per break (MPB) are set to 9 minutes, otherwise MPB is equal to TMB divided by 2, rounded to the nearest integer minute. The final calculation is for the break time as a percentage of the total work time. If TMB is greater than 34, the number of breaks (J) is set to 4 and J is compared to the maximum

```
10 DIM Z(99)
20 FOR I=50 TO 99: READ Z(I): NEXT I
30 INPUT"Initial starting value (A)=?";A
40 INPUT"Total number of minutes worked (NW)=?";NW
50 INPUT"Total number of minutes lunch (NL)=?";NL
60 INPUT"Discomfort score change per min. for work (MUW)=?";MUW
70 INPUT"Discomfort score change per min. for breaks (MUB)=?";MUB
80 INPUT"Discomfort score change per min. for lunch (MUL)=?";MUL
90 INPUT"Discomfort score threshold value (TV)=?";TV
100 INPUT"% of population below threshold value (%BTV)=?";PBTV
110 INPUT"Std. dev. of VDT operator discomfort scores (SD)=?";SD
120 INPUT"Maximum number of breaks (K)=?";K
130 TDV=TV-Z(PBTV)*SD
140 DSL=A+(NW*MUW)+(NL*MUL)
150 DELTAS=TDV-DSL
160 IF DELTAS>=0 THEN PRINT"No breaks needed": END
170 TMB=ABS(DELTAS/(MUB-MUW))
180 IF TMB>34 THEN GOTO 260
190 J=2: MPB=INT(TMB/2+.5)
200 IF TMB<18 THEN MPB=9
210 BTPWT=(MPB*J/NW)*100
220 PRINT:PRINT"Number of breaks =";J
230 PRINT"Minutes per break =";MPB
240 PRINT"Break time as a percentage of work time =";BTPWT
250 END
260 J=4
270 IF J>K THEN PRINT"Not enough rest breaks": END
280 IF TMB<=J*17 THEN MPB=INT(TMB/J+.5): GOTO 210
290 J=J+2: GOTO 270
300 DATA 0,.028,.052,.075,.104,.126,.151,.176,.202
310 DATA .228,.253,.279,.305,.332,.358,.385,.412
320 DATA .440,.468,.486,.524,.553,.582,.613,.644
330 DATA .674,.705,.739,.772,.806,.842,.878,.915
340 DATA .954,.994,1.036,1.08,1.126,1.175,1.227
350 DATA 1.282,1.341,1.405,1.476,1.555,1.645,1.751
360 DATA 1.881,2.054,2.326
```

Figure 2. A BASIC Program to Calculate the W-R Schedule

number of breaks (K). If J>K the model reports there are not enough rest breaks, otherwise TMB is compared to J*17. If TMB>J then J is set equal to J+2 and TMB is again compared to J*17. This continues until either TMB is less than or equal to the value J*17 or until J>K. If TMB is less than or equal to J*17 then the number of rest breaks equals J, MPB is equal to the nearest integer value for the quantity TMB divided by J and the break time as a percentage of the total work time is calculated.

RESULTS

Figure 3 compares the observed discomfort score averages (4 musculoskeletal questions) from Study 1 and Study 2 with the calculated discomfort scores using the updated W-R Schedule Model. The discomfort scores at the beginning and end of each work session have been connected by straight lines, mainly to show the

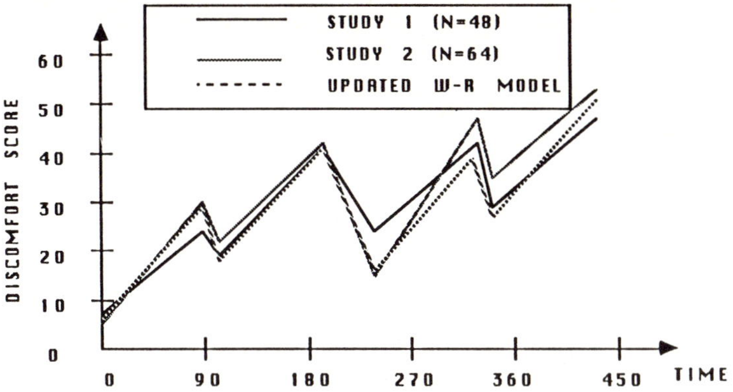

Figure 3. Comparison of Observed Discomfort Score Averages from Study 1 and from Study 2 with Calculated Discomfort Scores Using the Updated W-R Schedule Model.

trend over the workday. The scores from the two studies compare rather well with each other and the calculated scores.

Table 3 compares W-R schedules for the model presented in Zwahlen and Adams (1987) with those for the updated model. The W-R schedules are based on two 240 minute half workdays separated by a 60 minute lunch break. The updated model uses an initial starting value (A) of 6.3 and values for MUW, MUB and MUL of 0.25, -0.71 and -0.54, respectively. Table 3 indicates a very good agreement between the values based upon the previous model and those based upon the updated W-R Schedule Model. Using the updated Model and a rather high discomfort threshold value of 64 ("yes, very much so") and two rest breaks each of maximum length (17 minutes) only allows 56% of the population to have discomfort scores at the end of the workday below the threshold value. For 99% of the VDT operator population to be below the threshold value of 64, four breaks of 16 minutes would be required. Similarly, for a threshold value of 49, six 13 minute breaks would be required in order for 99% of the population to be below the threshold value. It should be noted that the values shown in Table 3 are most closely applicable to the conditions described in Zwahlen, et al. (1984a) and Zwahlen and Kothari (1986) and may or may not apply to other VDT work situations.

CONCLUSIONS

An easy to use updated Work-Rest Schedule Model, which is based upon subjective discomfort and presented in the form of a

TABLE 3. Summary of Selected Numbers of Rest Breaks, Lengths of Rest Breaks and Discomfort Threshold Value Combinations Indicating the Percentage of Break Time and the Percentage of the VDT Operator Population Below the Threshold Discomfort Score.

NO. OF BREAKS	MIN. PER BREAK PM	MIN. PER BREAK UM	THRESHOLD VALUE	% BREAKS PM	% BREAKS UM	% OPERATORS BELOW THRESHOLD PM	% OPERATORS BELOW THRESHOLD UM
2	14	16	64	5.8	6.7	50	50
	17	17	64	7.1	7.1	65	56
4	14	15	36	11.7	12.5	50	50
	17	17	36	14.2	14.2	75	66
	11	12	49	9.2	10.0	50	50
	17	17	49	14.2	14.2	95	91
	9	9	64	7.5	7.5	67	58
	15	16	64	12.5	13.3	99	99
6	11	11	36	13.8	13.8	80	71
	15	16	36	18.8	20.0	99	99
	11	11	49	13.8	13.8	96	93
	13	13	49	16.3	16.3	99	99

Threshold values: 36 - "somewhat"
49 - "quite a bit"
64 - "yes, very much so"
PM- Previous W-R Schedule Model (Zwahlen and Adams (1987))
UM- Updated W-R Schedule Model

BASIC program has been developed. The results from the updated Work-Rest Schedule Model should be regarded as tentative since they are based on subjective discomfort scores from two laboratory studies involving a total of 14 VDT operators.

REFERENCES

Zwahlen, H.T., Hartmann, A.L. and Rangarajulu, S.L., "Video Display Terminal Work with a Hard Copy - Screen and Split Screen Data Presentation", Ohio University, Dept. of Industrial and Systems Engineering Research Report, prepared for NIOSH, distributed by The Report Store, Lawrence, Kansas 66044, 375 pp., July 1984(a).

Zwahlen, H.T., Hartmann, A.L. and Rangarajulu, S.L., "Effects of Rest Breaks in Continuous VDT Work on Visual and Musculoskeletal Comfort/Discomfort and on Performance", paper presented at First USA-Japan Conference on Human-Computer Interaction, Honolulu, Hawaii, August 1984. Published in Salvendy, G., Ed.: Human-Computer Interaction, Elsevier Science Publishers B.V., Amsterdam, pp. 315-319, 1984(b).

Zwahlen, H.T. and Kothari, N.C., "The Effects of Dark and Light Character CRT Displays Upon VDT Operator Performance, Eye Scanning Behavior, Pupil Diameter and Subjective Comfort/Discomfort", Ohio University, Dept. of Industrial and Systems Engineering Research Report, distributed by The Report Store, Lawrence, Kansas 66044, 386 pp., June 1986.

Zwahlen, H.T. and Adams, C.C., "A Quantitative Work-Rest Model for VDT Work", paper presented at the Annual International Industrial Ergonomics and Safety Conference, University of Miami, Coral Gables, Florida, June, 1987.

Social, Ergonomic and Stress Aspects of Work with Computers,
edited by G. Salvendy, S.L. Sauter and J.J. Hurrell, Jr.
Elsevier Science Publishers B.V., Amsterdam, 1987 — Printed in The Netherlands

SOME ISSUES RELATED TO SEATED POSTURE AND WORKSTATION DESIGN

M. J. DAINOFF
Department of Psychology and Center for Ergonomic Research, Miami University, Oxford, Ohio, USA 45056

ABSTRACT

It is generally agreed that modern office workplaces require a certain degree of flexibility/adjustability of workstation components. However, there is considerable disagreement as to the rationale of, and extent of the need for, such adjustability. In an attempt to provide a framework within which such disagreements may be discussed and resolved, a theoretical model is proposed. The model assumes that sitting behavior is fundamentally a homeostatic activity reflecting conflicting demands on the body originating in the need for stability and support as well as relief from discomfort through movement. Evidence is presented in support of the model.

1. INTRODUCTION

As computer systems become the rule rather than the exception in workplaces, and "dedicated" operators, who must spend the majority of their workdays in fixed postures, more common, a fair amount of attention is now being paid to the musculoskeletal stress associated with such postures, and to the use of adjustable "ergonomic" chairs and workstation surfaces as possible aids in alleviating such stress. While the general principle of adjustability per se is widely accepted, there is considerable disagreement as to how this principle is to be applied for maximum effectiveness. The purpose of this paper is to provide a focus for discussing some issues surrounding this disagreement.

2. PSYCHOBIOLOGICAL MODEL FOR SEATED WORK POSTURE

2.1 Branton/Gibson Synthesis

The primary impetus for this model has been provided by the seminal paper of Branton (1976). Much of what follows is an elaboration of that work, but an elaboration which builds, as well, on the ecological insights of J.J. Gibson (1979) regarding the relationship between perception and action. Seating, according to Branton, is fundamentally homeostatic in nature, involving two demands which are (usually) inherently in opposition: the need for structural surfaces to provide stability to the body, and the need to dissipate pressure on bodily surfaces so as to counteract ischemia. In the specific case of seated work posture, the demand for stability is rather more crucial than it might be, for example, in Branton's own research on train passengers, or in reading a newspaper in an armchair. In VDT work, for instance, high speed keying/reading is likely to require a relatively precise localization of hands and eyes within fixed spatial regions. Other modern examples of seated work (microsurgery, microchip assembly) require even greater degrees of stability. However, those support surfaces which provide such stability will also exert some degree of pressure on body surfaces producing a tendency toward ischemia (reduced blood flow) which can only be relieved by some degree of motion on the part of the operator. If ischemia cannot be relieved, the presumed results (via the mechanism of lactic acid buildup) are sensations of discomfort (fatigue/pain). At some point, the discomfort itself, and/or the actions to deal with it (squirming/fidgeting), will begin to interfere with the ongoing task. Consequently, any comprehensive analysis of sitting behavior must simultaneously focus on attributes of the physical work environment together with the relevant aspects, physiological as well as behavioral, of the operator. The work of carrying out such analyses can be made markedly more efficient if an appropriate theory is available for guidance. It is the contention of this paper that the ecological perspective of J.J. Gibson (1979) provides such a theoretical framework. On the one hand, the Gibsonian concept of affordance focuses on invariant attributes of the environment explicitly relevant to a particular pattern of behavior. As recently formulated by Warren (1984), "...an affordance is the functional utility of the combination of environmental properties that support some activity for a particular animal." On the other hand, the complementary concept of effectivity or "...action capability of actor with respect to its environment..." (Mark and Vogele, In press) provides, within the same theoretical framework, a focus on the individual. Therefore, the ergonomic problem of "person-environment fit" can be restated in terms of: (a) specifying the characteristics of the workplace (structural aspects of chair and workstation) which simultaneously afford the operator support to efficiently carry out his/her work along with the capability to minimize discomfort (through reduction of ischemia); (b) understanding the capabilities and limitations, both physiological (biomechanical/anthropometric) and psychological, of the operator in effectively utilizing those structural characteristics of the workplace.

2.2 Outline of the model

Assume a person seated on the edge of a high, solid, flat surface (e.g., a physician's examination table.) This can be considered a limiting case of affordance for sitting, insofar as the sitter's effectivity with respect to sitting (weight supported primarily by the inverted pyramids of the ischial tuberosities with limited additional support of thighs) tends to place him in unstable equilibrium (Branton, 1967), resulting in the tendency for the trunk to rock about the tuberosities. Under these circumstances, virtually the only action the sitter can take to enhance stability is to form an "open-chain system of links" (Dempster, 1955) by bracing his hands and arms against the table edge. If, however, the person now moves to a conventional armchair, the affordance for sitting is improved in the sense that additional support surfaces (seat pan cushioning, backrest, floor) become available. Further improvement in affordance can be achieved by adding flexibility, as in modern adjustable task chairs and workstations. Given the dual demands of stability plus avoidance of discomfort, it can be assumed that the individual will seek to achieve a biomechanical/ homeostatically optimum posture in which the total effort exerted by muscle contraction, together with the combined load on bodily structures, is at some minimum level (Branton, 1976; Pheasant, 1986, Ch. 8). Within the existing boundary conditions established by the total set of affordances, it can be assumed that the operator will attempt to regulate his postural orientation homeostatically around this optimum. Kugler and Turvey (1987), working within the Gibsonian framework, describe such an optimum as an attractor state. However, it is important to assert that such attractor states are by no means fixed or stable, but must vary across different affordance conditions (workplace environments) as well as different effectivity conditions (human somatotypes and/or personality differences). In particular, consider the same operator working under two different sets of affordances--one with a high degree of flexibility ("good" ergonomic design), the second with limited flexibility. In the former case, the operator, assuming she is knowledgeable about the affordances available, can achieve a reasonable level of biomechanical optimality with relative ease by manipulating chair and workstation controls. In the latter case, the limitations in adjustability preclude achievement of the same degree of optimization; the operator's capability for homeostatic regulation is restricted to local bodily movement (fidgeting) or leaving the worksite! Consequently, it can be argued that the degree of affordance may be directly translated into operator comfort and work efficiency. In a direct comparison of workplaces with differences in affordances of the kind described above (good vs. less good ergonomic design), Dainoff (1983) demonstrated increased performance and decreased discomfort symptoms in operators performing the same tasks under both conditions. Mark, Vogele, and Dainoff (1985), in analyzing video records from the previous study, demonstrated lesser amounts of postural shifting in the more optimal workstation. In other experiments, Pustinger, Dainoff and Smith (1985), and Bhatnager, Drury, and Schiro (1985) discovered increased levels of postural shifting/fidgeting to be associated with decreased work performance.

3. SPECIFIC AFFORDANCES AND EFFECTIVITIES IN VDT WORKPLACES

3.1 Spinal Load

Traditionally, seated work has implied a straight-backed chair with a flat seat pan. However, as understanding of muscular and spinal dynamics grew, it became clear that, in such chairs, the act of sitting resulted in a backward rotation of the pelvis with a consequent tendency towards reversal of the lumbar curve of the spine from its normal (standing) inward curve or lordosis, to a flat or outward curving (kyphotic) posture (Chafin and Andersson, 1984, Ch. 9). Such kyphotic postures were, in turn, associated with increased pressures on spinal discs, leading to the possibility of long term back disorder (Nachemson, 1975), not to mention short term pain. Thus, ergonomic guidelines and recommendations began to discuss the importance of backrests with built-in support to the lumbar region which would oppose the kyphotic reversal and relieve pressure on disks (Corlett and Eklund, 1984).

As backrests which could be angled backwards became available, it seemed that, given a choice, workers would opt for a backwards leaning posture over an upright posture--even with a lumbar support (Grandjean,

Hunting, and Pidermann, 1983). Moreover, it was argued that backward leaning made good physiological sense in that a significant amount of load on the lumbar spine was relieved (Chafin and Andersson, 1984).

An alternative perspective, equally challenging to the standard upright posture, came from Mandal (1981,1986). He argued that, from the point of view of spinal load, it would be better for the seatpan to tilt forward. In this way, the pelvis would rotate forward--closer to the standing posture--and the lordosis would be restored. His research evidence indicated that people would be quite willing to work in this posture, and that individuals with existing lower back pain found that forward tilt would minimize such pain.

Recent work in our laboratory has attempted to reconcile both of these positions (Dainoff and Mark, 1986; Dainoff, Mark, Daley and Moritz, 1986.) We have argued, in effect, that either posture may be effective, depending on the total set of affordances available in the environment. A particularly significant element has been the demands of the task. In our research, we have asked subjects to alternate between data entry (copy intensive with high keying demands) and editing (screen intensive with minimal keying demands). In the former task, given the smaller type size characteristic of paper copy, together with the need for fingers to be in continual contact with the keyboard, the attractor state afforded by a forward tilting seat pan was considered optimal and was, in fact, observed by subjects performing that task. If, on the other hand, only a backward tilted or vertical back support is available, the user tends to lean forward in order to see the text, putting considerable pressure on lumbar spine as well as internal organs (Chafin and Andersson, 1984). With the editing task, the screen character size is typically large enough such that leaning against a rearward tilted backrest is now optimal.

3.2 Additional Sources of Load

While much attention has been focused on the problems of the lower spine, prolonged sitting can generate discomfort signals from other body systems which must be integrated into the overall homeostatic search for optimality. We have already discussed the problems generated when the operator is required to lean forward while bending at the waist; load is placed not only on spine and back muscles, but also on head and neck (Pheasant, 1986). If the keyboard surface is too high relative to the elbow, hands and wrists may be strained. There is evidence, summarized by Lueder (1986), that, given a choice, people will adjust chairs higher than their popliteal heights, presumably in order to optimize working position of hands and arms. However, seatpans which are too high place excess pressure on the underside of the thigh, resulting in hampered blood circulation in the lower leg, and associated discomfort (Kroemer, 1987).

4. INTEGRATION

To reiterate the attractor state concept, the seated VDT worker will attempt to homestatically find a postural balance point which minimizes discomfort from all of the above sources (and, most likely, many others not mentioned). A workplace with optimal affordances for flexibility and adjustments will allow the achievement of such a balance point, given the realization that discomfort signals from different parts of the body will likely arise at different times and with different intensities. Thus, for example, in our current research, we might propose that subjects doing data entry work would initially minimize discomfort of lower back and arms through forward tilt, at the cost, however, of pressure on the lower legs. After some period of time, given the capability of the chair to shift its position easily, a new attractor state is chosen, perhaps a rearward tilt with an accompanying upward adjustment of the keyboard height. This posture, as well, will eventually generate discomfort signals, generating yet another shift. However, the extent of such changes will be less then the continual movement (fidgeting) resulting from a less flexible workplace.

In the extensive focus, thus far, on homeostatic principles, it is easy to concentrate on the "mechanical" nature of the proposed model. However, implicit in the kind of regulatory behavior attributed to the individual in this model is the assumption that that person has the knowledge, skills,and training necessary to effectively manipulate the five or more degrees of freedom of adjustment in a state of the art ergonomic chair, and in the three or more degrees of freedom in a state of the art ergonomic workstation (Dainoff et al., 1986; Dainoff and Dainoff, 1986). It is this issue, together with concern over exactly how much flexibility/adjustability a workstation should afford, given the population effectivities (anthropometric dimensions/ willingness to adjust/need for training) that is the focus of much of today's controversy. (See, for example, Kleeman, 1987.)

5. REFERENCES

Bhatnager, V., Drury, C.G. and Schiro, S.G., 1985. Posture, postural discomfort, and performance. Human Factors, 27:189-199.
Branton, P., 1976. Behavior, body mechanics, and discomfort. In E. Grandjean (Ed.), Sitting Posture. London: Taylor & Francis.
Chafin, D.B, and Andersson, G., 1984. Occupational Biomechanics. New York: Wiley.
Corlett, E.N. and Eklund, J.A.E., 1984 How does a backrest work? Applied Ergonomics, 15.2:111-114.
Dainoff, M., 1983. Ergonomics of office automation--a conceptual overview. In M. Matthews and R. Webb (Eds.) Proceedings of the 1983 International Conference on Occupational Ergonomics. Volume 2: Reviews. Human Factors Association of Canada, Toronto.
Dainoff, M. and Dainoff, M., 1986. People and Productivity: A Manger's Guide to Ergonomics in the Electronic Office. Toronto: Holt, Rinehart and Winston of Canada.
Dainoff, M. and Mark, L., 1986. Task and the adjustment of ergonomic chairs. Proceedings of the International Conference on Work with Display Units. Stockholm.
Dainoff, M., Mark, L.S., Daley, R. and Moritz, R., 1986. Seated posture in the dynamic mode--preliminary results. Proceedings of the 30th Annual Meeting of the Human Factors Society, Dayton, Ohio.
Dempster, W.T., 1955. The anthropometry of body action. Ann. N.Y. Acad. Sci., 63:559-585.
Gibson, J.J., 1979. The ecological approach to visual perception. Houghton-Mifflin, Boston.
Grandjean, E., Hunting, W., and Pidermann, M., 1983. VDT workstation design: preferred settings and their effects. Human Factors, 25:161-175.
Kleeman, W. R., 1986. A different way to use anthropometric data as a tool for computer terminal workstation design. Human Factors Society Bulletin, 30(2):4-6.
Kroemer, K.H.E., 1987. Reply to Lueder. Human Factors Society Bulletin, 30(1):6.
Kugler, P. N., & Turvey, M. T. (1987) Information, natural law and the self-assembly of rhythmic movements: Theoretical and experimental investigations. Hillsdale, New Jersey: Erlbaum Associates.
Lueder, R., 1986. Seat height revisited. Human Factors Society Bulletin, 29(11):4-5.
Mandal, A.C., 1981. The seated man (Homo Sedans), the seated work position, theory and practice. Applied Ergonomics, 1981,12,19-26.
Mandal, A.C., 1986. The influence of furniture height on back pain. Proceedings of the International Conference on Work with Display Units. Stockholm.
Mark, L. and Vogele, D. (In press) A biodynamic basis for perceived categories of action: a study of sitting and stair climbing. J. of Motor Beh.
Mark, L., Vogele, D., and Dainoff, M., 1985. Measuring movement at ergonomic workstations. Proceedings of the 2nd Midcentral Human Factors/Ergonomics Conference. Purdue University, East Lafayette, Ind.
Nachemson, A., 1975. Towards a better understanding of low-back pain: a review of the mechanics of the lumbar disc. Rheumatology and Rehabilitation, 14:129-143.

Pheasant, S., 1986, Bodyspace. Taylor & Francis, London.
Pustinger, C., Dainoff, M., and Smith, M., 1985. VDT workstation adjustability: effects on worker posture, productivity and health complaints. Proceedings of the 2nd Midcentral Human Factors/Ergonomics Conference. Purdue University, East Lafayette, Ind.
Warren, W.H., 1984. Perceiving affordances: visual guidance of stair climbing. J. of Exp'l. Psych.: Human Percention and Performance. 10:683-703.

Social, Ergonomic and Stress Aspects of Work with Computers,
edited by G. Salvendy, S.L. Sauter and J.J. Hurrell, Jr.
Elsevier Science Publishers B.V., Amsterdam, 1987 — Printed in The Netherlands

DEVELOPMENT AND EVALUATION OF A RISK BASED INTERACTIVE DECISION AID

JAMES P. JENKINS and JAMES C. GLYNN
U.S. Nuclear Regulatory Commission, Washington, D.C., U.S.A.

ABSTRACT

The Plan Risk Status Information Management System (PRISIM) is a personal computer program that presents probabilistic risk assessment (PRA) results and related information for use by decisionmakers. The body of PRA information contained in PRISIM has been configured in terms of plant operating models associated data so as to address plant operational decisions having safety implications. The major capabilities addressed by PRISIM are:

* Risk monitoring of commercial nuclear power plants to obtain continuous read-out of core melt frequency
* Evaluation of and insights about risk impacts of operating changes, start-up and testing activities
* Risk significance of formalized, mandatory inspection procedures.

Initial PRISIM programs were developed for use by the Nuclear Regulatory Commission power plant inspectors, but now are being expanded to provide greater analytical applications in project management activities. The system's database contains both preprocessed plant specific risk based information and a set of plant system models for use in the assessment of the degree of safety of a particular plant status. This capability provides to the PRISIM user the relevant information and an analytical tool to assess plant safety and to take appropriate actions to prevent unnecessary increases in plant risk. Rapid access by the user, via menu driven displays and interactive formats, allow both ease of access and minimum PRISIM programs have been developed and installed at three United States commercial nuclear power plants with additional applications at other plant sites under consideration.

1 INTRODUCTION

The application of a probabilistic risk assessment (PRA) methodology to commercial nuclear power plants began in 1975 (NRC, 1975) thereby providing to the nuclear industry for the first time a mechanism to systematically assess plant risk. The analytical process contained in a PRA quantifies the likelihood of a potential severe core damage event and estimates the consequences and risk of those accidents that proceed to core melt. The PRA risk statements can be either the likelihood of a reactor core melt event or its concommittant off-site consequences in terms of early and late health effects. Severe accident sequences are developed in the form of logic models called an event trees, whose function is to describe the particular plant systems and component failure conditions leading to a situation involving a complete loss of cooling to the reactor core. Human reliability performance of plant operators and maintenance personnel, described throug human error analysis, combined with information on faults associated with nuclear systems and components, are mathematically combined to obtain the resultant probabilities of each accident sequence.

Since 1975, both the NRC and other nuclear industry groups have performed over 25 risk assessments of nuclear power plants or major systems therein (NRC, 1987). Eager to capitalize on the usefulness of their logic structures, information about system failure probabilities, and human error analyses the Office of Research, NRC, began an investigation as to how the voluminous printed PRA results could be melded into the regulatory system, in particular the inspection process. Given the large number of operating plants in the U.S. (.100) and the need to assure the capability of inspectors or operators to evaluate the safety and reliability was adequate to minimize risk, the initial focus of the research program was to identify the responsibilities and actions of the NRC's resident inspectors and then to establish the relationships between plant risks and inspection decisions (NRC, 1985). Early in this study, it was established that the information necessary for an inspector to support a decision should be: (1) readily accessible, (2) free of PRA jargon and arcane descriptors, (3) readily recognizable to inspectors with varied technical backgrounds as to its usefulness, and (4) responsive to the types of decisions inherent in their inspection program. The PRISIM program was then developed using these established criteria for a useable and effective system.

2 SYSTEM DESCRIPTION

PRISIM is a decision-oriented, user-friendly, menu-driven program that contains data base management and interactive routines to aid reactor inspectors, or other decisionmakers, in evaluating the risk significance of plant status information, and planning the focus of their inspection activities.

The program was written for an IBM XT personal computer with special high resolution graphics and a 20 megabyte disc. Its controlling feature is a data base manager that performs two main functions: (1) it selects screen images from the PRISIM data base and displays them on a monitor, and (2) controls the PRISIM interactive routine - that portion of the program that calculates based on the probabilistic plant modesl the risk status of the plant at a particular time - and then displays the results of these calculations on the monitor, e.g., change in core melt frequency. The data base contains preprocessed information that is relevant, regardless of the plant's status, whereas the interactive routine provides the user with PRA information that has been updated to reflect the actual status of the plant at the moment and correlates specific inspection actions precisely to the prevailing plant safety conditions (Glynn, 1986).

2.1 Data Base

The PRISIM data base contains pre-processed or "canned" information that is relevant to inspection decisionmaking regardless of the plant's operating status, i.e., specific components assumed not to be out of service, via an interactive routine. In the pre-processed information, a considerable amount of data pertaining to the prioritization are provided, based on safety importance functions, for important systems, subsystems and components. Also included in this data base is information relating to support system interfaces and plant specific data on component failures and operating events in which emergency systems may have been called upon. In the interactive portion of PRISIM, the data base are dynamic in that prioritization rankings are computed in real time, based on the existing plant status. The specific ranking information is:

(1) Risk reduction - a ranking of the unfailed equipment according to their relative contribution to catastrophic risk;

(2) Risk response - a ranking of the failed equipment according to the benefit of restoring each to service; and,

(3) Safety assurance - a ranking of the unfailed equipment, i.e., #1 above, if it were taken out of service.

2.2 Principles of Operation

When the user enters the system, via a two-step keyboard operation, they are presented a series of display formats on the monitor which are menus. Two paths are available: the first is the inspection procedure related path which is designed to provide PRA related information that can be useful for a new PRISIM user, while the second path is a more direct way to access the same type of information for a more experienced user. Figure 1 illustrates the two paths and the menu hierarchy (JBF,1986).

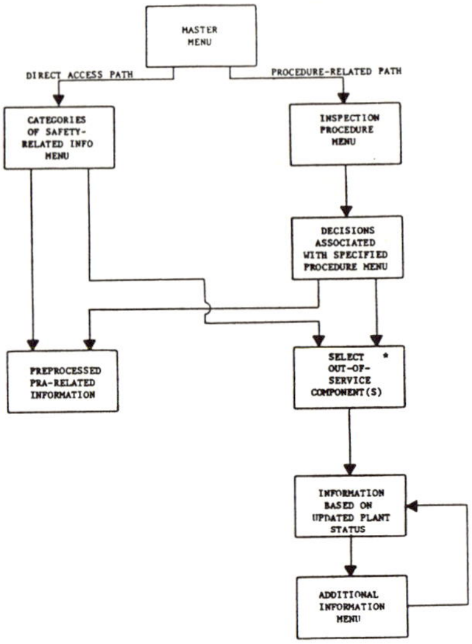

Fig. 1. PRISIM Menu Hierarchy Diagram

Subsequent formats provide information and decision options to allow more detailed examination of risk importance topics. For example, in the "Direct Path" referred to above, the user may select from over 1000 menu items, grouped into nine categories: (1) Risk Implications of the Current Plant Status; (2) Dominant Accident Sequences; (3) Safety Related Systems; (4) Safety Related Subsystems; (5) Safety Related Components; (6) Support System Interfaces; (7) Component Failure Data; (8) Human Interfaces; and, (9) Systems Testing/Survéillance. Discrete schematics of major systems can be accessed as well and components on them deleated so as to represent unavailable items.

3 APPLICATIONS

3.1 Test Plan

A prototype test bed was developed in April 1986 and installed at two facilities: (1) Arkansas Nuclear Unit-1 plant, and (2) NRC's Inspection and Enforcement offices, Arlington, Texas. Test and evaluation of PRISIM at these sites was described in a formalized test plan that contained the following:

(1) Objectives of the test program; (2) Description of the PRISIM system;

(3) Guidelines for test and evaluation which were to provide a list of concerns and questions which, in turn, would be used to make qualitative judgments on PRISIM usefulness and effectiveness; (4) Methods for data gathering; (5) Products from the test; (6) Training of user personnel; and (7) Schedule and milestones.

Installation and test began at both facilities in July 1986 with a similiar system developed and intalled at the Peach Bottom Atomic Power Station, Pennsylvania, in April 1987.

3.2 Scheduling of Inspection Activities

A major application of PRISIM is the planning and scheduling of an inspector's activities. Let us suppose that the plant has been operating normally and the inspector decides to carry out any number of mandatory plant inspections. PRISIM provides the inspector with a tool to assist him in judging what potential risk factors are inherent in a particular inspection module/oprocedure and how best he/she might perform this particular inspection from a risk perspective. On the other hand, if a system or component were to be taken out of service by the plant operator because of a casualty or due to scheduled maintenance, the inspector could determine the effect these new conditions have on the plant's overall risk index, as measured by the change in the core melt frequency, and focus his attention on those plant systems and components that have become most important for safety.

3.3 Management of Risk

Use of PRISIM as a risk management tool or sensitivity test of proposed changes in systems and components is another type of application. Let us assume that a major test is being planned while the plant continues on-line, i.e., the reactor is not scrammed. The Chernobyl accident originated from this type of test circumstances (NRC, 1987). PRISIM can be employed to assess the safety assurance associated with the systems and components which would be taken out of service if the test were conducted. In this case, the decisionmaker would want to find out the changes in the overall risk, the accident scenarios which would lead to an increase in core melt frequency, identification of the associated systems and components and their risk reduction contribution, and the operator actions. Based on the risk management values obtained from this sensitivity test the decisionmaker could arrive at a decision to proceed, i.e., accept the new risk, or schedule that particular activity during a shutdown period.

3.4 Performance Prediction

The PRISIM program is capable of being used to continuously monitor a plant's estimated probability of suffering a core melt accident. Such information was developed at one of the test sites and data were obtained over a one year period. As seen in Figure 2, the core melt frequency performance indicator revealed that major perturbations in the plant risk occurred roughly ever 2-3 months.

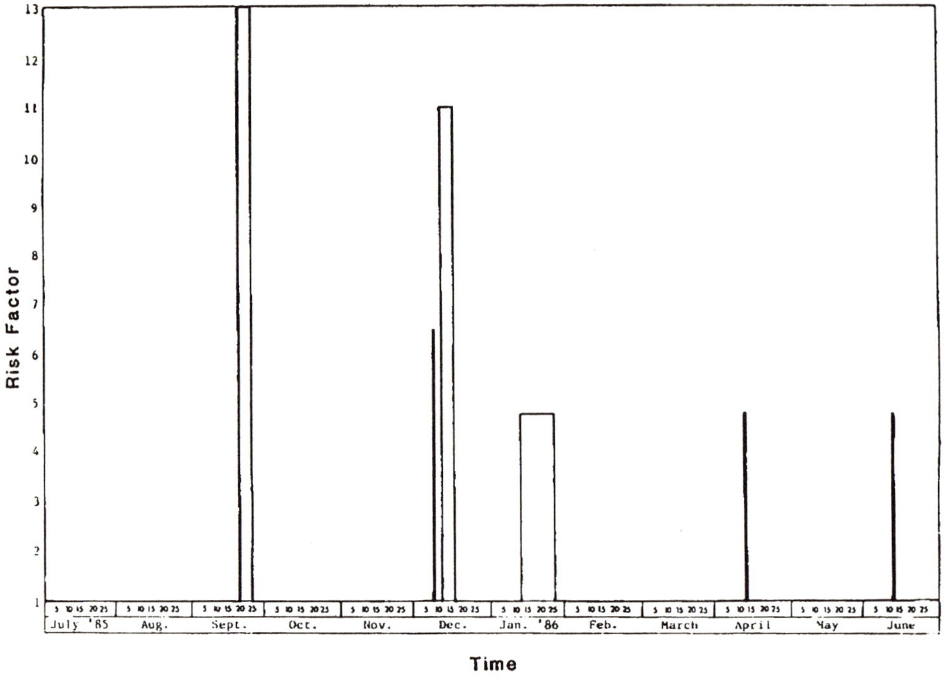

Fig. 2. Risk Factor Versus Time

The implications of these type data suggest the following: it is possible to make some judgments from objective data about the relative safety of a particular plant operation based on actual operating data and then compare the risk of the one plant versus another by: 1) determining from operating data the number of core melt frequency perturbations, or 2) determining the cumulative effect of the year's core melt perturbations above its nominal value, and 3) determining the seriousness of the the possible cause-effect relationships.

In addition to being able to provide such information as the daily core melt frequency of a plant, a sensitivity analysis can be performed to predict the potential effect of particular equipment malfunctions that may have been discovered at other plants. For example, a recent situation at a commercial reactor plant indicated serious problems with the operability of a type of motor operated valves. PRISIM results based on varying the nominal valve availability for these type of important motor operations were quickly obtained. Such data allowed an immediate assessment of the risk implications of a similar occurrence at other plants which relied on the same motor operated valve. Also, in the case where potentially unreliable valves were thought to exist, PRISIM provides a wealth of information pertaining to the identification of the most important remaining systems. Further, it provides identification of those transient indicators that pose the greatest risk to the plant. Finally, it provides a means to develop a course of action to address those contingencies most important to maintain the safety of the plant.

3.5 Assessment of PRISIM Performance

Both quantitative and qualitative measures of performance were obtained. The quantitative measures were manpower costs associated with hands-on usage of PRISIM by the two inspectors which were based on a Daily Log of frequency and duration of use. The Log also provided a qualitative reference to major uses of PRISIM, results and problems. Fourteen questions were developed which allowed the respondent to answer on a nominal scale (High, Medium, Low) as well as in a descriptive manner about the consequences of PRISIM on the inspector's functions and planning, as a decisionmaking tool.

4 TEST RESULTS

Two Reactor Inspectors at the Arkansas Nuclear Unit-1 site were primary sources of evaluation of PRISIM for the period August 1986 to March 1987 (when this report was written). A major finding was that PRISIM identified a new area of risk which had not been known prior to its application. Such information provided new insights into risk significant plant operating conditions and components. Because the plant was not in operation from September to December 1986, there was a limited basis for their evaluation and, as a consequence, the evaluation period was extended to May 1987. Use of PRISIM as a planning tool was limited because the plant did not suffer multiple casualties or have major safety components out of service. As the inspectors became more familiar with PRISIM and its degree of usefulness to daily operations of the plant they tended to use it only when a major change in plant status was envisioned. They rated it "High" in sensitivity to new problems which reflected additional sources of risk. The system was very easy to use and very reliable.

In conclusion, PRISIM type systems are designed to provide useful risk-based insights and information for decisionmaking by inspectors and other risk managers. Test results to date (March 1987) tend to support this finding.

5 REFERENCES

Glynn, J. C., 1986. A PRA based interactive system for planning reactor inspection activities. In: N. J. Holloway (Editor), Proceedings of the CSNI Workshop on Probabilistic Safety Assessment. United Kingdom Atomic Energy Authority, Warrenton, 15 pp.

_______, 1985. Risk Application to NRC inspection. Interim progress report JBF Associates, Inc. Knoxville, 15 pp.

_______, 1986. Plant risk status information management system. (PRISIM) user's guide. JBFA-103-86. JBF Associates, Inc., Knoxville, 69 pp.

_______, 1975. Reactor study-an assessment of accident risks in U.S. commercial power plants. U.S. Nuclear Regulatory Commission, WASH-1400 (NUREG-75/014).

_______, 1987. Reactor Risk Reference Document, U.S. Nuclear Regulatory Commission.

Social, Ergonomic and Stress Aspects of Work with Computers,
edited by G. Salvendy, S.L. Sauter and J.J. Hurrell, Jr.
Elsevier Science Publishers B.V., Amsterdam, 1987 — Printed in The Netherlands

Feedback content during complex skill acquisition

Jean McKendree, *MCC Human Interface Program,*
3500 West Balcones Center Dr, Austin, TX 78759

ABSTRACT

Systematic empirical study should be directed at developing a psychological theory of learning and skill acquisition which will then be a tool to guide designers of more usable systems. The level of expertise of the subjects and the types of feedback content were manipulated in three experiments using a tutor for geometry proof generation. The subjects were either algebra students (novices at geometry proofs) or geometry students (intermediate skill level students). The feedback involved various combinations of information about subgoals and about the reasons for the error. The results of the experiments suggest that feedback can facilitate the encoding of episodes in which the student engages in order to create the correct rules to be used in later problem solving. The vital information seems to be the decomposition into subgoals and the preconditions under which the rules apply. Implications for interface and training design are discussed.

1 INTRODUCTION

A major concern in interface design involves making learning of the system as short and painless as possible for a new user. Technology has made it possible to conceive of intelligent monitors and automated advisors existing in the near future, but the feasibility of creating ever more complex and sophisticated systems does not insure their usability. Carroll and McKendree (1987) in a review of the existing state of intelligent advising systems stress the importance of behavioral requirements on the design and assessment of systems. Systematic empirical study should be directed at developing a psychological theory of learning and skill acquisition which will then be a tool to guide designers of more usable systems.

The studies presented in this paper are an attempt to address one of many questions concerning the design of effective feedback for a tutoring system. Three experiments were run using a tutor for geometry proof generation. They investigated the information content provided after an error and the subsequent ability of the user to correct the problem. The results of the experiments are discussed in the context of the ACT* (Anderson, 1983) theory of skill acquisition and the implications for design of feedback are suggested.

The major questions addressed in the three studies discussed are these:

1. What information will be the most beneficial for learning after an error is made and why?

2. How should the presented information vary depending on the experience of the learner?

2 THE GEOMETRY EXPERIMENTS

2.1 General procedure and the tutor

Three geometry proof experiments were run using a Geometry tutor under development at Carnegie-Mellon University. The general structure of the tutor will be described followed by the specific changes made for each experiment. A more detailed description of the tutor is available in Boyle and Anderson (1984) and McKendree (1986).

The tutor was implemented in INTERLISP on XEROX DandeTiger machines. It was designed to monitor and structure the student's problem solving. For each inference, the student first specified the premises which were to be used by pointing to each and clicking a mouse button. After all the premises were selected, the student picked "Done selecting" from a menu. The system then prompted the student for the name of a rule which could be applied. The student typed in the name of the rule and then input the conclusion which could be drawn by selecting symbols from a menu or points from the diagram using the mouse. The diagram could also be selected as a premise in the case of rules such as *reflexive*. As the inferences were made, the tutor graphically built a proof tree of the inferences. An example of a simple complete graph tree is shown in Figure 1.

FIGURE 1: Complete graph tree on the Geometry Tutor

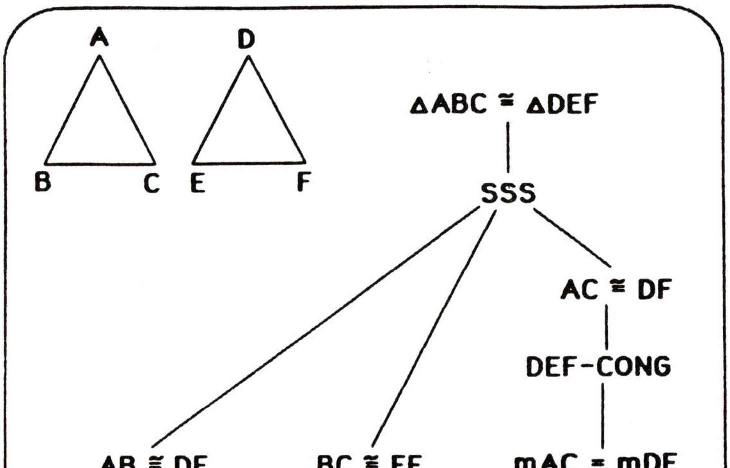

The students could make any legal inference without interruption. If an illegal move was made, the tutor immediately interrupted with a feedback message and allowed the student to try a second time. The feedback elicited after an error was the focus of the experimental manipulation. If the student made two errors in a row, the tutor provided the correct next step without an explicit request from the student.

2.2 Design of feedback

The tutor provided feedback when an error was made within an inference. Over all three experiments, four types of feedback were used: minimal, logical, goal and combined. The "minimal" feedback contained enough information to let the student know that the current move was incorrect. The "logical" feedback was a statement of the logical condition which had been violated in the attempted step. The "goal" statement was based on the conclusion of the next inference on the optimal proof path and stated the current correct subgoal. The "combined" feedback gave both the "logical" and the "goal" feedback.

An example of each type of feedback will be presented in the context of a typical proof. The first premise given indicates that the measure of one side of triangle PQR, side PR, is equal in measure to the corresponding side ST in triangle TUS. The other premises give that sides QR and PQ are congruent to US and TU, respectively. In this case, the *definition of congruence* must be applied to prove PR and ST congruent also. A very common error on this proof was to choose all three premises and attempt to use the *side-side-side* postulate to prove the triangles congruent.

Given: mPR = mST
QR ≅ US
PQ ≅ TU
Goal: Δ PQR ≅ Δ TUS

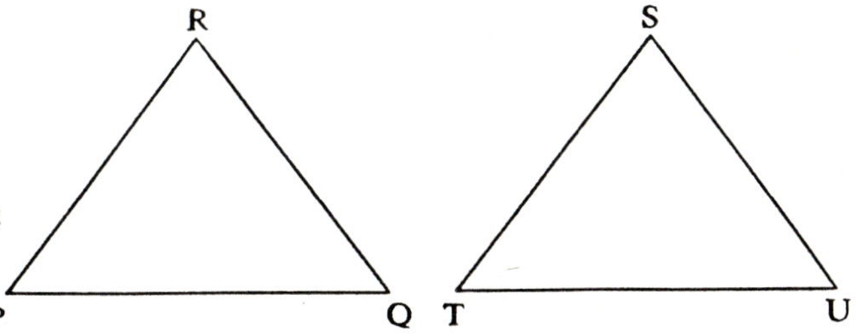

The following are examples of an interaction utilizing each type of feedback. The subject has chosen the three original premises and attempts to use the *SSS* postulate to prove the triangles congruent. The feedback message appears after the subject names the *SSS* postulate as the rule which applies.

MINIMAL FEEDBACK:
Tutorial response: "This rule does not apply with the premises you have chosen."

LOGICAL FEEDBACK:
Tutorial response: "SSS requires 3 CONGRUENT sides. You have chosen 2 CONGRUENT statements, but also 1 EQUAL statement."

GOAL FEEDBACK:
Tutorial response: "You should be trying to prove that PR is congruent to ST."

LOGICAL + GOAL FEEDBACK: "SSS requires 3 CONGRUENT statements. You have chosen 2 CONGRUENT statements, but also 1 EQUAL statement. You should be trying to prove that PR is congruent to ST."

2.3 The experimental manipulations

The three experiments differed in systematic ways which will be outlined along with a general statement of the main question being investigated in each. The first experiment involved the algebra students. This experiment was an attempt to assess the impact of different types of information on the performance of novice subjects. These students had not been introduced to proofs nor had they taken any geometry.

The materials covered two topics, congruent triangles and parallel lines. Each topic involved eight theorems and 15 proofs. All subjects worked through both topics and completed all proofs. Each subject saw the two types of feedback, one type for triangles and the other for parallel lines.

The second experiment was identical to the first except that the subjects were currently enrolled in geometry rather than algebra. This study was an attempt to find the impact of this feedback information on more experienced subjects.

The third experiment also used geometry students. There were four groups of subjects each given a single type of feedback: minimal, logical, goal or combined. The decision to move to a single type of feedback was an attempt to obtain more straightforward data which was not confounded by a subject learning from two types of feedback at once and to gain a clearer understanding of how each type of information effected performance. These subjects worked through several proofs on the first day after an introduction to the interface and the performance on these proofs was used as a pretest measure in the later analysis. Each subject then worked a series of 30 triangle proofs over the next two days. These proofs were all be from the topic of congruent triangles. They also worked a set of test problems on the final day on which all subjects saw "minimal" feedback.

2.4 Results and conclusions:
Novice versus intermediate students

The data from each subject were collapsed across problems according to feedback condition. These composite scores were used for the analyses. There were several measures calculated for each subject. However, only those which yielded significant differences will be reported.

The results of the first two experiments yield an interesting comparison. The subjects in the first experiment were complete novices at geometry and at proofs. The subjects in the second experiment had been in geometry for half a year and were more familiar with the geometric concepts and the fundamentals of proofs but were not experts. Though it is difficult to make a direct comparison between the two groups because of many uncontrolled factors, the differences in overall impact of the feedback types suggests some interesting conclusions.

Two measures seemed to most directly reflect the impact of feedback. The first was the number of times a feedback message led to a correct next step. The second measure of interest was the number of extra inferences generated during the test problems when the inference generation was not constrained. In the test problems, the students were allowed to complete any legal inference even if it was unnecessary or extraneous to the current proof. The time to complete a proof did not reflect a difference in the feedback types and will not be reported. This is due to the fact that the tutor provided the correct next step after two incorrect attempts by the student and therefore pushed everyone through the entire sequence of problems without allowing too much floundering.

Basically, in the case of the algebra students, none of the measures showed statistically significant differences. The "logical" information seemed to have little impact on the performance of these novice subjects. No evidence for a delayed impact of "logical" feedback was found in the test problems.

The second experiment was identical to the first except that the subject population, geometry students, was more experienced in the domain. The results from the second experiment with the geometry students showed more of an impact of the feedback types. The percentage of times that a "logical" feedback message led to a correct next step was 73.5%. The percentage for a "minimal" feedback message was 47.1%. This difference was significant ($F(1,7)=7.95$, $p<.02$).

The performance on the test problems was evaluated in the same way as in the first experiment. Again, only "minimal" feedback was provided after an error during the test. The number of extra inferences made in the test problems was 1.87 for the "logical" problems, 3.41 for the "minimal" problems and 2.61 for the combined problems. These differences were marginally significant ($F(2,12) = 2.91$, $p<.10$). A Neumann-Keuls test on the individual means showed that the difference was due to the "logical" condition being different from the other two conditions.

While the algebra students did not seem very able to utilize the "logical" information any more than the "minimal", the students with a more extensive background in geometry seemed to have a slightly better ability to integrate the information into their problem solving.

The geometry students did make fewer legal but unnecessary inferences in the test problems. This indicates that what they might have been learning was information concerning the goal structure of the proofs along with the logical conditions of application. Perhaps the information in the feedback aided the understanding of the applicability of the rules for the service of particular goals in the proof.

The within subjects design of these experiments made it very difficult to separate general learning from a differing impact of feedback. It is possible that the "logical" feedback was useful, but tended to increase ability on both types of proofs as the use of rules was better understood. It is also possible that "logical" information is not highly useful in either case. Therefore, the next study was designed as a between subjects study and sought to evaluate the impact of a wider variety of feedback types using a single type of proof.

2.5 Usefulness of feedback type for advanced students

According to current cognitive theories such as ACT* (Anderson, 1983), there are at least two types of information which are necessary for acquiring a skill such as proof generation. These are the current *goal* within the problem and the *conditions* under which a rule applies. The information about conditions corresponds to the "logical" information given in the earlier studies. Feedback information about the violation of conditions after an error allow the correct conditions to be acquired for subsequent applications. Information about the correct goal when given after an error can redirect the problem-solving episode to encourage the proper goal hierarchy to be learned.

It is still an open question just how these types of information should impact on learning. It is possible that both are necessary and will aid learning independently or according to the type of error. It may be that one type will be more easily assimilated or that one type of knowledge is necessary for using the other. Therefore, the final study chose to investigate these two types of information, goal and logical, and the combination of the two. A "minimal" group was also included as a baseline and because the previous experiments were not completely clear about the impact of this minimal information.

2.6 Results and Discussion

The primary measures of interest concern errors. Since feedback is given only after error, it is then that it will have a chance to effect learning. The first measure is the total number of errors over the training trials and in the posttest proofs. First, the minimal group was compared to the other three combined. The difference between the "minimal" feedback group with 2.22 errors per proof and the other three groups with a combined error rate of 1.37 per proof ($F(1,29)=2.96, p<.09$). The posttest problems also showed a significant advantage for the combined groups with .62 errors per proof over the "minimal" feedback group with an error rate of 1.77 errors per proof ($F(1,29)=4.16$, $p<.05$).

Contrasts comparing the "minimal" condition to each of the other three groups indicated that the "goal" group, with an adjusted error rate of .65 per proof, differed significantly from the "minimal" group which had an adjusted error rate of 2.98 for the training trials ($F(1,13)=7.50$, $p<.01$). The "minimal" and "goal" groups also differed significantly over the posttest trials with the "minimal" group making 1.73 errors per proof and the "goal" group making .37 errors ($F(1,13)=5.57$, $p<.03$).

The "minimal" group also differed marginally from the "combined" feedback group on the posttest. The "combined" group had an adjusted error rate of .37 per proof while the "minimal" error rate was 1.35 per proof ($F(1,12)=3.02, p<.1$). These groups did not differ significantly on the training trials though the trend was in the predicted direction which the "minimal" error rate being 2.40 and the "combined" group error rate of .83 ($F(1,12)=2.23, p<.16$).

While the total number of errors is an indication of differences in learning between groups, a more accurate measure of the immediate impact of feedback is the number of times the subject corrects an error on the next attempt after feedback. This error correction was calculated, as in the previous study, as the percentage of times the subject performed correctly on the next step taken. The "minimal" group had an overall correction rate of 24% and the "combined" group had a correction rate of 44% which was not statistically significant. However, the error correction by trial interaction was marginally significant ($F(21,630)=1.41$, $p<.1$). This interaction was primarily reflected in the early trials in which the "minimal" group had a higher correction rate which reversed in about 5 trials.

The most apparent impact and the one most easily predicted is the correction of errors involving a wrong conclusion. In this case, the "goal" and "combined" feedback actually states the conclusion which should be typed into the tutor. These two types of feedback resulted in over 70% correction of errors while the "logical" type showed 43% and the "minimal" showed 35%.

In summary, it seems that the "goal" and "combined" information types showed the greatest benefit for learners. This suggests that it is the "goal" information which is most effective in promoting learning of the skill. The impact of the "logical" information seems to be less, at least in training, though it shows slight advantage over simply indicating the presence of an error. Perhaps in proofs involving theorems with more complex conditions of application, this information would begin to have a differential impact. This general pattern of results is apparent in Table 1.

TABLE 1: Percentage of correct next moves according to feedback type

		Logical Feedback	
		PRESENT	NOT PRESENT
Goal Feedback	PRESENT	51	52
	NOT PRESENT	29	23

3 GENERAL CONCLUSIONS

The results of these studies indicate that skill acquisition can be enhanced by properly constructed feedback and that the type of information can impact on the effectiveness of the feedback. The major conclusion from all the studies is: Information about the nature of the error and the procedures for correct execution is more effective for learning and subsequent problem solving than simply being informed that an error has been made. This suggests that a very unconstrained "exploratory" type of instruction will not be as conducive to efficient learning.

This sort of tutorial design promotes the sort of contextual learning which cognitive theory suggests allows a quicker acquisition and better retention of a new skill (Anderson, Farrell and Sauers, 1984; Brown, Burton and deKleer, 1982). Generally, a problem is presented and the user can initiate action fairly freely. The responses of the student are compared to those of an "ideal student" model contained in the system and feedback is presented to shape the sequence of moves toward the model. The importance of the "ideal student" in this type of design suggests the importance of behavioral task analysis and the embodiment of knowledge in the system about the manner in which a competent student acquires the desired skill.

In summary, the results of the experiments suggest that feedback can facilitate the encoding of episodes in which the student engages in order to create the correct knowledge to be used in later problem solving. The vital information seems to be the decomposition into subgoals and the preconditions under which the rules apply. By emphasizing this information during the actual problem solving, instruction could be made more effective by minimizing the likelihood of an incorrect encoding and by guiding the examples acquired.

This sort of goal information probably could be obtained and implemented in a training sequence or an advising system much more easily than information involving diagnosing misconceptions and structuring an answer which will correct the problem without introducing others. This explicit "debugging" approach has been advocated by a number of researchers in the area of automated tutoring (Sleeman and Brown, 1982) but it has not been empirically shown to be a more effective strategy than providing the sort of goal information discussed here. These experiments indicate that there could be a real advantage in both effectiveness and effort required for implementation in concentrating on ideal user models and dynamic planners to provide novice users with correct plans for accomplishing the tasks they have when they approach the system.

Finally, the importance of empirical studies and testing of prototyped designs should not be underestimated. These sorts of studies not only provide concrete and immediate results which can be used, but also contribute to a better understanding of the cognitive issues involved in the design of effective systems and interfaces.

References

Anderson, J.R. (1983). *The Architecture of Cognition.* Cambridge, MA: Harvard Press.

Anderson, J.R., Farrell, R., and Sauers, R. (1984). Learning to program in LISP. *Cognitive Science, 8*, 87-129.

Boyle, C.F. and Anderson, J.R. (1984). Acquisition and automated instruction of geometry proof skills. Presented at the AERA *Skill and Strategy Acquisition in Reading, Writing, Math and Science Symposium.*

Brown, J.S., Burton, R.R. and deKleer, J. (1982). Pedagogical, natural language and knowledge engineering techniques in SOPHIE I, II, and III. In D. Sleeman and J.S. Brown (Eds.), *Intelligent Tutoring Systems.* New York: Academic Press.

Carroll, J.M. and McKendree, J. (1987). Interface design issues for advice-giving expert systems. *Communications of the Association of Computing Machinery,* January, Volume 30, Number 1, 14-31.

McKendree, J. (1986). Impact of feedback content during complex skill acquisition, Unpublished doctoral dissertation, Department of Psychology, Carnegie-Mellon University.

Sleeman, D. and Brown, J.S. (1982). *Intelligent tutoring systems.* New York: Academic Press.

Social, Ergonomic and Stress Aspects of Work with Computers,
edited by G. Salvendy, S.L. Sauter and J.J. Hurrell, Jr.
Elsevier Science Publishers B.V., Amsterdam, 1987 — Printed in The Netherlands

STUDY ON VDT DISPLAY COLOR AND READABILITY

S. FUKUZUMI[1], I. YAMAMOTO[2] and Y. HAYASHI[2]

1. Software Products Engineering Laboratory
NEC Corporation, 14-22, Shibaura 4-Chome, Minato-ku, Tokyo 108 (Japan)

2. Faculty of Science and Technology
KEIO University, 14-1, Hiyoshi 3-Chome, Kouhoku-ku, Yokohama 223 (Japan)

SUMMARY

This paper reports display color readability on a CRT screen. Three different dominant wavelengths (orange, green and purple blue) and three kinds of stimulus purity were examined under constant illuminance and luminance. Orange was found to be the most readable of the three, in regard to dominant wavelength unrelated to illuminance brightness. Small stimulus purity was found to be readable for the display color. Questionnaire results indicated that readability depended on favorite colors, viewer friendly (lack of eye strain) and viewing impression elegance.

INTRODUCTION

In recent years, work with Visual Display Terminals (VDTs) has become very popular work in general offices, due to the development of computers. Work with VDTs has brought about the consideration that the burden on workers' eyes is very intense, because computer display terminal viewers employ their visual functions concentratively.

Factors considered as special characteristics for VDTs that effect workers' eyes are gradation in the outline of display letters, flicker and glare on CRT displays and the display color.

Studies on the display color were reported. Yellow and green were compared by performance, visual acuity and pulse, as well as answers to a questionnaire. Results indicated that yellow was suitable to display (Haider et al., 1980). Green letters are more suitable than blue, as determined by measuring Small Fluctuations in Ciliary muscles, when using an Infra-red optometer (Iwasaki et al., 1984). In point of visual fatigue, yellow and green weakly caused visual fatigue by Critical Flicker Frequency, compared with red, blue and white (Osaka, 1985). Actual reading of the information depends on luminous contrast, not necessarily on color (Bouma, 1980). The brightness and the contrast, displayed by the symbols on the VDT screen, seem to be more important than the actual symbol colors. In all colors, yellow-green can be most recommended (Radl, 1980). As mentioned above, many results recommended yellow or green.

In the laboratory at KEIO university, display color has been studied. Concerning the combination of text color and background color, readability was found to depend on color difference (Namikawa, 1983, unpublished). Reading error during work with a VDT was considered to depend on a combination of text color and background color. However, it was reported that readability was effected by the amount of contrast between text color and background color (Nonaka, 1984, unpublished). Comparing white letters on a black background with black letters on a white background, the former was not considered more exhausting than the latter, under 300 lux (Fukuzumi et al., 1985).

These results were deemed important in order to study VDT work in the future. In spite of knowing that this relevancy is important (Frome, 1983), it has not been found yet. Therefore, this study was planned to clarify the relation between readability and attributes of colors, specially, psychological/physical quantity presented about colors (chromaticity coordinates of color, dominant wavelength, stimulus purity and so on (Durrett et al., 1982, Shneiderman, 1986)).

Display color is apt to be evaluated according to the workers' preference (Salvendy, 1986). However, many other factors in regard to display color evaluation have not yet been reported, for example, evaluation difference due to the time spent gazing at the screen, and a combination of a lot of colors. The object of this study was to determine effective factors to lighten a workers' burden and to increase productivity. Also, tests were made to determine what the color factor contributes to screen readability.

METHODS

Apparatus

This study paid attention to three dimensions of illuminant color, dominant wavelength (hue), luminance (brightness), and stimulus purity (chroma). Eleven different colors were used in this experiment. They were three kinds of stimulus purity for three kinds of dominant wavelengths (orange, green and purple blue), white letters on a black background, and black letters on a white background. (Table 1 shows each dominant wavelength, stimulus purity and background color with letters of each color, and shows X Y chromaticity coordinates for colors used in this experiment.)

Illumination in the laboratory was only ceiling lights, which illuminated the screen indirectly. This experiment was performed using two levels of vertical illuminance on the screen, 700 lux and 300 lux. These values are the upper and the lower limits presented in the VDU Guideline from the

TABLE 1
Dominant wavelength ,stimulus purity and X Y chromaticity coordinates of colors used in this experiment.
(Added number for each color is greater, chroma for each color is smaller.)

Items	Text Colors		Background Colors	Chromaticity	
	Dominant wavelength(nm)	Stimulus purity		X	Y
Green 1	542	.541	black	0.298	0.541
Green 2	542	.319	black	0.324	0.455
Green 3	542	.125	black	0.331	0.369
Orange 1	583	.594	black	0.474	0.416
Orange 2	583	.545	black	0.448	0.410
Orange 3	583	.224	black	0.399	0.359
Purple blue 1	461	.768	black	0.177	0.086
Purple blue 2	461	.602	black	0.226	0.155
Purple blue 3	461	.406	black	0.258	0.198
White	---	0	black	0.341	0.324
Black	---	---	white		

National Institute for Occupational Safety and Health (NIOSH). Display letters luminance was 44 cd/m^2 and background luminance was 23 cd/m^2 under 700 lux. Display letters luminance was 6 cd/m^2 and background luminance was 0.5 cd/m^2 under 300 lux. Text and background luminances were set up by the preliminary experiment. That is, screen luminance was adjusted according to the subjects' favorite luminance. From these values, experimental luminance was decided upon.

The letters were presented on NEC N5913 CRT Display. Personal computer NEC PC-9801V was used in this experiment.

<u>Subjects</u>

Twenty men and seven women were examined under 700 lux. Nine men and one woman were examined under 300 lux. All of them had normal color vision.

<u>Procedure</u>

Three line sentences on an average were displayed on the screen. Its color was one of the eleven categories shown in Table 1. After the subjects read the sentences, the same sentences in other colors were displayed on the screen, and the subjects read the sentences again. Then, the subjects evaluated which color was more readable for themselves. All of the combinations of eleven colors were evaluated.

To diminish of the fatigue and practice effect, one experiment was about ten

TABLE 2

Questionnaire about evaluation for each color.

	Very	Fairly	A little		A little	Fairly	Very	
Beautiful								ugly
Elegant								vulgar
Showy								quiet
Bright								dark
Tiring								not tiring
Favorite								unfavorite
Visible								not visible
Calm								irritating
Smart								boorish
Friendly								standoffish
Closed								frank
Convenient								inconvenient
Temporary								permanent

minutes long. Experiments were made several times for one subject.

After the experiment on all the combinations of eleven colors was finished, the subjects filled out the questionnaire shown in Table 2.

Analysis

The results of this experiment were analyzed, by Thurstone analysis for paired comparisons method, SD method and factor analysis.

RESULTS

The results of this experiment, by paired comparison method, are shown in Fig. 1 as the relation between stimulus purity and scaled value. From this result, in spite of illuminant condition, purple blue 1 (which has the largest

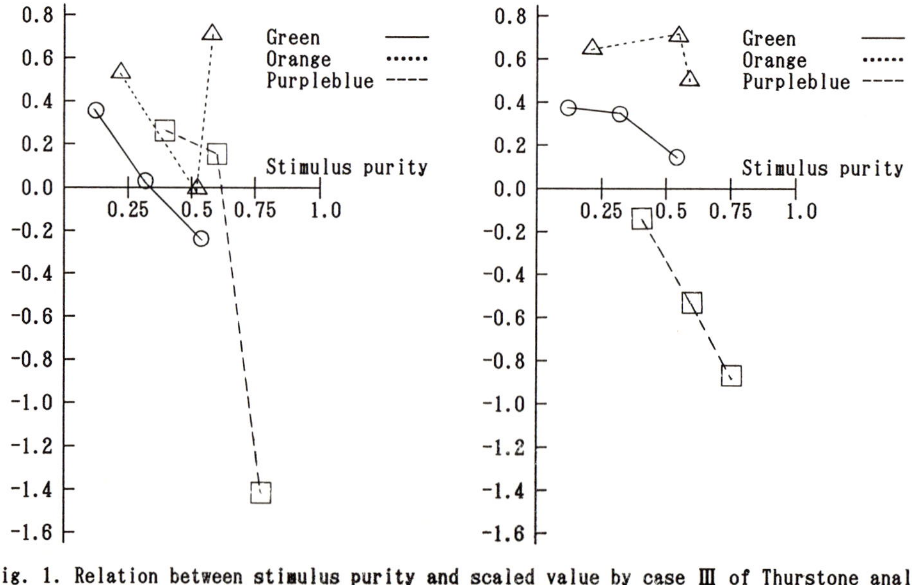

Fig. 1. Relation between stimulus purity and scaled value by case III of Thurstone analysis for paired comparisons method. (Left is under 700 lux and right is under 300 lux.)

stimulus purity of all purple blue colors) was evaluated very low. When stimulus purity is larger, the evaluation value for any dominant wavelength was larger, except for orange, but a contrary result was reported (Durrett et al.,1982). Under 700 lux, orange 3 (which has the smallest stimulus purity of all orange colors) had the highest evaluation. Under 300 lux, orange 2 (which has medium stimulus purity of all orange colors) had the highest evaluation. Except for these two cases, the purity evaluation tendency was similar under two conditions.

Compared with the evaluation value on a screen displayed between white letters on a black background and black letters on a white background, under 700 lux, both screens had similar value. However, under 300 lux, white letters on a black background had higher value. This result is considered by scotopic vision. This supports a previous report about it (Fukuzumi et al., 1985).

Compared with the evaluation value between men and women, women tended to receive a higher evaluation for lower stimulus purity.

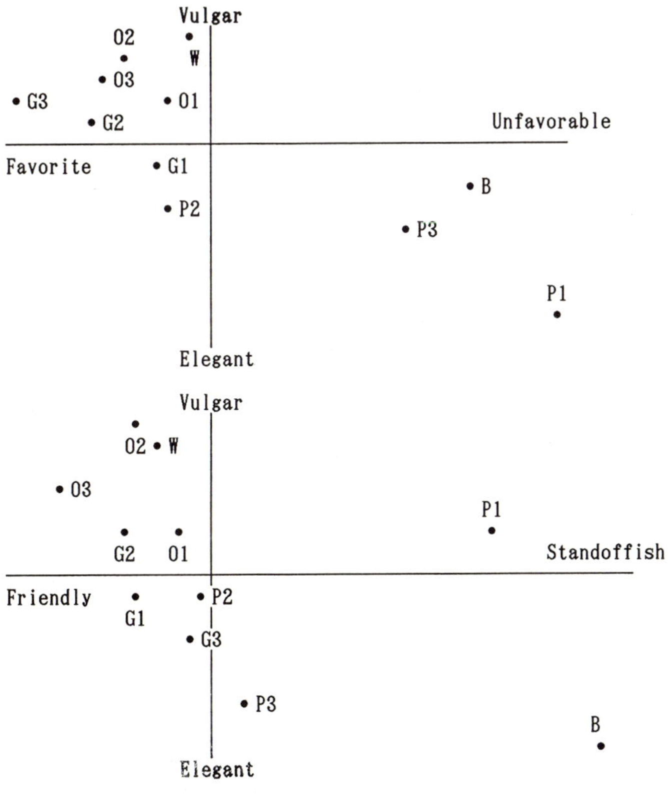

Fig. 2. Factor analysis results, plotted by first factor and second factor for each display color. (Upper is under 700 lux and lower is under 300 lux.)

The questionnaire results are shown in Fig. 2 as the results of factor analysis. Under 700 lux, dominant wavelengths are favorite in green, orange and purple blue sequence. This result is different from paired comparison. The reason is considered to be that the subjects are evaluated not only by readability, but also by visibility (U.S. N.R.C, 1980). Under 300 lux, dominant wavelengths were on the first axis in orange, green and purple blue sequence. Thus, this factor axis is considered to relate with scaled value for paired comparison.

Table 3 shows a correlation coefficient between the evaluation values and questionnaire results. From this, except for the second factor under 300 lux, each correlation coefficient is high. This result supports the results obtained from the paired comparison method.

TABLE 3

Coefficient of correlation between factor scores and evaluation value.

Factor score	Coefficient of correlation	
	700 lux	300 lux
First factor	* -0.6614	* -0.8505
Second factor	* 0.6162	0.3961

* Significant level 5%
** Significant level 1%

DISCUSSION

Figure 1 shows that purple blue 1 has the lowest evaluation. The reason is considered to be that colored lights are seen more brightly than white colored lights, when stimulus purity are greater. This tendency was strong near blue (wavelength was about 450 nm), and became very weak over or under 450 nm (Uchikawa et al., 1985). So, purple blue 1 is considered to have much brighter visibility than the other colors. From this result, optimal luminance is considered to differ as colors differ, and optimal luminance of blue is considered to be lower.

Orange 2 was evaluated lower than orange 1 and orange 3 under 700 lux. On the contrary, it was higher under 300 lux. Luminance for letters was established as being constant, but luminance for orange 2 was a little lower than the others. Thus, the contrast became low, where orange 2 is considered to be unreadable under high illuminance and to be readable under low illuminance.

As a result of data shown in Fig. 1, small stimulus purity gave a high evaluation value, as regards readability. The reason for this can be explained as follows;. A human being is considered to use chromatism for accommodation. Therefore, the wider the spectral energy distribution is, the easier the accommodation is, resulting in stable picture. It is reported that stimulus purity is small, if spectral energy distribution is wide (Miki et al., 1985, 1986). From this, small stimulus purity for letters on a screen are readable. However, from the evaluation on white, the optimum stimulus purity isn't the smallest. Therefore, it is considered that there is an optimum point somewhere on the axis of stimulus purity. This experiment didn't show where the optimum point was, due to few subject and small amount of data.

As a result of answers to the questionnaire, the color sequence on the axis of elegance was not always arranged in stimulus purity order. The reason is considered as being that the difference in the elegance rating was evaluated by a difference in stimulus purity, and that a difference in stimulus purity was added by a delicate difference in luminance.

CONCLUSION

The results of this study suggests that readability of display color depends on effective factors. These factors do not determine immediately which color is readable, but they may be useful in lightening a workers' burden and increasing productivity in the future.

The conclusions resulting from this study are as follow:

(1)The readability for color of letters displayed on a black background has a tendency to be long-wavelength (near 580 nm) as a dominant wavelength. Readability has a tendency to have a small value, as stimulus purity is unrelated to illuminance brightness.

(2)Comparing white letters on a black background and black letters on a white background, both evaluations are similar under 700 lux, but black letters on a white background had low evaluation under 300 lux.

(3)Two axes of evaluation were selected under each condition. Under 700 lux, one is the favorite axis, the other is favored in regard to elegance. Under 300 lux, one is favored in the friendly axis, the other is favored in regard to elegance.

REFERENCES

Bouma. H, Visual reading process and the quality of text display, Ergonomics Aspects of Visual Display Terminals, pp101-114.

Durrett. J and Trezona. Judi, How to Use Color Displays Effectively, BYTE, (April 1982), pp50-53.

Frome. F.S, Improving Color CAD Systems for Users: Some Suggestions from Human Factors Studies, the ACM/IEEE 20th Design Automation Conference proceedings, June 1983, pp18-27.

Fukuzumi. S, Shinya. T, Hayashi. Y, Effect on Visual Function by Visual Work with Self-illuminant Display, Proceeding of the conference on the Kanto branch of Japan Ergonomics Association, 1985, pp9-10, in Japanese.

Haider. M, Kundi. M, Weiβnbock. M, Worker strain related to VDUs with differently colored characters, Ergonomics Aspects of Visual Display Terminals, pp53-64.
Iwasaki. T, Kurimoto. S, Nomura. T, Ohkubo. R, Noro. K, Yamamoto S, Visual Fatigue by VDT Works, Clinical Ophthalmology, 38(5), 1984, pp483-487, in Japanese.
Miki. Y and Tamura. H, The Accommodation Response as a Function of Target Light Source and Spatial frequency, Proceeding of the second Symposium on Human Interface, 1986, pp151-156, in Japanese.
Miki. Y, Tamura. H, Accomodative Response of lens for many kinds of illuminant lights, Technical and Investigation Report of The Institute of Electronics and Communication Engineering of Japan, Medical Electronics and Bioengineering, vol.84, No.332, pp47-52, in Japanese.
Osaka. N, The effect of VDU color on visual fatigue in the fovea and periphery of the visual field, DISPLAYS, (JULY 1985), pp138-140.
Radl. G.W, Experimental investigations for optimal presentation mode and colors of symbols on the CRT-screen, Ergonomics Aspects of Visual Display Terminals, pp127-135.
Salvendy. G, Handbook of Human Factors: G.Salvendy (Ed.), Wiley, 1986, pp1364
Shneiderman. B, Seven plus or minus two central issues in human-computer interaction, CHI'86 Proceeding, (April 1986),pp343-349.
Uchikawa. K and Uchikawa. H, A Model for Luminance-Saturation Function of Equally Bright Colored-Lights, Optical Society of Japan, 14(6), 1985, pp486-471, in Japanese.
U.S. Nuclear Regulatory Commission, Guidelines for Control Room Design Reviews, 1980, pp6.7.16-6.7.42.

Social, Ergonomic and Stress Aspects of Work with Computers,
edited by G. Salvendy, S.L. Sauter and J.J. Hurrell, Jr.
Elsevier Science Publishers B.V., Amsterdam, 1987 — Printed in The Netherlands

QUANTITATIVE EVALUATION OF VDT OPERATIONS THROUGH THE ANALYSIS OF SPATIO-TEMPORAL CHARACTERISTICS OF EYE MOVEMENTS

SUSUMU SAITO

Department of Industrial Physiology, National Institute of Industrial Health, Nagao 6, Tama-ku, Kawasaki, 214 (Japan)

SUMMARY

The experimental survey to clarify the ergonomic problems concerned with VDT operations is described in this paper. Great emphases were put on the analysis of postural and eye movements characteristics of VDT operators. The usual jobs of system engineers were chosen as field research subjects. They were occupied with development of computer program during working hours. Eye movements were analyzed by computerized Eye-Camera. Both spatial and temporal momenta of eyeball in VDT operations were found to be about 2.5 times as much as those in other traditional visual works without CRTs.

INTRODUCTION

With so rapid progress of semiconductor or computer technology as background, the substance of labor is undergoing a change in various kinds of occupations. VDT operation in which CRT is used as most important information display unit is one of the typical jobs of newly appeared. However, as matters now stand, so many ergonomic problems still remain to be solved in (refs. 1-4). Additionally, it is also difficult to know how to use and locate adequately the new equipments such as VDTs in many offices and factories where the traditional jobs had been carried out for many years. The purposes of the present study were to quantitatively evaluate the VDT works and to compare the ergonomic problems of the work with those of other traditional visual works without VDTs. Particularly, on the eye movement characteristics in the works with and without VDTs were focused. It will be tried, in a word, to estimate human-computer interaction with some ergonomic and physiological scaling derived from analysis of eye movements.

METHOD

The subjects for field study of VDT works were five healthy employees who usually operated VDTs at their office as system engineers. The range of their age was 19-29 years and they had practical experiences in computer programing about two years. Their works normally involved both data entry and conversational types of VDT operations.

Measured items during VDT operations were eye movements detected by TV-type Eye-Camera (ref. 5), heart rate, respiratory movements. Operator's working postural images were continuously recorded on VTR through the working hours. CFF (Critical Flicker Frequency) was also intermittently measured at intervals of 30 minutes. The questionnaire for understanding the subjective feelings of fatigue was surveyed at the end of their daily work. Among these items, two peculiar aspects concerned with posture and ocular motility in VDT operations are mentioned with emphasis in the paper.

Postural data analyses

On the postural issue, viewing distances from eye to CRT were measured by frame by frame analysis of VTR image. Viewing distances during all through the working hours were numerically expressed as mean and standard deviation. Similarly, head movements of VDT operator were continuously measured by the video signal from semiconductor TV camera set on operator's forehead.

Ergonomic two indexes of eye movements

Eye movements during the work were detected by Eye-Camera and analysed by computer. The corneal reflex method employed in the equipment has about 0.5 degrees of spatial resolution (ref. 6), and 33 msec of temporal resolution. These specifications of the apparatus sufficiently satisfied the aim of the present study.

Some characteristics of eye movements were divided into two categories. Those were fixation and saccadic phases of eye movements. Fixation phase could

TABLE 1

Characteristics of eye movements in VDT operation. "FIXATION AREA" and "AVERAGED SACCADIC VELOCITY" were for ergonomic evaluation of the work.

-------ANALYSIS OF EYE MOVEMENTS-------

DATA. FILE: AMNI. DAT ANALYSIS WINDOW. : 1200 sec

FIXATION PHASE:
LOCATION : X-AXIS M= −5.68 SD=13.35(deg)
: Y-AXIS M= +2.78 SD=11.66(deg)
DURATION : XY-AXIS M=441.20 SD=493.08(msec)
TIME RATIO (FIXATION/TOTAL DATA) : 72.28 %

SACCADIC PHASE:
FREQUENCY : 1.638 /sec INTERVAL : 610.38 msec
AMPLITUDE : X−AXIS M= 9.18 SD= 10.12 (deg)
: Y−AXIS M= 8.29 SD= 10.68 (deg)
: XY−AXIS M=14.13 SD= 13.03 (deg)

GRAND TOTAL CHARACTERISTICS OF EYE MOVEMENTS
FIXATION AREA (X SD) × (Y SD) : 155.66 (deg×deg)
AVERAGED SACCADIC VELOCITY : 23.14 (deg / sec)

be statistically expressed in terms of its two-dimensional location and duration of glances as given for Table 1. Similarly, saccadic phase was represented by its frequency, amplitude and direction of motion. I have made a proposal that integrated characteristics of eye movements in the various visual works could be expressed in the following two ergonomic indexes. The first, fixation area for an arbitrary period of work can be obtained by numerical product of horizontal and vertical standard deviations of spatial distribution of fixation points. Secondarily, averaged saccadic velocity is computed as the product of saccadic frequency per second and mean amplitude of the eye trajectories. The former index represents the viewing extent and spatial characteristic of eye movements in the work. The latter value corresponds to the temporal aspects and is considered to stand for the amount of something looking around restlessly. Such analyses enable to describe objectively and also quantitatively the ergonomic aspects of various different kinds of works as mentioned follows.

RESULTS AND DISCUSSION

Decrease of CFF following VDT operations

Many number of physiological signs were investigated in connection with VDT workload or fatigue (refs. 1-4). The operators investigated in the present study showed marked decrease in CFF value following VDT work. The decreased rate was about 5 % as average of five operators.

Viewing distance and working posture

Constrained postures at VDT workplaces are associated with physical impairments and subjective complaints of fatigue in operators (refs. 1, 4). As

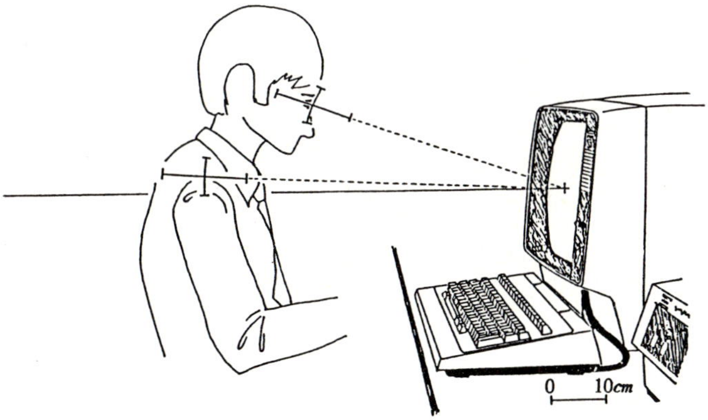

Fig. 1. The posture as grand average of 5 VDT operators through working hours.

closely related to the working posture, viewing geometry for VDTs has been recommended by several organizations (ref. 2). For example, the guideline has proposed by NIOSH (National Institute for Occupational Safety and Health, U.S.A.) that the viewing distance should be within 45 and 70 cm (ref. 7). In the present field survey, the working postures were analysed through a day. As average of five operators, mean and standard deviation of viewing distance between eye and CRT screen was 46.1 and 7.4 cm, respectively. The range of five operators was from 36.1 to 54.7 cm. These are illustrated in polar coordinates in Fig. 1 about positions of eyeball and shoulder. Coefficient of variance of viewing distance within working hours was 15 % in average of 5 operators. The distance was lengthened with the passing of working time in a day and was about 7 % longer than the value at beginning of work. The shorter length and small variance of CRT viewing distance can be indicated as peculiar problems to VDT operations. Besides, it was found by Laville that the distance eye to task was shorter as speed of execution was higher, and that the speed was also associated with a greater stability of the distance (ref. 8).

Relation between CRT viewing and head movements

It was able to be known by analysis of Eye-Camera image data that the operators were viewing what kind objects in their field of vision. They had been engaged in the occupations of computer programing as mentioned previously. Operators were looking at several different parts of work station. The objects viewed at the jobs were mainly restricted to CRT screen, keyboard and manuscript on the work station. In order, each time ratio spent in looking these objects was 64, 21 and 14 % as the grand average of five operators and seven working hours. The other viewing time was only 1 %. The CRT, as a matter of course, was the most frequently watched device in VDT operations.

By analysis of head movements in the work, the inverse correlation between CRT viewing ratio and head motility was found. The higher the CRT ratio, the less mobility of head. These mutual relationships were expressed by correlation coefficient of -.647. The CRT has the properties of being heavy and immobile structurally. These defects should be related to the less motility of head and small variance of viewing distance during VDT operations. These phenomena are considered to be associated with musculo-skeletal health complaints in the portions of neck, shoulders and back reported by several investigators (refs. 1. 4).

Characteristics of eye movements in various visual tasks

For quantitative evaluation of spatio-temporal characteristics of eye movements during visual works, the measuring system has been developed through the

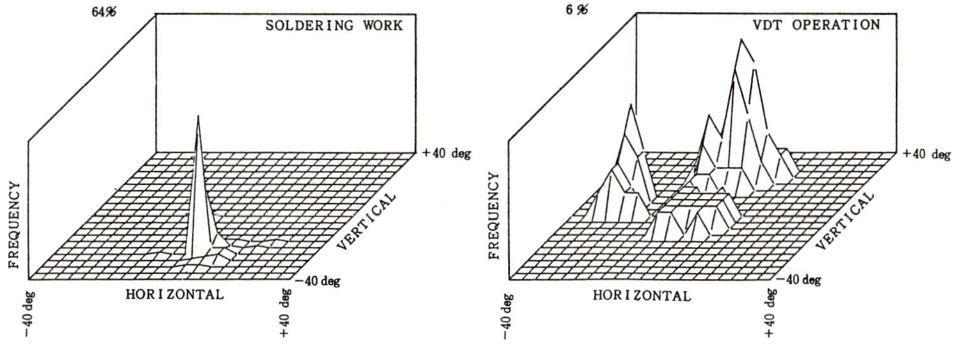

Fig. 2. Frequency distributions of fixation points in the work. The cases of electric soldering work (left) and VDT operation (right) are illustrated.

present study. Based on the eye movement properties, the ergonomic values specific to the work are able to be obtained by the system.

The fixation area in the work was found to depend on the types of jobs. The frequency distributions of fixation points in the case of electric soldering work and of VDT operation are three dimensionally illustrated in Fig. 2. Narrow and concentrated distributions are clearly shown in the case of soldering work. In VDT work, by contrast, distributions are divided into three locations. Those correspond to CRT, keyboard and manuscript. Fixation area in the cases of various kinds of visual tasks was compared to each other. The value of fixation area in VDT operations, which is numerical product of horizontal and vertical extent of fixation points, was 136.6 ± 34.6 deg x deg as mean and standard deviation of five operators through the working hours. This fixation area in VDT works was about 2.5 times as wide as averaged area in traditional visual works without CRTs such as text translation, document reading, drawing and so on. As well as wide extent of viewing area, upward glance (8 degree angle as average) should be pointed out as distinctive feature of VDT operations. Almost the traditional visual works were usually conducted with downward gaze at about 25 degree from the primary position of the eye. These were the spatial characteristics of eye movements in VDT operations.

Temporal aspects of eye movements were also evaluated as averaged saccadic velocity given by numerical product of mean amplitude and frequency of saccades. In the VDT works, saccadic frequency was 1.78 ± 0.16 /sec and averaged saccadic velocity was 22.76 ± 2.23 deg/sec as grand mean of five operators through their working hours. It was found that the eye velocity, in this meaning, in the VDT works was approximately 2.5 times as high as those averaged velocity in other visual works without CRTs. The value shown in VDT works was about 49 % of the maximum velocity obtained in the other eye tracking experiments. When separation between the visible two target was 18 degree the maximum eye tracking frequency was 2.67 ± 0.25 /sec as average of six young subjects and the value considered to be physiological bounds of saccadic motilities of eyeball. As mentioned above, it can be pointed out that one of the most remarkable phases of VDT operations is high velocity of eye movements. This property was due to both the high frequency and large amplitude of eye movements in the works. In usual traditional works without CRTs, it was experimentally confirmed that there was an inverse correlation between frequency and size. It was most common that larger saccade inevitably followed lower the frequency of eye movements.

ACKNOWLEDGEMENTS

The author would like to acknowledge Prof. Takao Ohkubo for his helpful advice. Thanks are due to my many colleagues with whom I have discussed this problem.

REFERENCES

Bergman, T., 1980. Health Effects of Video Display Terminals. Occupational Health and Safety, 49, 6: 24-28, 53-55.

Grandjean, E. and Vigliani. E. (Editors), 1980. Ergonomic Aspects of Visual Display Terminals. Taylor & Francis, London.

Grandjean, E. (Editor), 1984. Ergonomics and Health in Modern Offices. Taylor & Francis, London.

Laville, A., 1980. Postural Reactions Related to Activities on VDU. In :Grandjean, E. and Vigliani, E. (Editors), Ergonomic Aspects of Visual Display Terminals. Taylor & Francis, London, pp. 167-174.

Leigh, R.J. and Zee, D.S. (Editors), 1983. The Neurology of Eye Movement. F.A. Davis, Philadelphia.

National Research Council, 1983. Video Displays, Work, and Vision. National Academy Press, Washington, D.C.

Pearce, B.G. (Editor), 1985. Health Hazards of VDTs ? John Wiley & Sons, Chichester.

Saito, S., 1984. Binocular and Two-dimensional Eye Movements Analyzer for Physiological Evaluation of VDT Operations. In: Salvendy, G. (Editor), Human-Computer Interaction. Elsevier, Amsterdam, pp. 307-313.

Social, Ergonomic and Stress Aspects of Work with Computers,
edited by G. Salvendy, S.L. Sauter and J.J. Hurrell, Jr.
Elsevier Science Publishers B.V., Amsterdam, 1987 — Printed in The Netherlands

NOISE PROBLEM IN JAPANESE AUTOMATED OFFICE

YOSHIO TOMINAGA AND NAOKI MAEHARA

Institute for Science of Labour, 2-8-14 Sugao Miyamae-ku,Kawasaki City 213 Japan

SUMMARY

Workers' complaints of fan noise in office automation systems were surveyed. Noise environment of sixty six rooms in thirty six local taxation offices were monitored. Below 45 dBA the fan noise was evaluated as unnoticeable or undistracting by half of the workers. The authers recomended 45 dBA as the aim of noise control by source moving or using of the low partition for the fan noise. After the recomendation workers' complaints decreased.

NOISE ENVIRONMENT IN JAPANESE OFFICE

Generaly Japanese offices are described as being "open planed" as opposed to "closed planned " private office system . In a large unpartitioned room, desks are located in close proximity with the office staff seated face to face . Space allotted for workers is minimal. For instance, the Japanese governmental office area standard defines a minimum of (3.63 m^2 x number of workers) . It is only half the area of Australian standard .

In such a crouded office, noise environment is composed by two components . Primary one is noise generated by fellow workers in close proximity, such as that from verbal communication, movement, manual operation, foot steps etc . These are intermittent noise . Secondary component is a mixture of noise of human origin from persons seated afar, ventilation noise, outdoor traffic noise and so on . This can be defined as the continuous background noise. Although the primary component should be dominant when the room noise is monitored, the secondary component is an important factor in determining negative effect .

A sample of noise level through a whole day in a typical office

Fig 1 shows the noise level distribution through a whole day in a typical local government office's room . In this layout desk area averages 0.7 m² (1.1 m x 0.65 m) . Located 1 m behind is another worker . Noise at a disignated desk in this room was monitored A-weighted noise level was sampled each 0.1 sec through 6 hours.

In this distribution the component from neighbouring workers had levels around 60 dBA and more. Background noise was in 50 - 60 dBA. Equivalent Level of this noise was 61.7 dBA . Workers in this room did not complaint of the noise environment .

This case is a typical sample of a Japanese office . In many other Japanese offices noise environment are similar to the example in Fig 1 .

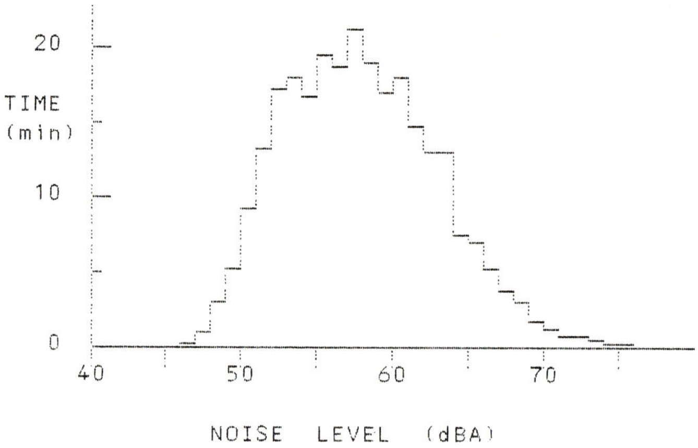

Fig 1 Noise level distribution in a office room through a whole day

Though Japanese office workers have been quite tolerant of the room noise, the introduction of office automation machines created the problem of complaints frequently. Plans to counter such complaints are obviously much more difficult in a Japanese office layout than in an ideal private office system work environment .

NOISE SURVEY IN AUTOMATED OFFICE

Some office workers complaints of the fan noise from terminal controllers were surveyed in 36 local taxation offices at different locations in Tokyo with similar rooms and office area . Each office contained a terminal controller and 8 VDTs fed from a central office source . The dissimilarity was the desk position and location with in each individual office from the noise source.

We tested the room characteristics in relation to the sound propagation, the level of the fan noise at each of the desks, background noise level at the center of rooms and subjective responce of those 384 workers .

Characteristics of source and sound propagation in room

The frequency components of the noise from fans were superior around 500 Hz . The level of those noise were 57 dBA at 1 m .

In many offices the controllers were located near desks. Distance between the noise source and the nearest desk was within 4 m in 23 offices . The maximum level of those fan noise at the nearest desk reached to 57 dBA .

The sound propagation was about 4 dB decay for doubling distance in all rooms .

A typical room layout and location of the noise sorce is shown in Fig 2 . Noise level at each desk is also shown .

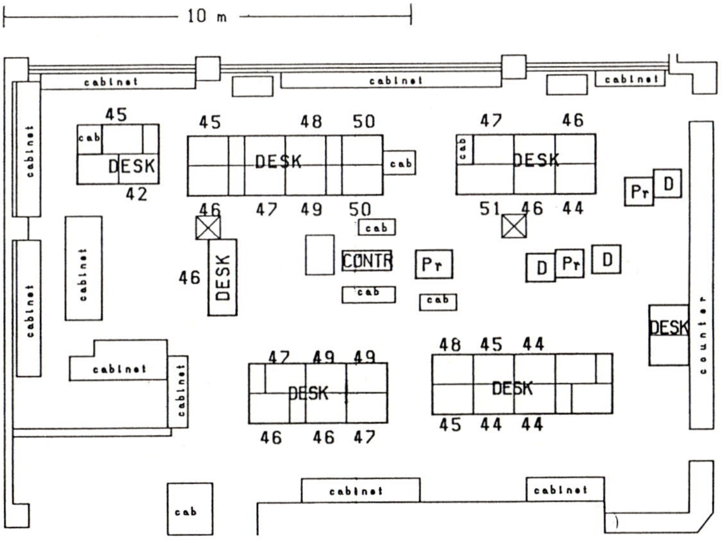

Fig 2 Location of desks and distribution of noise (dBA) from source .

In this example the noise sorce (CONTR) is at the center of the room . Thirty three desks are located within 8 m from the source . Noise levels were less than 45 dBA at only 9 desks .

Workers response

Workers response to the noise at their desks were shown in Fig 3 . Of workers subjected to noise levels below 40 dBA, 82 % of them did not notice or answered that the noise was noticeable but not distracting . At the desks recieving noise levels about 45 dBA, 50 % of workers answered the noise was unnoticeable or undistracting, but the other half complainted the noise was distracting or annoying at times .

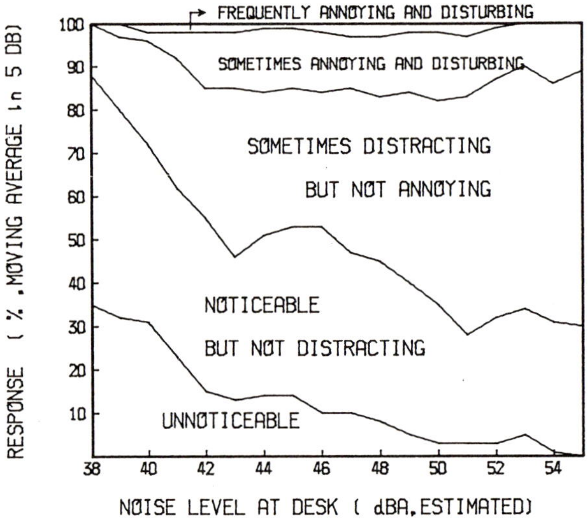

Fig 3 Subjective response of workers to the fan noise

Table 1 shows the responses by two groops, workers subjected the noise 45 dBA or more and less than 45 dBA. The response to room noise of 316 workers in those offices are also shown in Table 1

In this table the room noise were the background noise due to conversation, ventilation and outdoor noise . Noise conditions of those rooms were similar to the example illustrated in Fig 1 .

Difference among those response in Table 1 are significant . Of the annoyance from those noise the following relationship were found statisticaly .

fan noise 45 dBA or more > room noise
room noise > fan noise less than 45 dB

TABLE 1 Response to fan noise and room noise

response	percentage of subject		
	fan noise		room noise
	≧ 45 dBA	<45 dBA	
unnoticeable	7.5	34.1	0.0
noticeable but not distracting	39.1	34.1	51.9
sometimes distracting but not annoying	37.3	26.8	32.5
sometimes annoying and disturbing	14.3	2.4	12.4
frequently annoying and disturbing	1.9	2.4	3.2
number of subject	41	161	314

The level 45 dBA was the recomended aim of noise reduction for the terminal controller noise .

To keep such fan noise under 45 dBA, desks should be located at least 7 meters away from the noise source. If this is not possible due to space limitations low partitions to dampen noise may be an alternative . Dimensions and other charactaristics needed for the low partition were determined through an experiment with 1/4 scale model .

The effect of the noise reduction

After one year the noise survey for the same offices were conducted again. Except two offices terminal controllers were moved away from working area or enlosed in low partitions. Number of desks subjected noise above 45 dBA was greatly decreased from 220 to 26 and workers negative responce was correspondingly decreased (TABLE 2) . Because of having near future reconstruction plan or space limitation, any noise controle had not conducted in two offices .

TABLE 2 Response to fan noise and room noise after noise reduction

response	percentage of subject		
	fan noise		room noise
	≧ 45 dBA	<45 dBA	
unnoticeable	0.0	23.5	0.0
noticeable but not distracting	33.3	47.8	58.1
sometimes distracting but not annoying	27.8	24.3	31.3
sometimes annoying and disturbing	33.3	3.4	9.4
frequently annoying and disturbing	5.5	0.9	1.3
number of subject	18	115	384

* No significant difference is between the data of the fan noise and those of room noise .

CONTENTS

Social, Ergonomic and Stress Aspects of Work with Computers,
edited by G. Salvendy, S.L. Sauter and J.J. Hurrell, Jr.
Elsevier Science Publishers B.V., Amsterdam, 1987 — Printed in The Netherlands

OCCUPATIONAL HEALTH AND THE COMPUTER-MEDIATION OF INFORMATION WORK: RESEARCH NEEDS

Steven L. Sauter and Joseph J. Hurrell, Jr.

National Institute for Occupational Safety and Health,
Cincinnati, OH 45226

ABSTRACT

Employment statistics indicate that jobs involving computer and information work are among the fastest growing occupations. The present paper identifies deficits in understanding potential stress and health effects of work in these types of jobs. It is suggested that epidemiologic studies are needed to determine the potential for frank health disorders, and that experimental interventions are needed to evaluate ergonomic and organizational reforms. Exploration of little-studied job content factors such as cognitive demands is also recommended.

1 INTRODUCTION

According to Bureau of Labor Statistics (BLS) projections (Silvestri and Lukasiewicz, 1985), two of the three fastest growing occupations for the period 1984-1995 contain the adjective "computer" in the job title. Five of the ten fastest growing occupations involve non-manufacturing applications of computer technology. New jobs in these categories contribute to the growth in information work which, along with other service work, now outstrips manufacturing as a source of employment in this country. Projections cited by Bezold, Carlson, and Peck (1986) suggest that nearly 40 million white-collar jobs will be "affected" by computers and robots.

Concern over potential health problems associated with these employment trends has focussed with particular intensity on the consequences of work with video display terminals (VDTs), the most visible symbol of the advancing information industry (e.g., see Bergquist, 1984).[1] The purpose of this paper is to draw attention to some notable deficits in our understanding of health problems in relation to structural characteristics of computer-mediated information work, and VDT work in particular. One pressing issue in this regard involves the potential and basis for chronic somatic disorders, especially cumulative trauma injuries, in VDT work. A second concern involves the effects of job content changes brought about by the growing information sector. Of special interest is the increasing cognitive nature of jobs in this area, a subject that has received little attention in considerations of the potential health impact of computer/information

[1] A broader but less visible public health concern which should not be ignored centers around the longer-term implications of changes in employment and the nature of health care delivery associated with a growing information sector (e.g., see Bezold et al., 1986; Office of Technology Assessment, 1985; Pedersen, 1986).

work.

2 SOMATIC DISORDERS: PREVALENCE, CAUSES AND CONTROL MEASURES

Giuliano (1982) has conceptualized evolutionary changes in the organization of office/information work. In his view, the "information-age office," as distinguished from the "industrial office," exploits new technology (computerization) to avoid the negative aspects of the industrial office (production-line approach with narrow, fragmented tasks). Benefits in terms of increased productivity and satisfaction are predicted for transaction-intensive tasks, as well as management, research, and administrative activities.

From the perspective of somatic health and exposure to ergonomic stressors, however, this scenario is less appealing. The literature shows uniformly high prevalence rates for somatic complaints among VDT operators (Schleifer and Sauter, 1987), and it is not clear that (at least in the near future) such computerization offers much potential for relief of ergonomic stressors common to industrial-age information or office work. Quite to the contrary, the concern has been that computerization may exacerbate these problems, although the data are equivocal on this issue.

Many information workers (e.g., professionals), whose jobs were performed in an almost "craft-like" manner and who were largely immune to the ergonomic stressors associated with computer terminal work, become exposed to these stressors in the information-age office. For workers performing very basic and stereotypic information processing tasks such as data entry, replacement of electromechanical with microelectronic systems offers little in the way of near-term job reform and relief from ergonomic stressors. And while computerization may benefit general secretarial workers in terms of improved resources and satisfaction, these individuals are still subject to salient ergonomic stressors such as long spells of keyboard and display work. Importantly, absolute employment in this capacity is projected to increase (although growth rate will decline in relation to other information or service work). BLS statistics suggest that the number of clerical jobs will increase by almost two million (almost 300 thousand new secretaries) to 20.5 million in 1995, with clerical work remaining the largest occupational grouping (Silvestri and Lukasiewicz, 1985).

2.1 Knowledge Gaps and Research needs

(i) Potential for frank musculoskeletal disorders. It is curious that reports of relatively high rates of musculoskeletal complaints among VDT operators have generated only limited epidemiologic study of frank pathology or injury in this population. At this point in time, for example, there seems to be no reliable information in the U.S. on prevalence or incidence rates for carpal tunnel syndrome or other distinct biomechanical disorders among VDT users (or, for that matter, among their VDT non-user counterparts).[2] Australian data on "repetition strain injury" (RSI) rates among VDT operators suggest a strong potential for disabling musculoskeletal disorders in this type of work (see Hocking, 1986, Public Service Board, 1985). However, RSI refers to a broad classification of rheumatic disorders, including chronic pain syndromes which lack sound understanding in terms of etiology and pathology. Occupational prevelance rates (including VDT-type work) and occupational differentials for musculoskeletal disorders are also found in the Japanese literature (e.g., Ohara, Itani, and Aoyama, 1982). However, as with RSI, the classification tends to be broad and diffuse, defying clear attribution of causal factors and pathological basis.

(ii) Injury mechanisms. Leaving aside the question of potential for frank musculoskeletal disorders in VDT work, empirical data pointing to the proximal causes of such disorders in terms of individual factors or job demands are scarce. While Australian investigators have sought associations of RSI with factors such as age, sex, keystroke rate, etc., with only a few exceptions (e.g., Sauter, Schleifer, and Knutson, 1987; Hunting, Laubli, and Grandjean, 1980), have studies probed for fine-grain associations of individual characteristics and workplace ergonomic factors with musculoskeletal disturbances. Apparently, case-control type studies involving injured or symptomatic workers, which might be especially revealing, are completely lacking.

(iii) Ergonomic controls. Although studies yielding injury or injury mechanism data have not been forthcoming, the last decade has witnessed a dramatic response by industry in terms of ergonomic enhancement of products for computer terminal work. However, until recently (e.g., Bendix, Winkel, and Jessen, 1985; Nakaseko et al., 1985) there have been few studies (at least in the public domain) of the health or performance benefits of specialized or redesigned products for computer terminal work. And with only a few exceptions (e.g., Springer, 1982; Dainoff et al., 1982; Shute and Starr, 1984), there seems to have been little systematic study of the payoffs of generalized ergonomic reforms in the information-age workplace. At this time there apparently exist no controlled, prospective studies of the effects of ergonomic interventions in an actual workplace where VDTs are used.[3]

(iv) Organizational controls. Similarly, control technology information in terms of organizational aspects of computer-mediated information processing work is sparse. While there has been considerable discussion of improved work-rest schedules, exercise regimens, job enlargement, training, etc., to mitigate stress experienced in this type of

---

[2]According to the National Commission on Compensation Insurance (1985), worker compensation claims among clerical workers accounted for 2%-3% of all losses each year over the period of 1981-1984. And in reference to the question of increasing health risk with information age work, these data show no trend toward a relative increase in losses over this period. However, the Commission does not differentiate these losses by injury, nor discuss potential methodological problems in use of worker compensation data for such research/surveillance purposes.

[3]Alternately, it may simply be that this information is not readily accessible. Springer (1986) cited numerous instances of gains realized from ergonomic improvements in office design. However, much of this information was gleaned from consultative experiences or reports in trade journals, and thus its scientific merit is difficult to judge.

work, research has been limited to a few, relatively small-scale, laboratory studies (e.g., Sundelin et al., 1986; Zwhalen et al, 1986; Lee and Waiker, 1985).

In general, laboratory studies investigating both task-organizational and ergonomic prevention measures have utilized young, healthy subjects (for ethical and experimental control purposes). Ecological validity might be an important concern in these studies considering the prevalence of impairment (especially musculoskeletal and visual) in the working population, and the potentially increased sensitivity of affected individuals to both job stressors and control measures.

3 JOB CONTENT ISSUES

Giuliano's (1982) analysis acknowledges the interaction of computerization with the content and organization of information/office work. In terms of invoking new stresses, this process has been discussed with particular lucidity by Turner and Karasek (l984), and investigated in numerous studies over the past several years. However, salient aspects of information-age work have received insufficient attention.

3.1 Knowledge gaps and research needs

(i) Effects of cognitive demands. By definition, trends towards increasing computerization and information work involve a shift toward occupational tasks which load more heavily on mental or cognitive components, although such demands are likely to be highly differentiated as a function of task or occupation. Emerging data (e.g., Jennings, 1983) now suggest connections between cognitive demands and health parameters. However, the cognitive demands construct, which includes aspects of attention, memory, decision-making, response requirements, etc., has tended to escape recognition, characterization and investigation in the occupational stress literature.[4] Indeed, standard psychometric indices for assessing features of cognitive demands do not exist. With regard to contemporary information processing or computer tasks, we are unaware of any stress-related investigations incorporating this construct in a systematic fashion.

(ii) Effects of worker control. A second important construct which has not escaped attention, but requires much further conceptual treatment, is "controllability." Information-age work involving interaction with intelligent machine systems may influence allocation of control over task functions between the operator and the machine, with the potential for displacement of control to the machine. This effect is indicated in two studies showing significantly lower ratings of job autonomy among VDT users in comprison to their non-user counterparts (Sauter, Gottlieb, Jones, Dodson, and Rohrer, 1983; Smith, Cohen, Stammerjohn and Happ, 1981), although other studies are

---

[4]Mental workload and cognitive demands has been a subject of inquiry in the psychological and human factors literature for some time (e.g., see Moray, 1979). However, the construct has never been systematically addressed in relation to stress and health. For a perspective on such a connection, refer to the paper by Ganster, Hurrell, and Thomas in this volume.

more equivocal on this issue (see Turner and Karasek [1984] for a review).[5] Such control loss may have significant adverse health implications. Expansive Swedish and U.S. studies have linked low intellectual discretion and personal freedom at work to affective and somatic disorders, including cardiovascular disease (Karasek, 1979; Baker, Marxer, Ahlbom, and Theorell, 1981).[6] Ganster (1986) points out, however, that "control at work" is a multidimensional construct and that little scientific concurrence exists on operational definitions of control or the importance of different domains of control. These methodological issues have received minimal attention in the occupational stress literature, and none whatsoever in the context of computerized information work.

4 SUMMARY OF RESEARCH NEEDS

Research is needed to resolve basic questions on risks and control measures for health problems in computer-mediated information processing work. With regard to somatic health concerns, studies yielding prevalence/incidence rates for frank musculoskeletal disorders are needed to determine injury risks. Such studies would benefit from task analysis and workstation evaluation to help isolate causal factors and suggest control measures. Additionally, experimental studies are needed to examine the efficacy of ergonomic and organizational reforms designed to relieve physical and psychosocial stress in computer-mediated information work. Especially relevant in this regard are prospective intervention studies in actual work settings. In general, investigations to isolate ergonomic stressors and to define solutions may benefit from a focus on symptomatic workers.

At a more exploratory level, research on psychosocial stressors in information processing work needs to begin to consider the nature of cognitive demands and their health implications. Similarly, efforts need to be devoted to dimensionalizing and operationalizing controllability, and toward exploration of suggested occupational health impacts.

---

[5]Mixed findings on the influence of computerization on control allocation suggests the effect may be highly dependent upon the nature of the task.

[6]It should be noted that Haynes (1985) reported evidence of a link between VDT work and chest pain suggestive of angina, and that the chest pain was more common among VDT users with lower perceived job control. However, an independent evaluation of this study and data reanalysis (Schnorr et al., 1987) point out a number of significant methodological limitations which detract from the reliability and interpretability of the data.

5 REFERENCES

Bergquist, U.O. (1984). Video display terminals and health: A technical and medical appraisal of the state of the art. Scandinavian Journal of Work, Environment and Health, 10(suppl. 2), 1-87.

Bendix, T., Winkel, J., and Jessen, F. (1985). Comparison of office chairs with fixed forward or backwards inclining, or tiltable seats. European Journal of Applied Physiology, 54, 378-385.

Bezold, C., Carlson, R.J. and Peck, J.C. (1986). The Future of Work and Health. Dover, MA: Auburn House Publishing Co.

Dainoff, M., Fraser, L. and Taylor, B.J. (1982). Visual, musculoskeletal, and performance differences between good and poor VDT workstations: Preliminary Findings. Proceedings of the Human Factors Society 26th Annual Meeting, Seattle.

Ganster, D.C. (1986). Exploring the role of control in occupational stress. In Methodological Issues in Occupational Stress Research. Cincinnati, OH: National Institute for Occupational Safety and Health.

Giuliano, V.E. (1982). The mechanization of office work. Scientific American, 247(3),148-165.

Haynes, S. (1985, January). Presentation at the 58th Scientific Session of the American Heart Association, Washington, DC.

Hocking, B. (1986). Epidemiology of R.S.I. in Telecom (OH&S Report no. H86-10). Melbourne: Telecom Australia.

Hunting, W., Laubli, Th., and Grandjean, E. (1980). Constrained postures of VDU operator. In E. Grandjean and E. Vigliani (Eds.), Ergonomic Aspects of Visual Display Terminals (pp. 175-184), London: Taylor and Francis.

Jennings, J. R. (1983). Attention and coronary heart disease. In D. Krantz, A. Baum and J. Singer (Eds.), Handbook of Psychology and Health: Vol. III. Cardiovascular Disorders and Behavior. Hillsdale, NJ: Lawrence Erlbaum.

Karasek, R.A. (1979). Job Demands, job decision latitude, and mentals strain: Implications for job redesign. Administrative Science Quarterly, June, 285-307.

Karasek, R.A., Baker, D., Marxer, F., Ahlbom., A., and Theorell, T. (1981). Job decision latitude, job demands, and cardiovascular disease: A prospective study of Swedish men. American Journal of Public Health, July, 494-705.

Lee, K.S. and Waikar, A.M. Reduction of physical stress experienced by VDT operators. Proceedings of the International Scientific Conference: Work with Display Units, Part I (pp. 507-510). Stockholm.

Moray, N. (1979) Models and measurement of mental workload. In N. Moray (Ed.), Mental Workload: Theory and Measurement, New York: Plenum.

Nakeseko, M., Grandjean, E., Hunting, W. and Gierer, R. (1985). Studies on ergonomically designed alphanumeric keyboards. Human Factors, 27(4), 175-188.

National Council on Compensation Insurance: (1985). Emotional Stress in the Workplace - New Legal Rights in the Eighties. New York: Author.

Office of Technology Assessment. (1985). Automation of America's Offices (OTA-CIT-287). Washington, DC: U.S. Government Printing Office.

Ohara, H., Itani, T., and Aoyama, H. (1982). Prevalence of occupational cervicobrachial disorder among different occupational groups in Japan. Journal of Human Ergology, 11, 55-63.

Pedersen, D., Sieber, W., and Sundin, D. (1986). Health care trends in United States Industry. The Changing Nature of Work and Workforce: Proceedings of the Third Joint U.S. - Finnish Science Symposium (pp.27-30). Cincinnati, OH: National Institute for Occupational Safety and Health.

Public Service Board. (1985). Census of Repetition Strain Injury in the Australian Public Service (Statistical Bulletin, December 1985). Canberra.17.

Sauter, S., Schleifer, L., and Knutson, S. Musculoskeletal discomfort and ergonomic predictors in VDT data-entry work. In S.S. Asfour (Ed.), Proceedings of the 1987 Annual International Industrial and Ergonomics and Safety Conference, Amsterdam: Elsevier (in press).

Sauter, S.L., Gottlieb , M.S., Jones, K.C., Dodson, V.N. and Rohrer, K.M. (1983). Job and health implications of VDT use: Initial results of the Wisconsin-NIOSH study. Communications of the Association for Computing Machinery, 26, 284-294.

Schleifer, M. and Sauter, S. (1987). A questionnaire survey to assess ratings of physical workplace conditions, somatic discomfort, and work inefficiency among VDT users. In L. Mark, J. Warm, and R. Huston (Eds.), Ergonomics and Human Factors: Recent Research, Springer-Verlag, (in press).

Schnorr, T.M., Thun, M.J. and Halperin, W.E. (1987). Chest pain in users of video display terminals. Journal of the American Medical Association (letter), 257(5), 627.

Silvestri, G. T. and Lukasiewicz. (1985). Occupational employment projections: The 1984-1985 outlook. Monthly Labor Review, Nov. 1987, 42-57.

Smith, M.J., Cohen, B., Stammerjohn, L.W. and Happ, A. (1981). An investigation of health complaints and job stress in video display operations. Human Factors, 23(4), 387-399.

Springer, T.J. (1986, June). Presentation at a NEOCON 18 seminar (Productivity: A dialogue for the future). Chicago.

Springer, T.J. (1982). VDT workstations: A comparative evaluation of alternative. Applied Ergonomics, 13(3), 211-212.

Shute, S.J. and Starr, S. (1984). Effects of adjustable furniture on VDT users. Human Factors, 26(2), 157-170.

Sundelin, G., Hagberg, M. and Hammarstrom, U. (1986). The effects of pauses on muscular load and perceived discomfort when working at a VDT word processor. Proceedings of the International Scientific Conference: Work with Display Units, Part I (pp. 501-502). Stockholm.

Turner, J.A. and Karasek, R.A. (1984). Software ergonomics: Effects of computer application design parameters on operator task performance and health. Ergonomics, 27(6), 663-690.

Zwahlen, H.T., Hartmann, A.L. and Kothari, N. (1986). How much do rest breaks help to alleviate VDT operator subjective ocular and musculoskeletal discomfort? Proceedings of the International Scientific Conference: Work with Display Units, Part I (pp. 503-506). Stockholm.

Social, Ergonomic and Stress Aspects of Work with Computers, 219
edited by G. Salvendy, S.L. Sauter and J.J. Hurrell, Jr.
Elsevier Science Publishers B.V., Amsterdam, 1987 — Printed in The Netherlands

PROBLEMS OF A NOVICE: SOME INTROSPECTIVE OBSERVATIONS

A. F. SANDERS

Institut für Psychologie, RWTH Aachen, Jägerstr. 17-19, 5100 Aachen, F.R.G.

ABSTRACT

A central issue in the field of human-computer interaction concerns the question how non-experts in computer systems can acquire sufficient expertise to use the machine for their specific aims. Usually prospective users cannot be supposed to go through extensive training programs. Hence operating the machine should require minimal new knowledge. This means that the machine properties should be as compatible as possible with existing human perceptual and cognitive schemas. Preferably, the remaining acquisition should be mediated through texts that are suitable for self-paced study. Similar requirements obviously apply to programming languages.

The paper discusses the experiences of the author on getting acqainted with some essential operations on two current personal computers. One - an Apple Macintosh Plus - has mouse and menus as a standard means of control. The other system - a Toshiba 1500 - is an IBM compatible machine. I used it with an exclusive keyboard control. The experiences are accompanied by some ergonomical recommendations, and discussed in the context of current notions on human-computer interaction.

1 INTRODUCTION

I am a novice. Until recently I had never worked with either one of the present generation personal computers; nor did I have any operational command of systems, such as the PDP family, which are still current in psychological laboratories. My programming experience stemmed from the Fortran of the sixties. The underlying reason was clearly that I had had no clear needs. With regard to experimentation I was so lucky to have excellent technical and programming assistance. There was also always the secretary to type my papers. Hence there was no real impetus until I received an invitation for this conference. I realised that preparing a paper on Human-Computer Interaction without knowing anything about this interaction from personal experience would not be satisfactory. Thus I started working with the idea in mind that a report on my proceedings "from novice to somewhat less of a novice", together with observations on the ergonomics might be a first exercise.

How to start? As I said there are various people in my environment who are real experts and they made quite helpful recommendations. Thus, a first task was to experience the situation of a manager who has a newly acquired PC. Usually managers have no clear notions what to do with the device other than that they expect that the machine can rapidly provide data that are relevant for policy decisions, such as financial reviews, sales and their development over time, and even economic trends for the branch. The manager understands that acquainting himself with the ins and outs of the machine takes some time, but he is not willing to spend more than a couple of hours on mastering elementary skills, such as how to open and close a file, how to copy a document, etc. My primer on these essentials consisted of getting acqainted with an *Apple Macintosh Plus*. This configuration is commonly considered as quite "user-friendly", due to the concrete "desk top"metaphor. The availability of a mouse in combination with menus and the principle of direct manipulation is an essential element in this mode of control.

The second task was related to the first one and consisted of mastering the *textprocessing* skill on the same machine, since everybody tells me that it is great to type your own papers. Again, this is a reasonably simple skill, building upon the primer, and with the further advantage that it is commonly found in real life tasks.

A third task brought me to a *Toshiba T1500* under MS-DOS 2.11. It is obviously useful to learn the secrets of how to operate another machine in order to enable comparisons. It should be clear that these comparisons do not say anything about the qualities of the machines as such but rather about the typical configurations in actual use. For example, our Apple has a mouse which was not the case for our Toshiba. This does of course not mean that a mouse-type configuration cannot be obtained for the Toshiba. Rather than stressing the differences between the machines it is better, therefore, to emphasize the differences between mouse and keyboard operations in acquiring the essentials. It is probably still better to say that even mouse and keyboard are not essential but rather the underlying software facilities. In addition the nature and qualities of instruction aids are relevant. I had a handbook for the Apple, and a couple of interactive teaching programmes for the Toshiba.

The core of the paper consists of some observations on how these tasks were accomplished. During the practice sessions a fairly detailed protocol was kept, which will be summarised in the following sections, completed with brief comments. Final sections will be devoted to a summary of the observations and some more general comments.

2 ACQUIRING THE ESSENTIALS: THE MOUSE CONFIGURATION

In the first session, I found myself armed with a Macintosh, a system disk, and a handbook. It must be said that the handbook did well in teaching various trivial but essential preliminaries such as starting, inserting a system-disc, and using the mouse in a number of essential operations. The use of the mouse in operations such as activating a file, displaying the menu of a file and manouvring the contents along the display are simple and straightforward. In standard form, the displays consist of icons as well as labels, which has the danger of overloading the available space. In turn this means that not all information can be displayed at once, at least not when several files are open at the same time. This can be counteracted by omitting either the label or the symbol,which, then, leads to a sizable increase in the index of difficulty when moving the mouse to a desired target. The basic problem is that the display is fairly small.

Two more _ secondary but still noteworthy _ problems with the display are worth mentioning:The first concerns its upright position. If the machine is located on a real desk _ as is probably the case for many managers _ the consequence is that there is a suboptimal viewing angle. In particular people wearing bifocals ,as I do, find themselves with too elevated a head-neck angle.

The second problem with the display is related to the operations involved in scrolling the window. First, I found the scrolling sensitivity to be is insufficient, in particular when working with various menu-listings at the same time. Second, I found the relation between the direction of scrolling and its corresponding signal _ i.e. an upward and a downward arrow _ to suffer from a lack of compatibility. The human prejudice is that an upward arrow corresponds to an upward movement of the text, or, in other words, to presenting later text. The opposite is the case, which leads to unnecessary errors.

Returning to my selftaught course I made the observation that mastering the first tricks took me some 30 minutes. As said this was mainly due to the good initial didactic quality of the handbook, as well as to the ease and naturalness of most operations. Unfortunately the high standard of the handbook was not maintained at the point somewhat more complex operations were introduced; after some attempts, I found myself exploring the further possibilities by trial-and-error ! This implied that, although I was under the impression that I was progressing much faster, I ended up with overlooking, and failing to master, various aspects of the operating system. I must admit, though, that the very fact that I could progress and did not get stuck altogether, was a main advantage.

The major problem of the handbook could well be that its authors anticipated impatience from the side of the users. Hence they proceeded at too early a stage from explaining how to operate the system to providing an example, i. c. text-processing. This occurred at the cost of discussing essential concepts such as the option of having an hierarchical file structure, how to get rid of dated files or documents, and copying files on other discs. In addition, the meaning of a number of operations remained unclear. At the same time, the introduction to text processing was fairly scant and limited to deleting and inserting bits of text. Another problem, which is found with technical handbooks in general, is that the information does not fully apply to the system at use. For example, at first reading , I was under the impression that the "Finder" program, so essential for all applications, was on a seperate disc. In addition some combinations of mouse and keyboard responses were indicated which were no longer valid.

About eight hours of further practice were sufficient to master by trial and error further essentials such as learning how to open, close and delete new files, acquiring the meaning of most menu items of the operating system and, inadvertently, getting acqainted with elementary aspects of the "Macwrite" program, which sufficed to type and edit this paper. Then the major disadvantage of trial-and-error became clear, namely that I limited the communication with the machine to what I had learned sofar while neglecting remaining possibilities. Therefore I listed all menu items which I did not yet understand and returned to the chapter of the hand- book, which listed all actions and functions. The main problem I faced was that I did not know what further options of the machine to expect. There- fore I asked an expert on the machine to prepare a "progress-test",in the hope to learn about remaining gaps. Without going into details it turned out that I could correctly solve about 70 percent of the problems. The remaining ones were either fully "terra incognita" or were approached by way of clumsy solutions.

Yet it should be noted that the eight hours were well invested. One main property, adding to this success, concerned the highly corresponding relations between commands and actions. Most options such as "Delete" and "Insert" as well as various formatting and print options were in clear language, requiring minimal acquisition time. In addition the mouse-operated analog fashion of issuing commands adds a clear dimension of control to the verbal one. In particular in text processing _ a highly verbal activity _ it is quite relevant to perform non-verbal elements in an analog fashion. In the final section of this paper I will return to this issue.

Another interesting feature which greatly aids the beginner is that commands which are at odds with a given state - for instance opening a new document while being engaged with another one, is simply impossible. This is indicated on the screen by a reduced brightness contrast of the impossible commands which automatically changes when the command fits the situation.

With regard to the handbook my conclusion is that without a properly written text, I will never master various fine details. It is interesting that the person, who had composed the progress test, reported that initially he had met the same problems, and still, after two years of practice,was unaware of some of the machine's more detailed potential.

3 ACQUIRING THE ESSENTIALS: THE KEYBOARD CONFIGURATION

The Toshiba is an I.B.M. compatible machine. As a first introduction I took an IBM "interactive" teaching programme course on the DOS operating system,written under SEF authoring system. I got the recommendation to do so since written materials, although present,were not considered as optimally didactic by the institute's experts. The computer had no Mouse option so that all instructions had to be entered by keyboard. This is of course still the usual practice for most present day personal computers and, hence, the acquisition of this skill seemed a useful exercise.

As a first comment I would like to remark that, as is often the case,the programmed instruction course suffered from the problem that I experienced it as trivially slow at some instances - for example it explained the principle of computer interaction at some length - while at other moments I could have used many more practice trials.

On the first day of practice I followed the computer course for about three hours. At the end I noted three major problems which seemed to hamper progress. The major one had to do with *compatibility.* It refers to the common observation that one has to master the relation between system functions and their corresponding keyboard actions by almost sheer paired-associate learning. For example, at the left side of the keyboard, there are 10 function keys, labelled F1 to F10. Some of these function keys played a role in the teaching program, like creating the possibility to quit, to sollicit for the correct solution, etc. Lateron, however, I was told that they all play a role in textediting actions such as copying, deleting parts of a line etc. Hence the function of a function key appears to depend on the defining program. This is fair, of course, but not immediately selfevident to the novice. Once the novice has discovered this principle he wonders about the generality of the specific functions, which are taught later in the teaching program. Yet I felt that I should try to master the rules but I suffered from heavy interference, so that upon completing the programmed course I had not the proper knowledge. At this point,the instruction program might have inserted Study-Test trials until reaching perfect recall _ following the best traditions of paired associate learning paradigms _ but it did not. Too late I realised that I should have taken notes in order to memorize the associates afterwards. However, I had started from the idea that, in the absence of explicit instructions to do so, this is not in accordance with the philosophy of programmed instruction. I met similar problems - and perhaps even more pronounced - when I was introduced to actions, the initiation of which required a combination of keys. There were some attempts to provide meaningful codes for the function keys, such as DEL(ete), and INS(ert).They were quite helpful and belonged to the few things which I could properly handle the next day. There were other keys, such as NumLock and PrtScr, the underlying meaning of which remained obscure for some time.

The second main problem which I experienced in my novice status on the keyboard system was that several concepts were introduced, the meaning of which remained initially either unclear or could not be put in operation. For instance, I was told in a relatively early stage that the DOS system _ Disc Operating System - has four main elements i.e. BOOT, IBMBIO.COM, IBMDOS.COM and COMMAND. DOS with different main functions, such as loading other programs in the main memory, controlling files, and accepting commands. Obviously the abundant use of abbreviations introduces another compatibility problem. This observation may be considered as a commonplace. Yet it cannot be denied that the practice of abbreviations in communication sciences has traditionally led to unreadable documents, up to the point where it is impossible for even experienced programmers to read eachother`s work. And this is not only a matter of improper documentation. Of course I realize that limits in wordlength are one reason underlying this state of affairs, and that improving on this would reduce the possibilities of the machine otherwise. In other words it is a matter of priority in a clear case where a human factor competes with machine and ,possibly , organisational factors.

Apart from the abbreviation problem, however, I suffered from the fact that the precise operational meaning of the functions remained unclear. For example the declarative knowledge that the function of IBMDOS.COM concerns the control of files and that COMMAND.COM accepts instructions and subsequently executes the corresponding programs, is not useful if it is not immediately accompanied by procedural activities. Perhaps this is a consequence of a computerized instruction course. It is hard or even impossible to take the course and at the same time "play" with the machine,unless one has two machines available ! There are practice trials in the course but they are usually not very convincing simulations.However, another possibility is that knowledge about the functions of IBMDOS.COM and other parts of DOS may be not of primary importance to the novice.

It is usually taken for granted that the contents of handbooks and of programmed instruction courses should be composed by experts. Yet,it is gradually becoming clear that expertise can also exert a negative effect on knowledge transfer since the expert cannot escape very easily from his own internal representation,and has little feeling for acquisition of knowledge by the novice. An essential further requirement is that the laws of knowledge acquisition and transfer are obeyed. Selfevident, as these statements may be, they should be repeated until they are seriously considered in the practice of composing instruction materials. Preparing a proper instruction is not an "easy" task, and of secondary relevance. In fact, it determines the acceptance of any technical system.

Thus, my main problem was that, after the first two hours of practice, I had not yet the faintest notion what I could do with the machine, and that I still failed to know various elementary operations. Thus, when quiting that first day I had to ask how to turn the machine off!

The second day I repeated parts of the course and wrote down everything I would have to memorize. This was followed by desperate attempts to master the contents which was hampered by the fact that I still was not aware of the operational uses of the various commands. For instance, I had learned that pressing a combination of Ctrl and Num would have the effect ofinterrupting the "current system operation", but I had only vague ideas what that might mean. It was only some time later that I understood the relation with,say, stopping a scrolling screen ! The programmed course claims to introduce the user to the system in three hours. Strictly speaking this was correct: It took me about that time. The question whether I had learned enough to do anything with the machine must be answered in the negative. The course cannot be recommended to starting managers !

The next possibility was to take an _ also programmed _ course on text processing as a perhaps more adequate introduction for a beginner.

I had a set of tutoring programs for some textprocessing facility, consisting of seven programmed instruction sessions, of which I took the first two in order to get the feel. I should add that the facility is quite commonin textprocessing sothat I cannot be blamed of indulging in a dated system.

Again, a main problem of mastering this skill consisted of the many highly incompatible,and partially conflicting, command-effect relations. There are only a couple of instances where the relations were compatible ,and this immediately supported learning. One compatible instance was how to move the cursor in single steps, for which a diamond of keys is assigned in which "up", "down","left", and "right" indeed refer to moving the cursor one step into the same direction. On the other hand there were many instances where no obvious attempts towards compatible solutions could be noticed. For example one might still argue, as is done in the Opening Menu, that a "P" for Print is a reasonable solution. Whether this has been intentional may be doubted when inspecting some of the remaining File commands, which I met: "E" for Renaming, "O" for Copy, and "Y" for Delete. What is even more complicating is the lack of homogeneity at the different levels of the Menu. Thus, at the Main menu "Del"stands for deleting a character and "Ctrl-T" for deleting a word. In addition I was instructed at the Preliminary commands of the Opening menu that keying "Ctrl-S" would delete a character at that level. The system had also a number of options aiming at increasing the possibilities and efficiency. Among others there is a Quick submenu, to be activated by "Ctrl-Q",with a number of options for generalizing certain commands to larger sections of text. Although as such these commands are quite useful, they are again connected to a number of unrelated keying instructions.

There has been a clear attempt to forego these problems by displaying the (sub)menus on the screen. During the course the novice is repeatedly told that there is no need to memorize the various paired associates since they will remain on display. Learning is supposed to occur by doing. This is undoubtedly the case, but the amount of time which needs to be invested over and over again by any new novice should not be underrated. I was certainly not impressed by the lay-out of the display, which suffered again from abbreviations,and, in addition, from a deficient organisation and a tendency towards overload. Lack of space prevented me from showing an example but they can be observed anywhere where this type of the system is used.

A final remark concerns the way special elements _ such as change of print, lay-out and general formatting etc _ are accomplished. In the system any such element is accompanied by a keying instruction, and sometimes even by double keying instructions, to indicate the beginning and the end of, say, a section in italics. Although, again, such operations can be undoubtedly mastered they are certainly not optimally user-friendly.

4 GENERAL COMMENTS AND CONCLUSIONS

I would like to limit my comments to three main issues: First the role of verbal vs. spatial codes or resources; second ,the relevance of S-R compatibility; and third, the general notion of acquiring cognitive schemata or internal representations in handling a machine.

Verbal vs. spatial codes: Research on human information processing during the fifties and sixties has been largely centered around processing verbal material. This was common in research on memory (Murdock 1974), as well as on selective attention (Broadbent 1958, 1971). They were almost exclusively concerned with verbal material to the extent that, for some time, it was believed that non-verbal features of signals were rapidly forgotten unless they could be translated in a verbal articulatory code (Conrad 1964). In the literature on selective attention stimulus encoding was thought to be completed when it activated its corresponding dictionary unit (Treisman

1964), a concept with at least a strong verbal connotation.

Towards the end of the sixties this picture gradually changed. This was due, first, to discoveries on hemispheric differences (Trevarthen 1968) with regard to an ambient, or spatial, and a focal, or discrete, mode of neural processing, Second, there were the important developments in the area of mental imagery (Paivio 1971), and, more generally, there was the recognition of the pronounced role of analogue internal representations in information processing. It is interesting to note the relations between imagery and spatial internal representations; recalling a group of spatially arranged stimuli is selectively interfered by a spatially distracting task, while a non-spatial interfering task has more effect when the stimuli are also not spatially arranged. (Janssen 1976). A third development has been the accumulating evidence that a combination of a predominantly spatial and verbal task shows considerably less inter- ference. This applies to tests of short-term memory tasks, (e.g. Brooks 1968), as well as to dual task performance (Kinsbourne and Hicks 1978) although, in the latter case, considerable interference may still occur, due to converging requirements on processing mechanisms that are unrelated to spatial vs. verbal processing . (Sanders 1979)

The results of this research are of considerable relevance for models of the human operator, extending from display principles to more general aspects of handling machines by man. Thus, a major consequence is that in man-machine system design the correspondence between the "behavior" of the machine and the nature of existing human internal representations is quite relevant. For example the developments in the design of altimeters for aircrafts have led to a moving pointer vertical analog display (Roscoe *et al.* 1981) rather than to a moving scale vertical display, in sharp contrast with a digital display and - obviously - the dated circular three-pointer display. The advantages of the moving pointer display are primarily that they correspond well with the human internal representation in which "up" refers to a higher altitude, and, moreover, an upward or downward movement of the pointer refers to ascending and descending.

Spatial codes have also found their way in the design of navigational maps (Thorndyke, 1980), and in many other instances of display design. A particularly striking example is found in the recent work of Gopher (1984) on keyboard design for typewriting. He found a highly positive effect of vision-based imagery on a transcription skill. His keyboard consisted of two 3x2 matrices of keys, which were operated by either hand. Each Hebrew letter was represented by a chord, bearing strong resemblance to the shape of a letter. Gopher found much faster acquisition of the typing skill in comparison to standard letter keyboards. Similar results were obtained with roman letters. The need of keying chords rather than single keys affected the asymptotic level of performance to some extent but much less than might have been intuitively expected. (See also Gopher, this volume)

From the above observations on getting aquainted with the "Mouse" and "Keyboard" operated computer configurations the relevance of taking account of such evidence has become abundantly clear In summary, the advantages of the spatial aspects of a Mouse operated configuration concern (a) the absence of *verbal* keying instructions to accomplish spatial aims, (b) the direct spatial correspondence between movements and aimed result, and ,(c)the division between verbal and spatial properties in one and the same task; this allows for allocating *both* verbal and spatial resources (Wickens,1984) instead of verbal resource only.

It might be added that the quality of spatial representations poses a major problem in various present-day computer applications. From my more recent practice in Ergonomics I may mention two instances. The first concerned the control of a railway station together with some railroad junctions. Here the traditional geographical "map-like" display on a wall was replaced by two visual display units, each covering about half of the original configuration. The problem was of course that each individual screen was to small to contain the full configuration, displayed at a minimally visible detail. On the other hand, dissecting the display into two parts had the clear disadvantage of ruining the relation between display and internal map. Yet, in this example it was still attempted to maintain a spatial configuration. In another case _ concerned with the occupancy of berths in a large harbour system _ one of the original ideas was to eliminate the geographics altogether and to replace it by simple categorical listings, or, in other words, by an exclusive verbal system.

S-R Compatibility. : I can be brief about the role of Stimulus-Response Compatibility in Human-Computer interaction. This well documented performance principle, first described by Fitts and Seeger (1953), was particularly ignored in the many abbreviations and in the seemingly random relations between stimuli and responses as found in the keyboard operated configuration. I think that I have demonstrated that, despite the fact that the principle of compatibility is widely known and intuitively appealing, it is still often not properly taken into account. I may add that measures towards excluding the possibility of choosing an improper command is usually not subsumed under S-R Compatibility but can be well considered in these terms. Anyway it prevents many errors.

Acquiring cognitive schemata: Operating a computor _ or any system _ is often said to require an integrated set of cognitive schemata or an "internal model" of the man-machine interaction. The internal model is a hypothetical construct, the main function of which is to provide a detailed overview over the possibilities and the modes of interaction between man and machine in a given system. Thus, it is a framework of knowledge, declarative as well as procedural,which guides plans of action, the interpretation of inputs, and the choice of adequate methods to accomplish subgoals that are part of the total plan (e.g. Moray 1981, Rasmussen 1979). In the area of human-computer interaction the GOMS model aims at specifying the internal model of the user.A hierarchy of "Goals" is formulated; there are a number of "Operators" or perceptual-motor acts; a set of "Methods" describes procedures for accomplishing the (sub)goals; and finally, there are "Selection rules" with respect to the methods and operations. (Card, Moran, and Newell 1983) A main task for research is to describe the detailed structures of the internal model in terms of the GOMS. In addition, it is clear that the internal model has a variety of features, depending on the nature of the machine as well as on the level of acquisition and the already available internal models of the human operator. The more a novice can capitalize on already available internal models, the less there is a need for learning, and the more the system is user-friendly. This remains a trivial statement as long as is not specified where the gains can be obtained. The complexity of the "Operators" seems a prime candidate for savings: They are essential for accomplishing the Goals but basically uninteresting. The methods can be conceived of as subroutines which use the Operators, but at the same time add more general knowledge about the system's possibilities. Methods and Selection rules must be learned and we need properly written texts or programmed courses for this. For a more detailed discussion of the relationship between mental models and metaphors, see Streitz (1987 a).

As already mentioned, preparing a text or even an instruction poses many ergonomical problems. One of the shortcomings of the Macintosh text was that goals were too widely introduced before operations and methods were sufficiently available. The IBM courses introduced concepts on too high a hierarchical level to be useful at that particular state of knowledge.

One of the major issues is that a proper text should be "client-centered"; that is it should be based on a model of knowledge acquisition. In a sense this was attempted in the programmed courses; there the main restriction, was the lack of flexibility and of possibilities to "play" with the new knowledge. The basic problem is the expert's lack of understanding of the novice's needs. Detailed task analysis and statement of accompanying knowledge requirements (Drury *et al* 1986) is needed to avoid at least some of the more common errors in instruction.

As a final, perhaps superfluous remark, I would like to say that the work on human-computer interaction at our Aachen Institute is usually not introspective but based upon sound behavioral experimentation. (e.g. Streitz, 1987 b, this volume) Yet on occasion, and in particular in early stages of thinking, as well as in taskanalysis, introspection has its merits as a way of inventarization of a problem.

5 REFERENCES

Broadbent, D.E.(1958) *Perception and Communication.* Pergamon, London.

Broadbent, D.E.(1971) *Decision and Stress.* Academic Press, London.

Brooks, L.R.(1968) Spatial and verbal components in the act of recall. *Canadian Journal of Psychology,* 22 , 349-368.

Card, S.K., Moran, T.P. and Newell, A.(1983) *The Psychology of Human Computer Interaction.* Erlbaum, Hillsdale.

Conrad, R. (1964) Acoustic comparisons in immediate memory. *British Journal of Psychology,* 55, 75-84.

Drury, C.G., Paramore, B., Van Cott, H.P., Grey, S.M., and Corlett, E.M.(1986) Task Analysis. In: G. Salvendy (Ed) *Handbook of Human Factors,* Wiley, New York.

Fitts, P.M. and Seeger, C.M.(1953) S-R compatibility: spatial charac- teristics of stimulus and response codes. *Journal of Experimental Psychology,* 46, 199-210.

Gopher, D (1984) The contribution of vision-based imagery to the acquisition and operation of a transcription skill. In: W. Prinz and A. F. Sanders (Eds.) *Cognition and Motor Processes.* Springer, Heidelberg.

Gopher, D (1987) Cognition at the tip of your fingers: anew look at the design of data entry keyboards. This volume: G. Salvendy (Ed.), *Proceedings of the 2nd International Conference on Human-Computer Interaction*. Amsterdam: Elsevier Publ.

Janssen, W. (1976) *On the nature of the mental image.* Dissertation, Academische pers, Amsterdam.

Kinsbourne, M. and Hicks, R. (1978) Functional cerebral space. In: J. Requin (Ed) *Attention and performance 7,* Erlbaum, Hillsdale.

Moray, N. (1981) The role of attention in the detection of errors and the diagnosis of errors in man-machine systems In: J. Rasmussen and W. Rouse (Eds) *Human detection and diagnosis of system failures.* Plenum, New York.

Murdock, B.B.(1974) *Human Memory: Theory and Data.* Erlbaum, Hillsdale.

Paivio,A (1971) *Imagery and Verbal Processes.* Holt, New York.

Rasmussen, J (1979) Reflections on the concept of operator workload. In: N. Moray (Ed), *Mental workload.* Plenum, New York.

Roscoe, S.N.(1981) *Aviation Psychology.* University of Iowa Press.

Sanders, A.F. (1981) Some Remarks on Mental Load. In: N. Moray (Ed), *Mental Workload,* Plenum, London.

Streitz, N.A. (1987 a) Mental models and metaphors: implications for the design of adaptive user-system interfaces. In: H. Mandl and A. Lesgold (Eds) *Learning Issues for intelligent Tutoring Systems*. Springer, New York.

Streitz, N.A. (1987b) Cognitive compatibility as a central issue in human-computer interaction: Theoretical framework and empirical findings. this volume: G. Salvendy (Ed.), *Proceedings of the 2nd International Conference on Human-Computer Interaction*. Amsterdam: Elsevier Publ.

Thorndyke, P.W.(1980) Performance models for spatial and locational cognition. *Technical report R-2676-ONR* Washington D.C.

Treisman, A.M.(1964) Selective attention in man. *British Medical Bulletin,* 20, 12-16.

Trevarthen, C.B.(1968) Two mechanisms in vision in primates. *Psychologische Forschung,* 1968, 31, 299-337.

Wickens, C. (1984) *Engineering Psychology and Human Performance.* Merrill, Columbus,Ohio.

Social, Ergonomic and Stress Aspects of Work with Computers,
edited by G. Salvendy, S.L. Sauter and J.J. Hurrell, Jr.
Elsevier Science Publishers B.V., Amsterdam, 1987 — Printed in The Netherlands

DEVELOPMENT OF A SCALE TO ASSESS OCCUPATIONAL COGNITIVE DEMANDS

DANIEL C. GANSTER
Department of Management, University of Nebraska, Lincoln, NE, U.S.A.

JOSEPH J. HURRELL, JR.[1]
National Institute for Occupational Safety and Health, Cincinnati, OH, U.S.A.

LINDA T. THOMAS
Department of Management, University of Nebraska, Lincoln, NE, U.S.A.

ABSTRACT

A psychometric instrument to measure aspects of occupational cognitive demands was developed and empirically examined. Data were obtained from 512 employees representing diverse occupations. Analyses focused on the factor structure of the cognitive demands measure, the reliability of the subscales, and the relationships between cognitive demands and psychological and health outcomes. Results suggest that the measure provides a reliable assessment of three cognitive demand factors representing (1) information characteristics, (2) memory requirements, and (3) uncertainty. These demands, in turn, are predictive of somatic health outcomes and psychological stress.

1 INTRODUCTION

It is apparent that the nature of work in industrialized societies is changing. Efforts to accomplish large-scale automation have caused and will continue to cause major changes in job design and work structure. These changes are expected to lead to an increased reliance on cognitive rather than physical effort on the part of the worker. Unfortunately, little is known about the physical and psychological health consequences of cognitive work demands.

---

[1]Dr. Hurrell is to be recognized for his contribution in developing the item pool reported upon in this study. The findings, opinions, and conclusions expressed herein are not necessarily those of the National Institute for Occupational Safety and Health.

It is clear that cognitive demands produce physiological changes. Laboratory studies have shown, for example, that cardiovascular parameters respond to changes in cognitive demand. These parameters include both blood pressure (Ettema, 1969) and heart rate (Hitchen, Brodie, and Harness, 1980). The importance of such cognitively induced changes in the disease process is currently unknown; however, there is mounting evidence for a coronary heart disease linkage (Jennings, 1983).

In the occupational realm there is evidence to suggest that there is a disproportionate frequency of both physical and mental health problems among occupants of cognitively demanding jobs. Karasek and his colleagues (Karasek, Theorell, Schwarts, Pieper, and Alfredsson, 1982; Alfredsson, Karasek, and Theorell, 1982), for example, have reported significant variation among occupations in terms of "psychological demands" which predict occupational level psychological strains and risk of myocardial infarction. Likewise, epidemiological studies (e.g., Cobb and Rose, 1973; Rose, Jenkins, and Hurst, 1978) have shown that air traffic controllers (whose job, according to Finkelman and Kirschner (1980) "requires more attention to stimuli for longer periods than most other occupations") suffer from a high incidence of hypertension and gastrointestinal disorders.

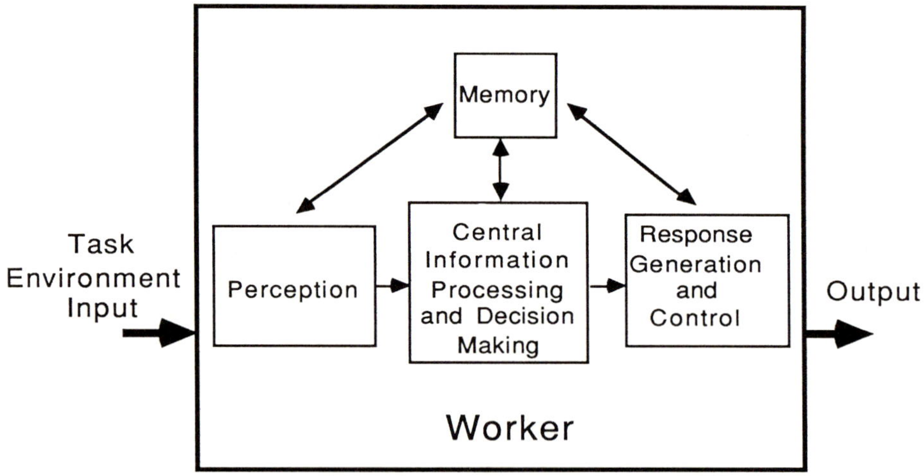

Fig.1. Human Information Processing Model

Perhaps the greatest single reason for lack of knowledge concerning the relationship between job mental demands and health consequences is the lack of valid, reliable, and practical methods for assessing cognitive demands in the workplace. As Moray (1982) noted in a review of the literature, it is astounding that there has been so little research into what leads to excessive mental workload. In addition to the paucity of research, ambiguity concerning the nature of the cognitive demand construct has resulted in its measurement by many methods, including heart rate, heart rate variability, primary and secondary task performance, subjective rating, and pupillary responses (see Wierwille, Rahimi, and Casali, 1985). The relationships among these methods are not well established. Indeed, there is no consensus on the nature of the cognitive demand construct and its assessment.

The research described here is developmental in nature and seeks to provide a self-report psychometric instrument for assessing cognitive task demands. Once developed, this instrument can be used to further examine the effects of cognitive work demands on health outcomes.

2 METHODS

2.1 Development of mental demands measure

An item pool was developed using a comprehensive conceptualization of human information processing depicted in Figure 1. In this conceptualization, properties of the task performed by the worker have the potential of contributing to overall cognitive load. For example, task environments that present impoverished information or which present information rapidly have a greater likelihood of imposing load than do tasks that present in a clear and reasonably paced manner. Tasks that require the worker to attend to such informational input make the perception and interpretation of information more difficult, and thus impose more cognitive load. Secondly, mental structures and processes that are required by tasks that the worker performs affect cognitive load. Here consideration was given to such factors as whether the worker is required to perform two or more tasks simultaneously, whether the memory requirements of the task are substantial or relatively inconsequential, whether the worker must perform certain operations while holding certain pieces of information temporarily, and whether the worker has achieved a level of automaticity with respect to the job such that a level of competency has been gained that reduces the cognitive demand initially experienced. Lastly, questions were generated that assess response generation requirements such as the extent to which competing responses are required during the performance of the task, the extent of response pacing demands, and the availability of feedback regarding performance. In short, an attempt was made to write items that reflected the major processes in the information processing model of Figure 1. In all 62 items were developed to tap these dimensions.

2.2 Subjects and Procedures

Workers from a variety of occupations were recruited as respondents. In all, 512 workers provided complete data on the cognitive demands measure. Over 50 distinct job titles were represented in the sample. Occupations included blue-collar jobs such as press operator and grocery stocker, clerical and service jobs such as data entry personnel and store clerks, supervisory jobs such as grocery department managers, professional and technical jobs such as newspaper editors, and middle management jobs. Respondent occupations were chosen in a non-representative fashion with the objective of maximizing the variance in cognitive demands across the entire sample. Respondents from each occupation were surveyed on a voluntary basis at their place of employment or in training sessions. Mean respondent age was 32.7 with a range from 18 to 66 years. The average educational attainment was "some college," and 52% of the sample was female.

As noted above, 62 items were written to reflect dimensions of cognitive demands. Each item employed a 4-point Likert response format (Strongly Agree to Strongly Disagree). These questions were combined in a questionnaire with scales measuring 17 different somatic complaints (e.g., headaches, dizziness, tense muscles, loss of sleep), items assessing overall subjective stress, mental and physical exhaustion, and demographic and occupational characteristics.

3 ANALYSES AND PSYCHOMETRIC RESULTS

3.1 Factor analyses

All 62 items were factor analyzed using a principal axis factoring extraction method. This analysis produced 18 factors with eigenvalues of 1.0 or higher. Based upon inspection of the Scree plot of eigenvalues by factor, the first three factors (accounting for approximately 25% of the variance) indicated a logical break point. The analysis was then recalculated constrained to a three-factor solution. The resulting factor loading matrix was then rotated using the varimax criterion.

3.2 Subscale construction

Based upon the rotated factor loadings individual questions were combined into three cognitive demands subscales of 26 items (factor 1), 24 items (factor 2), and 12 items (factor 3). The total sample was divided into two parts, using alternating cases in order to form a validation sample (N=257) and a hold-out cross-validation sample (N=255). A reliability and item analysis was made of these three factor-based subscales using cases in the validation sample. Using the corrected item-total correlations as criteria, items were eliminated from each of the three scales with the goal of maximally shortening the scales while maintaining adequate reliability. While the item-total correlations were used as criteria for item selection, the cut-off criterion that was used varied for each of the subscales. Our goal was to optimize the resultant reliability of each scale while minimizing the number of items retained. Since the average inter-item correlation differed for each of the three subscales, the cut-off values of the item-total correlation differed. For the first and second scales, all items with item-total correlations below .40 were dropped. For the third scale, items with item-total correlations below .25 were dropped. This item elimination step reduced the three subscales to lengths of 14, 11, and 9 items. Internal consistency reliability estimates (coefficients alpha) were recalculated on these reduced scales. The shortened subscales were then reassessed in the cross-validation sample. This procedure yielded three subscales with replicated reliabilites ranging from .70 to .83. These final subscales were thus retained for further analysis and interpretation, and the remaining 28 items were discarded.

3.3 Scale interpretation

The first subscale contains 14 items most of which reflect perceptual demands relating to the characteristics of the information input to which the worker must attend. These questions generally pertain to either the amount of information that must be processed, the speed of its presentation, or its clarity and organization. Sample items include: I can easily see or hear information I have to use in my job; The information I have to look at or listen to is presented too rapidly; Sometimes it is difficult to understand instructions that are given to me. Since these items refer mainly to characteristics of the informational input, we named the scale "information load." This scale had a replicated reliability of .83.

The second subscale contains 11 items most of which refer to memory requirements of the job. Sample items include: To do my job well, I have to do a lot of things mentally at the same time; My job makes heavy demands on me to recall information; The memory requirements of my job are heavy. We thus termed this scale "memory load." This scale had a replicated reliability of .77.

The third subscale consists of 9 items that appear to characterize the routineness or automaticity of procedures that are done on the job. Sample items include: The decisions that I make on the job require little thought; My job basically consists of one set of rules with little deviation from them; In my job there are set rules that I follow over and over again. Since these items are recoded so that high scores indicate lack of routineness, we termed the scale "uncertainty," and it showed a replicated reliability of .71.

TABLE 1

Correlations among cognitive demands scales and outcome variables.

	Information Load	Memory Load	Uncertainty
Age	-.12**	-.04	-.03
Education	.06	.02	.07
Sex[a]	.06	-.11*	-.08
Job Tenure	-.12**	.02	-.08
Hours work/wk.	.11*	.21**	.15**
Subjective Stress	.39**	.23**	.23**
Mental Exhaustion	.42**	.25**	.26**
Physical Exhaustion	.40**	.18**	.18**
Somatic Complaints	.29**	.14**	.12**

[a] Sex is coded 1=male, 2=female. Males thus have higher memory load than females.
* $p<.05$
** $p<.01$

3.4 <u>Relationships between Cognitive Demands and Health Outcomes</u>

As an initial attempt to examine criterion-related validity of the cognitive load subscales, correlations between them and the somatic complaints and psychological stress measures were examined. These correlations are reported in Table 1. As can be seen, each of the three aspects of cognitive demands relates significantly to overall prevalence of somatic complaints and subjective stress. In comparing the strength of the relationships of the cognitive demands factors, the information load factor generally appears to be the best predictor of the outcomes. The effects of memory load and uncertainty on the outcome variables are approximately equal and uniformly lower than those of information load. In all cases, however, the cognitive demand factors are positively associated with subjective stress, exhaustion, and somatic complaints.

The pattern of correlations between the cognitive demands and the set of demographic variables is mostly one of non-significance. All three of the demands factors correlate significantly with the number of hours worked per week, although the magnitude of these relationships is generally small. It would seem that relationships between cognitive demands and the outcome variables are not merely a function of the confounding effects of demographic factors.

4 CONCLUSIONS

The results of this study, while preliminary, provide some insight into the assessment of the cognitive demands of work. The scales developed are reliable and are predictive of subjective stress and somatic health outcomes. The findings suggest that the most taxing cognitive demands may stem from the nature of the information with which the worker must deal. Relative to the memory requirements and the uncertainty of decision criteria, the demands of attending to and deciphering unclear information, or information presented too rapidly, seem to have greater health consequences. It should be noted that while the items loading on these factors generally represent the aforementioned aspects of cognitive demand, the loadings of some items are not entirely consistent with these interpretations and factor labels. For example, while the first scale generally concerns informational characteristics, some items do not neatly fit with this categorization. This scale includes items such as: "I often feel mentally over-burdened on my job; I have plenty of time to make decisions; At times I experience mental blocks." Similarly, while the second scale generally focuses on memory requirements, it includes items such as the following: "I am fully responsible for the consequences of my decisions; My job often requires me to learn new procedures." The latter item might imply increased memory requirements, but it also implies a lack of routineness which is the focus of the third scale. Thus, as is often the case with the interpretation of factor-based scales, the labeling of the factors is somewhat subjective and open to other interpretations. Clearly, future work is needed to further refine the subscales.

5 REFERENCES

Alfredsson, L., Karasek, R., and Theorell, T. 1982. Myocardial infarction risk and psychosocial work environment characteristics: An analysis of the male Swedish work force. Soc. Sci and Med., 16: 463-467.

Cobb, S. and Rose, R. 1973. Hypertension, peptic ulcers, and diabetes in air traffic controllers. J. of Amer. Med. Assn., 224: 489-492.

Ettema, J. H. 1969. Blood pressure changes during mental load experiments in man. Psychother. Psychocom., 17: 191-195.

Finkleman, J. M. and Kirschner, C. 1980. An information processing interpretation of air traffic control stress. Human Factors, 22: 561-567.

Hitchen, M., Brodie, D. A., and Harness, J. B. 1980. Cardiac responses to demanding mental load. Ergonomics, 23: 379-385.

Jennings, J. R. 1983. Attention and coronary heart disease. In: D. Krantz, A. Baum, and J. Singer (Eds.), Handbook of Psychology and Health, Volume III, Cardiovascular Disorders and Behavior. Lawrence Erlbaum, Hillsdale, New Jersey.

Karasek, R., Theorell, T., Schwarts, J., and Alfredsson, A. 1982. Job, psychological factors, and coronary heart disease. Advanced Cardiol., 29: 62-76.

Moray, N. 1982. Models and measurement of mental workload. In: N. Moray (Editor), Mental Workload: Theory and Measurement. Plenum, New York.

Rose, R., Jenkins, C. D., and Hurst, M. W. 1978. Air traffic controller health change study. United States Department of Transportation, Federal Aviation Administration Report No. Am 78-39.

Wierwille, W. W., Rahimi, M., and Casali, J. G. 1985. Evaluation of 16 measures of mental workload using a simulated flight task emphasizing meditational activity. Human Factors, 27: 489-502.

Social, Ergonomic and Stress Aspects of Work with Computers,
edited by G. Salvendy, S.L. Sauter and J.J. Hurrell, Jr.
Elsevier Science Publishers B.V., Amsterdam, 1987 — Printed in The Netherlands

MOTOR FEEDBACK CONTROL OF HUMAN COGNITION - IMPLICATIONS FOR THE COGNITIVE INTERFACE

THOMAS J. SMITH, School of Kinesiology, Simon Fraser University, Burnaby, British Columbia, Canada V5A 1S6
KARL U. SMITH, Professor Emeritus, Behavioral Cybernetics Laboratory, University of Wisconsin-Madison, 1001 Tower Blvd., Lake Wales, FL 33853

ABSTRACT

The concept that human cognition is mediated by motor behavior dates back nearly a century. More recently however, the prevailing opinion has been that cognitive activity is a product of the substantive mind, embodied within the brain as an invisible, mental phenomenon divorced from motorsensory interaction with the external world. This concept has led to development of the idea of the human-computer cognitive interface, distinct from the physical interface, whose design features should be directed at mediating interaction of information processing activity between the human brain and the computer. We believe that the assumption by modern cognitive science that human cognition is somehow detached from its motorsensory substrates perpetuates a mythology of behavior that has preoccupied traditional psychology for decades. This paper first amplifies the historical experimental view of cognition as a motor coordinate process, in light of recent research findings from behavioral cybernetics and neurobiology supporting the view that cognitive behavior, like all other modes of behavior, is feedback integrated with motor activity. We then go on to address the implications of this analysis for understanding human-computer interaction. Because motor feedback control of perception, not information processing, guides cognitive behavior, the human factors design imperatives for the physical and the cognitive interface are one and the same. We evaluate these needs using the concept of human-computer interaction as a social cybernetic process, ideally involving mutual, multimodal control through social tracking of sensory feedback across the interface. From this perspective, modality and sensory feedback control limitations of current machines make them relatively impoverished social tracking targets for human partners. Research directions suggested by this social cybernetic analysis are then explored, which focus on analysis of the concrete, observable motorsensory substrates of cognitive activity and emphasize the need for objective human factors experimentation on optimal behavioral cybernetic design criteria for human-computer interactive systems.

1.0 INTRODUCTION

A major emerging theme in the scientific study of human-computer interactive systems is that of the cognitive interface (Nickerson, 1986). This term refers to the cognitive aspects of matching computer systems to users, in relation to, "relatively invisible variables that are critically important to an effective user-system interaction" (IBID, p. 112). The premises on which this concept of the cognitive interface is based may be summarized as follows: (1) cognition - the process of perception and knowing - is a product of higher mental functioning of the brain (Wickens, 1987); (2) the essence of such functioning is central information processing, entailing perceptual recognition, memory storage and recall, decision making, and response selection by the brain (Foley and Moray, 1987; Wickens, 1987; Knight, 1987); and (3) the cognitive interface represents an interaction between two information structures, one residing in the computer and the other in the user's head (Nickerson, 1986).

Explicit in this conceptual understanding of cognition and the cognitive interace are two assumptions: (1) cognitive activity is embedded within the brain as an abstract, intangible, higher order mental phenomenon somewhat divorced from motorsensory interaction with the external world; and (2) because it services interaction between human and computer information processing operations, the cognitive interface differs in its design and performance requirements from the physical interface.

Our objectives in this paper are threefold: (1) to challenge the information processing model of cognitive behavior outlined above, in favor of a behavioral cybernetic theory of cognition as a feedback controlled motor coordinate process; (2) to present experimental evidence in support of this theory; and (3) to discuss the implications of cognitive cybernetics for understanding human-computer interaction. Details of our approach are provided in two recent publications (Smith and Smith, 1985, 1987a), plus a forthcoming monograph (Smith and Smith, 1988).

The rationale for discussing cognitive mechanisms in the context of this conference is straightforward. We believe that general disregard for the cybernetic properties of cognition, in favor of preoccupation with the information processing model, has had two unfortunate consequences for furthering scientific understanding of human-computer interaction. First, analytic emphasis on the putative information processing features of the cognitive interface has detracted from objective experimental analysis of its more critical motorsensory control features. This in turn has engendered misplaced priorities and erroneous conclusions regarding the human factors design imperatives of human-computer interaction.

2.0 MOTOR COORDINATE FEEDBACK CONTROL OF COGNITIVE BEHAVIOR

2.1 Statement of Theory

The basic postulate of behavioral cybernetic theory is that behaviour is a closed-loop, self-regulated process mediated by motor feedback control (Smith, 1972; Smith and Smith, 1987a, 1985, 1966, 1962). As applied to cognition, the theory assumes that skilled motor coordinate activity directly and continuously feedback controls and organizes the development and expression of all cognitive skills, manifest in cognitively-directed social communication and interaction, individual thoughts (memory, problem-solving, creative thinking), and thought-guided machine operations. Our basic postulate is that cognition, like all other known biological processes, is subject to feedback control (Adolph, 1982).

The behavioral cybernetic principles upon which this motor feedback control theory of cognition is based may be summarized as follows.

1. Perception represents an active process, based on an integrated series of motor coordinate mechanisms which feedback control receptor processes and environmental stimuli. This principle is illustrated in Figure 1 in terms of six major ways in which such motor feedback control is achieved (Smith and Smith, 1987a; 1985; Dewey, 1896): (1) self-stimulation of receptors by movement; (2) modulation of physical properties of stimuli; (3) control of sources of stimulation; (4) control of transmission of stimuli; (5) control of orientation and

interaction of receptors; and (6) control of receptor sensitivity. The term sensory feedback control encompasses all of these mechanisms.

2. Cognition emerges in childhood as a series of individual and communicative perceptual, memory-based, specialized skills - perceptual judgement, speech, graphic behavior, writing, numerical/mathematical performance, overt and symbolic problem solving, artistic performance and creativity, operational machine skills, and social interaction - all of which have a motor coordinate basis (Smith and Smith, 1987b).

3. Adult thought emerges through maturation and refinement of these skills, in relation to individual coordinate cognitive behavior, active feedback control of diverse environmental conditions, social communication, and integration of the physiological substrates of cognitive activity (Smith and Smith, 1988).

4. Social cognitive behavior and communication operate in terms of a great variety of interpersonal and group modes of mutual interactive control of sensory feedback, all of which have a motor coordinate basis (Smith and Smith, 1987a, 1985).

5. Primary machine operations require the integration of three parameters of motor coordinate control of sensory feedback and cognition, related to interaction with the control, display, and actuator components of the machine (Smith and Smith, 1987a, 1966).

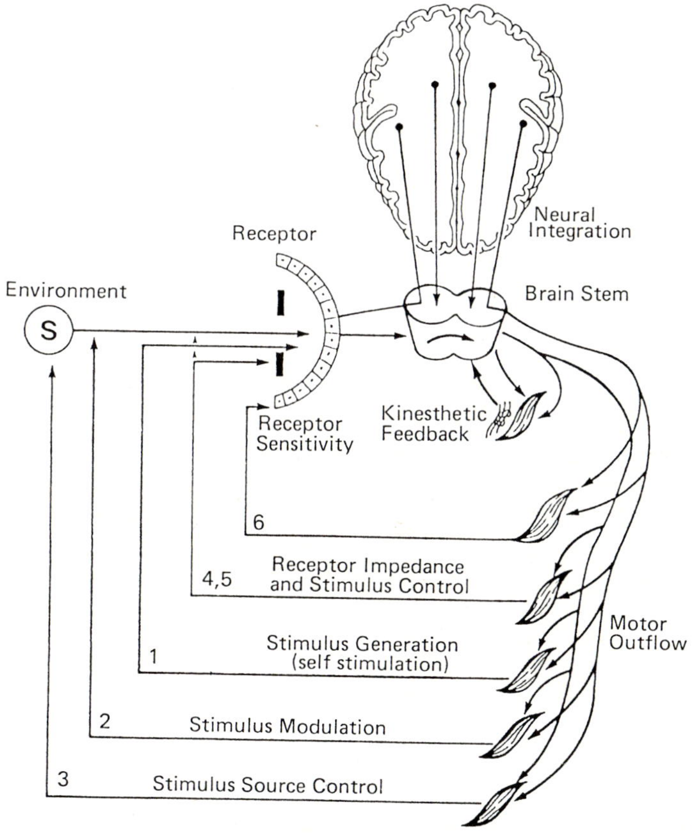

Figure 1. Modes of sensory feedback control by motor behavior.

2.2 Historical/Evolutionary Perspectives

The historical origins of a behavioral feedback theory of cognition go back to some of the most distinguished early theorists in psychology, namely James (1894, 1890), Dewey (1896), Munsterberg (1889), and Delebarre (1891, 1898), all of whom viewed the mind as an active, tangible process subject to motor control. They questioned the idea of the substantive mind and the psychophysical view of perception as a direct, open-loop product of sensation, challenges which could be extended to contemporary models of the brain as a body-detached information processing system.

There is compelling evidence (Smith and Smith, 1988) that from the earliest human origins, the evolution of all modes of cognitive behavior - graphic and verbal communication, time perception and control, creation and invention, and so forth - has been guided by motor coordinate use, manufacture, and integration of tools, implements, and machines. In other words, the entire process of human evolution, including particularly the emergence of cognitive skill, represents a motor controlled feedback product of human-machine interaction. Human-computer systems of today, with their cognitive demands, represent only the latest stage in a human factored evolutionary process dating back millions of years.

2.3 Summary of Experimental Evidence

Experimental demonstration that cognitive and motor behaviors are mutually integrated through reciprocal feedback control mechanisms rests upon four major lines of evidence, delineated in the following subsections. Collectively, this evidence strongly supports the cybernetic conclusion that perceptual cognitive skills, like other physical motor skills, arise out of specialization in motorsensory feedback control.

2.3.1 Motor Concomitants of Cognitive Function

Various findings summarized below indicate that motor activity is a dominant feature of cognition. This suggests (but does not prove) a motor control influence.

1. Jacobson (1932) demonstrated consistent, specialized electromyographic components of different cognitive operations.

2. Stetson (1951) recorded muscular manifestations of all of the phonetic components of speech.

3. The numerous distinct motions produced in writing and graphic behavior are specialized in terms of compliance between the spatial patterns of body motions and those of sensory input (Smith and Wargo, 1963).

4. At least 71 % of the time devoted to a cognitive task (problem-solving) directly requires motor behavior, for the encoding and responding phases (Sternberg, 1986).

5. Although the general population is predominantly right-faced, almost all highly talented musical composers and performers are or were left-faced. Facedness refers to lateral assymmetry in facial muscular movements: facial muscles on the dominant side are more animated and are used preferentially for vowel-consonant articulation and coarticulation during speech (Smith and Smith, 1988; Smith, 1984).

These observations support the following general conclusions about the role of motor activity in cognition: (1) motor control demands are prominent in the expression of various specialized cognitive skills, such as speech, writing, graphic behavior, and problem-solving; (2) thought-guided use of language in speech, a type of cognitive behavior which most observers agree clearly distinguishes humans from lower animals, is a motor coordinate skill (Stetson, 1951); and (3) musical creativity, a sublime example of human cognitive achievement, exhibits a direct link between motor and cognitive specialization.

2.3.2 Sensory Feedback Control of Perception and Cognition

A variety of experimental evidence indicates that perception and cognitive behavior are directly dependent on control of sensory feedback by the motor system (Fig. 1). This evidence is derived primarily from studies demonstrating the perceptual and cognitive effects arising out of perturbation of the spatial and/or temporal properties of sensory feedback. In particular, the performance of every cognitive skill so far examined, including the use of language in speech and writing, visual perception, thought-guided visual-manual tracking and machine operation, graphic behavior, social cognitive behavior, learning, and memory, is substantially and often profoundly degraded by such perturbation (Smith, 1972, 1962; Smith and Smith, 1987a, 1985, 1966, 1962). Specific examples are summarized below.

1. Visual perception depends on eye movements, as Helmholtz first postulated over a century ago (Smith and Smith, 1987a). Direct demonstration of the role of ocular movements in visual perception has now been provided by a series of studies showing that: (a) with a stabilized retinal image, normal vision disappears (Ditchburn and Ginsborg, 1952; Riggs et al.,. 1953); (b) spatial inversion, reversal, or displacement of the visual image perturbs normal vision and visual-manual tracking (Gould and Smith, 1962; Smith, 1962); (c) real time spatial reversal of eye movement control of the retinal image severely impairs normal vision and depth perception (Schmidt et al., 1974; Smith, 1970; Smith and Molitor, 1969); and (d) real time temporal delay in eye movement control of the retinal image severely impairs eye tracking (Smith, Putz, and Molitor, 1969). Generally, this research confirms that the retinal image is under direct feedback control of at least eight distinct ocular movement patterns, which determine visual perception of movement, movement direction, depth, form, pattern, texture, size, and color.

2. Temporal delay in the auditory feedback of speech sounds produces marked disturbances in all aspects of speech control, manifested as stuttering, blocking, pitch elevation, increased loudness, and language errors in grammar and syntax (Lee, 1950). Based on the experiences of the senior author as a regular subject in such studies (Smith et al., 1964), these speech disturbances may persist for months.

3. Spatial or temporal perturbation of visual feedback degrades both handwriting and graphic reading skills (Smith, Kao, and Knutson, 1969; Smith and Smith, 1966, 1962).

4. Memory in a steering task is impaired by delayed sensory feedback (Sussman and Smith, 1970, 1969).

5. Learning of all cognitive skills so far examined is never as rapid nor as complete with perturbed sensory feedback relative to normal feedback conditions (Smith and Smith, 1966).

6. A comprehensive body of evidence unquivocally indicates that spatial or temporal perturbations in sensory feedback severely degrade tracking performance in thought-guided machine operation tasks (Knight, 1987; Smith and Smith, 1987a, 1966).

2.3.3 Motor Feedback Control of Social Cognitive Skills

The cybernetic nature of cognitive behavior reaches its fullest expression in the social interaction between two or more individuals. This is because the individual in a social context must feedback control through motorsensory integration the behavioral and physiological mechanisms required to meet not only intrinsic regulatory demands, but also those created by interacting with one or more social targets. Every other individual in a social setting faces the same challenge, underscoring the reciprocal feedback properties of social interaction. The study of these feedback relationships represents the focus of the new field of social cybernetics, directed towards delineating the closed-loop, behavioral-physiological manifestations and properties of social interaction and cognition (Sauter and Smith, 1971; Smith and Smith, 1987a, 1985; Smith, 1972).

The term social tracking is used to describe the feedback-controlled process by which an individual follows or tracks a social target. In cybernetic terms the activities of one person in a social tracking process produce behavioral-physiological effects in others, which in turn feedback influence the ongoing behavior and physiology of the first. These effects arise as a consequence of control by each participant of sensory feedback generated by the other social tracking partners.

Social tracking is one of the essential ingredients of the human condition. All group behavior and interpersonal interaction involve social tracking. The acquisition of the entire constellation of social cognitive skills, like language communication and learning, is based on social tracking (Cherry and Sayers, 1956; Smith and Smith, 1987b, 1966). In addition to verbal communication, social tracking may entail at least seven distinct modes of nonverbal communication which are critical to social cognitive behavior (Smith and Smith, 1987b, 1973).

Experimental feedback research (Sauter and Smith, 1971; Smith and Arndt, 1969; Ting, Smith, and Smith, 1972) has established five general cybernetic principles of social tracking which strongly suggest that cognitive skills at the social as well as the individual level are based upon motor feedback control: (1) social tracking is most effective under conditions of mutual sensory feedback control; (2) social tracking is central to all modes of cognitive behavior and represents the principal determinant of effective learning of cognitive skills; (3) nonverbal motor behavioral modes of social tracking often are more effective than verbal modes in promoting effective learning and communication; (4) competency in social tracking is age-dependent, correlated with the maturation of physical motor skills; and (5) like individual movement integration, social tracking performance also is degraded severely by temporal and/or spatial perturbations in sensory feedback control.

2.3.4 Motorsensory Organization of Neuronal Manifestations of Cognition

Results from perturbed sensory feedback research described above indicate that cognitive processes of learning and memory are critically influenced by the degree of compliance between the spatial and temporal properties of sensory feedback and those of the motor behavioral mechanisms that control that feedback. In a broader sense, the implication is that the central neuronal manifestations of cognitive function are organized in a feedback manner by peripheral motorsensory activity.

A variety of direct and indirect evidence supports this view, as outlined by Smith and Smith (1987a) and by other studies cited below.

1. During development, neuronal integration is guided by the maturation of peripheral motor and sensory systems, in accord with the spatial organization of these systems.

2. Peripheral motor and sensory systems must undergo normal functional activity during maturation to ensure functional maturation of central neuronal centers subserving these systems (Jacobson, 1967).

3. Under conditions of sensory deprivation, expecially during critical stages of early development, neuronal mechanisms become disorganized both functionally and structurally (Barnes, 1986).

4. As might be predicted from the functional validation and sensory deprivation findings, efferent neurons in the primate motor cortex are somatotopically organized, not purely in relation to muscle anatomy (as was believed until about 1975), but more closely in relation to different movements which these neurons evoke (Humphrey, 1986). Consequently, a single muscle (because it participates in a variety of movements) may have multiple representation in the cortex. During voluntary movement, ensembles of motor cortical neurons display activity along a three-dimensional vectorial axis which is spatially congruent with the direction of movement (Georgopoulos, Schwartz, and Kettner, 1986).

5. Memory is widely distributed in the mammalian brain, and localized to different cortical and subcortical regions in accord with sensory modality (John et al., 1986; Thompson, 1986).

These observations support the general conclusion that both anatomically and functionally, the organization of central cortical and subcortical neurons is established and maintained through direct feedback links with peripheral motorsensory activity. Recent memory research compels rejection of a computer-like model of the brain, with information stored in discrete registers, in favor of a conceptual model of cognitive behavior based on functional interaction of huge ensembles of neurons acting cooperatively as an integrated system (John et al., 1986). Overall, current evidence strongly supports the proposition that motorsensory feedback, generated by motor behavior, organizes in a continuous fashion the coordinated integration of neuronal activity underlying cognition.

2.4 Behavioral Cybernetics of Cognition - Summary

Our purpose with the foregoing cybernetic analysis is to raise doubts that decades of research in traditional psychology, linguistics, and artificial intelligence have generated a science of human cognition which goes beyond the realm of pure speculation. Cognitive man is not made in the computer's image (Jordan, 1963). The purpose of human cognitive behavior is to control perception through control of sensory input. This process is mediated by motor activity, which levels continuous, dynamic sensory feedback control over every feature and modality of environmental stimulation and sensory input (Fig. 1). Perception therefore is a feedback product of the real time and real space properties of motor activity. What we know is defined by our motor system, not by any known pure information processing or mental factors. Consequently, as suggested in Figure 2, every manifestation of human cognition - thinking, learning, memory, creativity, language, communication, and so forth - is feedback integrated with the motor system and motor behavior. The foregoing evidence indicates that all three stages of motor cognitive feedback control designated in Figure 2 - on afferent, central, and efferent traffic - are not merely symbolic but represent tangible regulatory influences. Decrements in human cognitive performance, without exception in our opinion, arise out of spatial and/or temporal perturbations in sensory feedback control by the motor system, with concomitant degradation in cognitive feedback control. From this perspective, the human factors requirements of the cognitive and physical interface are one and the same.

3.0 IMPLICATIONS FOR HUMAN-COMPUTER INTERACTION

Over the past three decades the influence of computer technology on cognitive psychology has been unmistakable; some observers attribute the rise of modern cognitive science in the mid-fifties to the advent of information theory and high-speed computers (Kessen and Cahan, 1986). In a sense then, modern cognitive science itself is a product of human-computer interaction. The preoccupation of cognitive theorists in drawing parallels between the workings of brains and computers is dramatized by such current cognitive scientific concepts as information processing, information structures, storage and retrieval of memory items, and the cognitive interface.

Not surprisingly, these theoretic parallels have been carried over into interface design. From a human factors standpoint, the results are not reassuring. As Tanner and Wein (1985) note, a common designer assumption has been that a computer user has no feet, only one hand, and no ears. More specifically, there are at least five major themes in interface design which reflect prevailing designer assumptions that the interface exists to

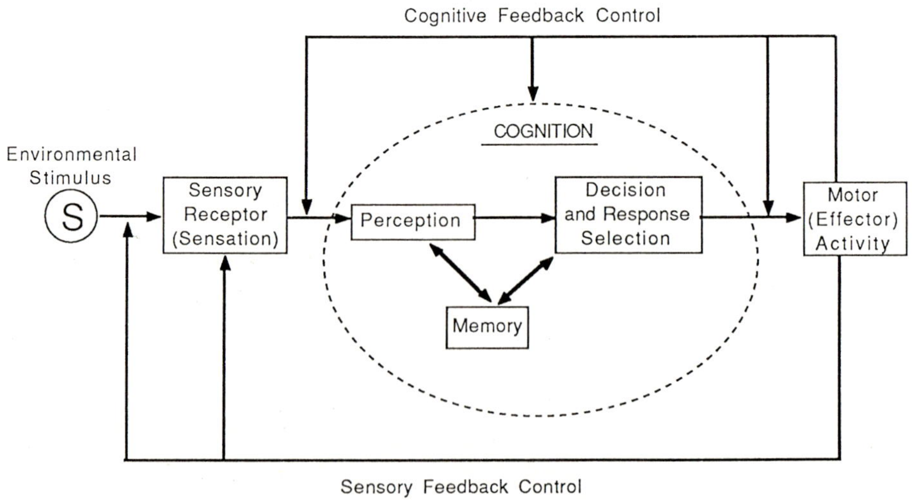

Figure 2. Role of the motor system in sensory and cognitive feedback control.

facilitate human-machine information processing exchange (Wickens, 1987; Nickerson, 1986): (1) emphasis on written verbal communication; (2) static visual display; (3) discrete user input via keyboard; (4) open-loop computerized pacing of performance; and (5) minimal computer control of sensory feedback from its own and user actions. Despite recent advances in automated speech production and recognition, and in non-keyboard manual input technology (Simpson et al., 1987; McCauley, 1984; Greenstein and Arnaut, 1987), these themes continue to dominate interactive system design motifs.

From a behavioral cybernetic perspective, human interaction with a computer may be compared to interaction with any other machine or tool, in that optimal performance is achieved only if a high degree of spatial and temporal compliance exists between the reactive, instrumental, and operational sources of sensory feedback from the control (input device), actuator (stored program), and display elements of the machine, and the sensory feedback control capabilities of the human operator (Smith and Smith, 1987a). Interaction with a computer differs from that with other machines primarily in terms of greater emphasis on use of graphic, verbal, and nonverbal mathematical cognitive skills in machine operation.

As with other machines however, feedback research indicates that computer interaction is severely degraded under conditions of perturbed sensory feedback. Delays in auditory feedback during cognitive keyboard performance tasks severely degrade all aspects of task performance (Ansell and Smith, 1966). Spatial perturbation of visual sensory feedback likewise degrades cognitive performance in CRT display reading tasks (Smith and Smith, 1969; Smith, Cambria, and Stephens, 1964). These findings suggest that operational cognitive proficiency in human-computer interaction is integrally dependent on optimal motor control of sensory feedback from the computer.

This cybernetic analysis leads directly to three major human factors criticisms of the information processing model, in terms of its prevailing influence on interface design of interactive systems.

1. It has promoted subjective speculation and modelling of central information processing as the key feature of cognitive interaction with computers, to the detriment of hard experimental analysis of the more meaningful motor feedback control mechanisms required for cognitive proficiency in computer use.

2. As a result, it has discounted the concrete, observable spatial and temporal constraints of sensory feedback control in cognitive behavior, and their role in determining operational skill in cognitive interaction.

3. The overall consequence has been conceptual disregard for the basic cybernetic properties of human-computer interaction, particularly as they define the critical human factors requirements of interface design.

The jury already is in on the human factors validity of these criticisms. The VDT, whose interface design generally embodies the information processing approach, has aroused a greater volume and variety of user problems and complaints, with broader geographic and cultural distribution, than has any other new technologic product in recorded history (Dainoff, 1982; Helander, Billingsley, and Schurick, 1984). To us, the VDT story represents a clear indictment of the shortcomings of the information processing model of interface design in adequately addressing the human factors requirements of human-computer interaction. More generally, the same message may be inferred from the sad story of office automation in the United States. In recent years, office automation, again guided predominantly by the information processing approach, has been sweeping through American business at a prodigious rate. Office computers now account for a large percentage of total American industrial investment, and American offices typically are much more computerized than those of any foreign competitor. Yet in the face of this immense investment in capital and technology, American office productivity has tumbled (Baily, 1986; Thurow, 1986).

Debate about the nature of human cognition does not provide insight into the human factors requirements of interface design. Neither, apparently, does the information processing model. Accordingly, we now describe a social cybernetic model of human-computer interaction which we believe provides a comprehensive framework for developing human factors criteria and principles of interface design. The essence of this model is that man-machine interaction represents a social tracking process whose behavioral cybernetic features by their very nature define the human factors design requirements of the interface.

3.1 Human-Computer Interaction as a Social Cybernetic Process

The conceptual model of human-computer interaction as a social tracking process stems directly from the general theory of social cybernetics described earlier (Section 2.3.3). As illustrated in Figure 3, the model assumes that man and machine are linked as social

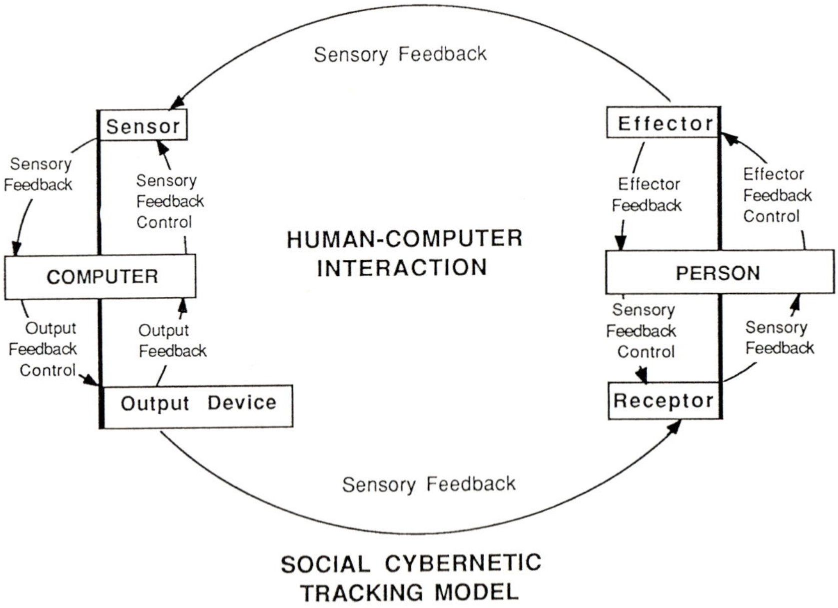

Figure 3. The social cybernetic model of human-computer interaction.

partners in a mutual social tracking process. This process ideally entails feedback control by each of the partners of sensory input derived not only from intrinsic and environmental stimulation (Fig. 1), but also from activity of the social target. The pattern and proficiency of social tracking are defined by the nature and the accuracy of this coupled sensory feedback control.

Examined from the social cybernetic perspective illustrated in Figure 3, today's interactive computer system stands as a relatively limited, impoverished social tracking target for the human partner. This point is driven home by Figure 4, which expands upon Figure 3 by comparing specific sensory and response capabilities of humans versus computers across the interface. The computer capabilities suggested in Figure 4 do not necessarily imply technological or research limits, but rather are meant to characterize operational design features in use today.

Several features of the comparison in Figure 4 command attention. There is no modality for which the computer is highly compatible with the human partner across all mutual classes of interaction. From the perspective of a motor feedback theory of cognition, the most serious social tracking deficiency revealed by Figure 4 is the imbalance in mutual sensory feedback of motor activity. Computer capabilities for kinesthesis and proprioception are limited, and machine detection of the three-dimensional coordinates of multiple movement patterns remains a formidable research problem. On the output side,

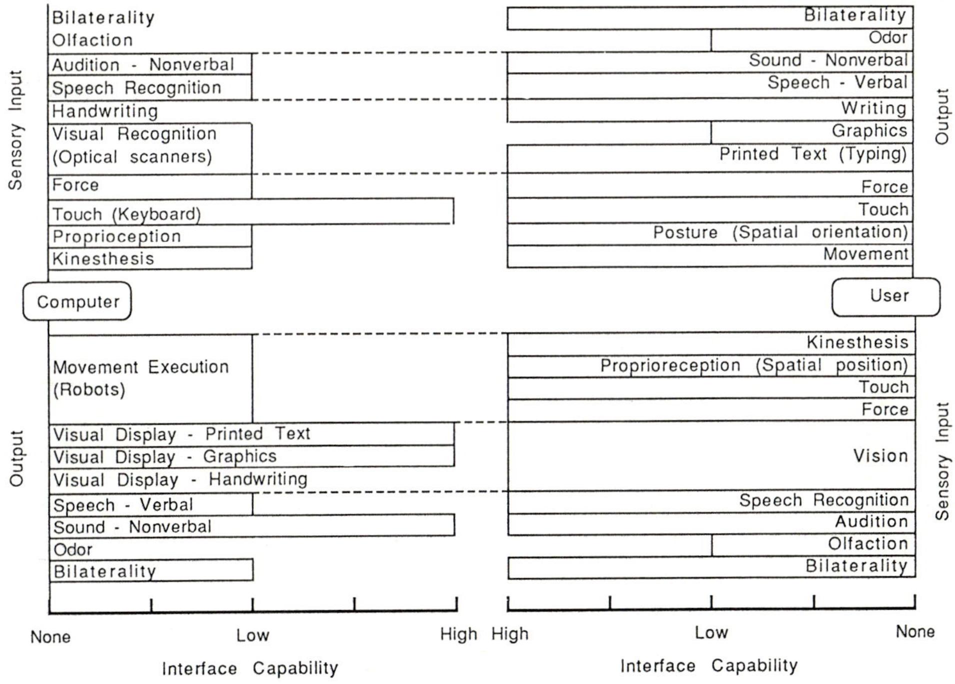

Figure 4. Comparison of sensory and output modality capabilities of the user versus the computer across the interface.

systems may provide visual feedback of movement (graphic animation), but typically do not move independently.

Further limiting social interaction is the restricted range of machine control over output and sensory feedback (Fig. 3). In the former case, examples of computer self-control over text processing output are numerous (i.e., automated spelling), but machine capabilities for self-modulation of other output modalities such as speech remain primitive. Machine capabilities for control of sensory feedback likewise are relatively primitive. Many interactive systems provide immediate feedback information to users regarding the quality of their performance, the condition of the system, and requisite steps to achieve a desired outcome (Williges, Williges, and Elkerton, 1987). This represents a form of computer sensory feedback control, in that it hopefully will influence subsequent user output sensed by the computer. On the other hand, automated feedback control of sensor properties themselves (i.e., dynamic range, sensitivity, timing, and/or spatial orientation), in accord with real time demands of the social tracking process, has not been introduced.

These machine social tracking limitations in turn create sensory feedback control problems for the human partner, by introducing sensory incompatibilities as well as spatial and temporal perturbations in machine sources of sensory feedback. For example, social cognitive behavior among humans is predicated in large part upon use of vision to detect sensory feedback, and of movement to both produce and control sensory feedback. With most interactive systems, the machine lacks both modalities, forcing substantial cognitive compensation by the human partner.

Further compensation typically is required to deal with perturbations in machine sensory feedback. Thus, feedback delays in computer response appear to be a ubiquitous feature of interactive dialogue and speech systems (Williges and Williges, 1984; McCauley, 1984). Of the manual input devices in common use today, only the graphic tablet, touch-sensitive screen, and light pen provide direct spatial compliance between user motion and sensory feedback to the computer (Greenstein and Arnaut, 1987). All other devices require some sort of spatial translation in output-input tracking between user and computer. Spatial perturbations in user sensory feedback also are created by the typical lack of computer vision and movement, forcing spatial transformations in user control of self- and computer-derived sensory feedback.

4.0 CONCLUSIONS AND RESEARCH IMPLICATIONS

We suggest that the social cybernetic model shown in Figure 3, with the modality limitations depicted in Figure 4, define the nature and challenge of interface design as a human factors problem. Human cognition is a feedback product of motor behavior, yet reduced modality compliance in current interface design, with concomitant restrictive and perturbing effects on the ability of the human partner to adequately maintain motor control of sensory feedback across the interface, inevitably degrades the quality of cognitive interaction. Effective social tracking and cognitive exchange among humans is intensely dependent upon sensory feedback control of nonverbal as well as verbal modalities, yet interface design has emphasized the latter to almost the complete exclusion of the former.

Human cognition is a feedback control process, yet interface design assumes it is an open-loop product of information processing. Given these fundamental cognitive and social tracking interface limitations, it is no wonder that most users find sustained interaction with a computer to be an unappealing proposition.

User friendliness means an enriched and robust social tracking environment. From a social cybernetic perspective, therefore, the only viable strategy for bringing about meaningful improvement in the cognitive interface is to substantially enhance the social tracking capabilities of the machine partner. Such a suggestion is not unreasonable, considering that in a long line of human-machine systems dating back to prehistory, the computer represents the first such partner which realistically can be adaptively configured for integrated, multimodel control of sensory feedback.

There are a large number of research directions that should be explored in support of this objective, aimed at enriching social cognitive interaction between user and machine. Four are summarized below.

1. Task Allocation. Traditional methodology in this area has relied upon functional allocation of tasks at the design stage, using tables of relative merit (Kantowitz and Sorkin, 1987). Substantial human factors inadequacies of this approach have led to proposals for alternative strategies, such as adaptive dynamic allocation based on real time evaluation of system behavior (IBID; Chignell and Hancock, 1985). In social cybernetic terms, adaptive task allocation represents a control problem of group dynamics, which can best be addressed by continuously evaluating sensory feedback control and social tracking capabilities of system participants relative to system demands. Research issues raised by this perspective include: (1) optimal social tracking modalities for adaptive task allocation in different contexts; (2) advantages of enhanced machine sensory feedback control in adaptive task allocation and performance; (3) evaluation of mental workload in terms of motor coordinate feedback control; (4) effects on human task performance of spatial-temporal sensory feedback control perturbations across the interface; (5) relative merits of behavioral versus physiological monitoring of human performance status during dynamic task allocation; and (6) application of parallel processing to adaptive task allocation.

2. Software Creation and Use. Software creation represents a prominent example of social intimacy between humans and computers; the product of the creative effort in turn largely defines the social environment for interaction between user and machine. Recent attention to human factors considerations in interactive software has addressed some important cybernetic guidelines for software creation and design (Williges, Williges, and Elkerton, 1987; Williges and Williges, 1984), but many research questions remain: (1) approaches to machine adaptive feedback control of output dialogue in relation to sensory input; (2) appropriate use of nonverbal communication (such as music) to augment interactive dialogue; (3) motorsensory concomitants of software creation as a behavioral cybernetic process, in relation to both conception and implementation; (4) optimal modality combinations for feedback indication of system status during software monitoring/debugging; and (5) interactive machine guidance of software development and implementation as a social cybernetic process of adaptive task allocation.

3. Computer-Based Instruction (CBI). Human learning is a behavioral cybernetic process mediated by motor behavior (Smith and Smith, 1966). There were 600,000 computers in U.S. schools in 1984, and CBI systems almost certainly will become a ubiquitous fixture in the classroom (Eberts and Brock, 1987, 1984). Given these facts, both the challenge and the future implications arguably are greater for CBI interface design than they are for any other interactive application in common use today. To transform the information processing design approach, which clearly characterizes present CBI systems (IBID), to a more appropriate and relevant social cybernetic emphasis, a number of research issues should be addressed: (1) improved machine detection and production of movement, to enrich motor coordinate social tracking with the student across the interface; (2) emphasis on machine

imitative behavior as in movement or speech imitative tracking - imitation is the sincerest form of flattery, and a machine which imitates student behavior becomes very friendly; (3) improved machine interactive tracking capabilities - learning occurs more rapidly, and at an earlier age, if children are introduced to skills like handwriting as tracking activities, with nonvisual tracking modalities (i.e., tactual) often superior to visual tracking for learning purposes (Smith and Smith, 1987b); (4) expanded use of nonverbal modes of communication; and (5) greater use of facial or face-like displays - for humans generally and children particularly, the face represents the most compelling visual social tracking target.

4. Computer Vision. This area has been struggling for three decades, with notable lack of success. Workers in the field seem to have forgotten that visual perception is a cybernetic product of motor behavior, involving at least eight distinct ocular movement patterns (Smith and Smith, 1987a). The research direction indicated is to explore feedback control of visual sensor movement as an integral feature of computer vision systems.

5.0 REFERENCES

Adolph, E.F., 1982. Physiological integrations in action. The Physiologist (Suppl.) 25(2):1-67.

Ansell, S., and Smith, K.U., 1966. Application of a hybrid computer system to human factors research. American Society of Mechanical Engineers Human Factors 69:1-8.

Baily, M.N., 1986. What has happened to productivity growth. Science 234:443-451.

Barnes, D.M., 1986. Brain architecture: Beyond genes. Science 233:155-156.

Cherry, C., and Sayers, B. McA., 1956. Experiments on total inhibition of stammering by external control and some clinical results. Journal of Psychosomatic Research 1:233-246.

Chignell, M.H., and Hancock, P.A., 1985. Knowledge-based load leveling and task allocation in human-machine systems. In Proceedings of the Twenty-First Annual Conference on Manual Control. Columbus, OH.

Dainoff, M.J., 1982. Occupational stress factors in visual display terminal (VDT) operation: A review of empirical research. Behavior and Information Technology 1(2):141-176.

Delabarre, E.B., 1898. A method of recording eye movements. American Journal of Psychology 9:572-574.

Delabarre, E.B., 1891. Uber Bewegungs-empfindugen. Inaugural Dissertation zur Erlangen der Doctorwurde. Freiburg, Germany: Alberts-Ludwigs Universitat zu Freiburg.

Dewey, J., 1896. The reflex arc concept in psychology. Psychological Review 3:357-370.

Ditchburn, R.W., and Ginsborg, B.L., 1952. Vision with stabilized image. Nature 170:36-37.

Eberts, R.E., and Brock, J.F., 1987. Computer-assisted and computer-managed instruction. In G. Salvendy (Ed.). Handbook of Human Factors. New York: Wiley, pp. 976-1011.

Eberts, R., and Brock, J.F., 1984. Computer applications to instruction. In F. A. Muckler (Ed.). Human Factors Review: 1984. Santa Monica, CA: The Human Factors Society, pp. 239-284.

Foley, P., and Moray, N., 1987. Sensation, perception, and systems design. In G. Salvendy (Ed.). Handbook of Human Factors. New York: Wiley, pp. 45-71.

Georgopoulos, A.P., Schwartz, A.B., and Kettner, R.E., 1986. Neuronal population coding of movement direction. Science 233:1416-1419.

Gould, J., and Smith, K.U., 1962. Angular displacement of visual feedback of motion. Science 137:619-620.

Greenstein, J.S., and Arnaut, L.Y., 1987. Human factors aspects of manual computer input devices. In G. Salvendy (Ed.). Handbook of Human Factors. New York: Wiley, pp. 1450-1489.

Helander, M.G., Billingsley, P.A., and Schurick, J.M., 1984. An evaluation of human factors research on visual display terminals in the workplace. In F. A. Muckler (Ed.). Human Factors Review: 1984. Santa Monica, CA: The Human Factors Society, pp. 55-129.

Humphrey, D.R., 1986. Representation of movements and muscles within the primate precentral motor cortex: Historical and current perspectives. Federation Proceedings 45(12):2687-2699.

Jacobson, E., 1932. The electrophysiology of mental activities. American Journal of Psychology 44:677-694.

Jacobson, M., 1967. Retinal ganglion cells: Specification of central connections in larval Xenopus laevis. Science 155:1106-1108.

James, W., 1894. What is an emotion. Mind 9:188-205.

James, W., 1890. Principles of Psychology (2 Volumes). New York: Holt.

John, E.R., Yang, Y., Brill, A.B., Young, R., Ono, K., 1986. Double-labeled metabolic maps of memory. Science 233:1167-1175.

Jordan, N., 1963. Allocation of functions between man and machines in automated systems. Journal of Applied Psychology 47:161-165.

Kantowitz, B.H., and Sorkin, R.D., 1987. Allocation of functions. In G. Salvendy (Ed.). Handbook of Human Factors. New York: Wiley, pp. 355-369.

Kessen, W., and Cahan, E.D., 1986. A century of psychology: From subject to object to agent. American Scientist 74(6):640-649.

Knight, J.L. Jr., 1987. Manual control and tracking. In G. Salvendy (Ed.). Handbook of Human Factors. New York: Wiley, pp. 182-218.

Lee, B.S., 1950. Effects of delayed speech feedback. Journal of the Acoustical Society of America 22:824-826.

McCauley, M.E., 1984. Human factors in voice technology. In F. A. Muckler (Ed.). Human Factors Review: 1984. Santa Monica, CA: The Human Factors Society, pp. 131-166.

Munsterberg, H., 1889. Beitrage zur Experimentellen Psychologie. Freiburg, Germany: Mohr.

Nickerson, R.S., 1986. Using Computers: The Human Factors of Information Systems. Cambrtidge, Massachusetts: The MIT Press.

Riggs, L., Ratliff, F., Cornsweet, J.C., and Cornsweet, T.N., 1953. The disappearance of steadily fixated test objects. Journal of the Optical Society of America 43:495-501.

Sauter, S., and Smith, K.U., 1971. Social feedback: Quantitative division of labor in social interactions. Journal of Cybernetics 1:80-93.

Schmidt, J., Gottlieb, M., Coleman, P., and Smith, K.U., 1974. Reversal of the retinal feedback of binocular eye motions in depth vision. American Journal of Optometry and Physiological Optics 51:382-399.

Simpson, C.A., McCauley, M.E., Roland, E.F., Ruth, J.C., and Williges, B.H., 1987. Speech controls and displays. In G. Salvendy (Ed.). Handbook of Human Factors. New York: Wiley, pp. 1490-1525.

Smith, K.U., 1984. 'Facedness' and its relation to musical talent. Journal of the Acoustical Society of America 75:1907-1908.

Smith, K.U., 1972. Cybernetic psychology. In R. N. Singer (Ed.). The Psychomotor Domain. New York: Lea and Febiger, pp. 285-348.

Smith, K.U., 1970. Inversion and delay of the retinal image: Feedback systems analysis of a classical problem in space perception. American Journal of Optometry and Physiological Optics 47:175-204.

Smith, K.U., 1962. Delayed Sensory Feedback and Behavior. Philadelphia: W. B. Saunders.

Smith, K.U., Ansell, S.D., Koehler, J., and Servos, G.H., 1964. Digital computer system for dynamic analysis of speech and sound feedback mechanisms. Journal of the Association for Computing Machinery 11(2):240-251.

Smith, K.U., and Arndt, R., 1969. Experimental computer automation in perceptual-motor and social behavioral research. Journal of Motor Behavior 1:11-28.

Smith, K.U., Cambria, R., and Steffan, J., 1964. Sensory feedback analysis of reading. Journal of Applied Psychology 48:275-286.

Smith, K.U., Kao, H., and Knutson, R., 1969. An experimental cybernetic analysis of handwriting and penpoint design. Ergonomics 12:453-458.

Smith, K.U., and Molitor, K., 1969. Adaptation to reversal of retinal feedback of eye-movements. Journal of Motor Behavior 1:69-87.

Smith, K.U., Putz, V., and Molitor, K., 1969. Eye-movement retina delayed feedback. Science 166:1542-1544.

Smith, K.U., and Smith, M.F., 1973. Psychology: Introduction to Behavior Science. Boston: Little Brown.

Smith, K.U., and Smith, M.F., 1966. Cybernetic Principles of Learning and Educational Design. New York: Holt, Rinehart and Winston.

Smith, K.U., and Smith, T.J., 1969. Systems theory of therapeutic and rehabilitative learning with television. Journal of Nervous and Mental Disease 148:386-429.

Smith, K.U., and Smith, W.M., 1962. Perception and Motion. An Analysis of Space-Structured Behavior. Philadelphia: W. B. Saunders.

Smith, K.U., and Wargo, L., 1963. Specialization of space-organized sensory feedback mechanisms and learning. Perceptual and Motor Skills 16:749-756.

Smith, T.J., and Smith, K.U., 1988. Behavioral Cybernetic Foundations for Human Factors/Ergonomic Science. Amsterdam: Elsevier (North-Holland). To be published.

Smith, T.J., and Smith, K.U., 1987a. Feedback-control mechanisms of human behavior. In G. Salvendy (Ed.). Handbook of Human Factors. New York: Wiley, pp. 251-293.

Smith, T.J., and Smith, K.U., 1987b. Behavioral-Physiological Foundations of Human Development. Burnaby, B.C.: Simon Fraser University School of Kinesiology.

Smith, T.J., and Smith, K.U., 1985. Cybernetic factors in motor performance and development. In D. Goodman, R.B. Wilberg, and I.M. Franks (Eds.). Differing Perspectives in Motor Learning, Memory, and Control. Amsterdam: Elsevier (North-Holland), pp. 239-283.

Sternberg, R.J., 1986. Inside intelligence. American Scientist. 74(2):137-143.

Stetson, R.H., 1951. Motor Phonetics: A Study of Speech Movements in Action. Amsterdam: North-Holland.

Sussman, H., and Smith, K.U., 1970. Sensory feedback modes as determinants of learning and memory. Journal of Educational Research 64:64-66.

Sussman, H., and Smith, K.U., 1969. Analysis of memory as a feedforward control mechanism. Journal of Motor Behavior 2:101-117.

Tanner, P.P., and Wein, M., 1985. Parallel input in computer-human interaction. Proceedings of the 18th Annual Conference of the Human Factors Association of Canada. Rexdale, Ontario: Human Factors Association of Canada, pp. 141-144.

Thompson, R.F., 1986. The neurobiology of learning and memory. Science 233:941-947.

Thurow, L.C., 1986. A positive-sum strategy for productivity that does not quite add up. Review of R. Landau and N. Rosenberg (Eds.). The Positive Sum Strategy: Harnessing Technology for Economic Growth (Washington, D.C.: National Academy Press, 1986). Scientific American 255(3):24-31.

Ting, T., Smith, M., and Smith, K.U., 1972. Social feedback factors in rehabilitative processes and learning. American Journal of Physical Medicine 51(2):86-101.

Wickens, C.D., 1987. Information processing, decision-making, and cognition. In G. Salvendy (Ed.). Handbook of Human Factors. New York: Wiley, pp. 72-107.

Williges, B.H., and Williges, R.C., 1984. Dialogue design considerations for interactive computer systems. In F. A. Muckler (Ed.). Human Factors Review: 1984. Santa Monica, CA: The Human Factors Society, pp. 167-208.

Williges, R.C., Williges, B.H., and Elkerton, J. Software interface design. In G. Salvendy (Ed.). Handbook of Human Factors. New York: Wiley, pp. 1416-1449.

Social, Ergonomic and Stress Aspects of Work with Computers, 255
edited by G. Salvendy, S.L. Sauter and J.J. Hurrell, Jr.
Elsevier Science Publishers B.V., Amsterdam, 1987 — Printed in The Netherlands

CUMULATIVE FATIGUE SYMPTOMS OF DIALOGUE AND DATA ENTRY VDU OPERATORS

T. MAREK and C. NOWOROL
Department of Work Psychology, Institute of Psychology, Jagiellonian University, Golebia 13, 31-007 Cracow (Poland)

ABSTRACT
The aim of our study was to find out if the different types of workload of dialogue and data entry VDU operators effect the intensity and configuration of cumulative fatigue symptoms, and if the investigated groups of operators are homogeneous. The questionnaire used consists of 59 items related to symptoms of cumulative fatigue. The symptoms were grouped into six categories: (A) general physical fatigue, (B) weakened vitality, (C) irritability reflecting conditions of loading, (D) physical breakdown, (E) anxiety, (F) failing morale. Thirty-nine dialogue operators and forty-four data entry operators aged 21 to 33 were examined. The group of dialogue operators rates higher in categories (B) and (C) than the group of data entry operators. In categories (A), (E) and (F) the reverse occurs. It was proven that in both groups of operators higher ratings in all categories are characteristic for those who have a short work period.

1 PROBLEM

Video display unit operators deal with many various tasks. The most common groups of VDU operators are dialogue and data entry. Both types of clerical work, dialogue and data entry, may be classified as very strenous. Clerical work with VDU's seems clean and safe but the nature of this technology has many subtle, cumulative and chronic health and safety problems associated with it. One of the most difficult things to understand about the health and safety problems associatead with clerical work is the interactions of many small irritations which may result in some major health problems due to the exposure to chronic stress.

Recently many authors have pointed out that data entry operators reported a significantly higher amount of visual, postural and mental symptoms of strain, and lower job satisfaction than dialogue operators (Elias et al., 1980; Johansson Aronsson, 1980; Smith et al., 1981). Besides important environmental, mental and social factors affecting the stress experienced by data entry and dialogue VDU operators, these authors often mention work duration. From the point of view of health problems due to long work duration the problem of cumulative (chronic) fatigue is one of the most important for both groups considered here. Long-term chronic fatigue may result in behavioral responses such as burn-out, absenteeism or damage to the organism manifested through illness.

The aim of our study is to find out if the different types of workload for dialogue and data entry VDU operators have an impact on the kind, structure and intensity of cumulative fatigue. The first question is whether cumulative fatigue causes significant differences between the groups of dialogue and data entry operators. The second important question of our study is whether cumulative fatigue, from the point of view of type and intensity, is statisticly the same for each operator within the dialogue group as within the data entry group. In other words the question is whether the investigated groups are homogeneous.

2 METHODS

2.1 Subjects

Thirty-nine dialogue VDU operators and forty-four data entry VDU operators aged 21 to 33 years were examined. All operators were female. The work interval was from two months to six years. None of these suffered from any illness.

2.2 Investigative method

According to the aim of our study the cumulative fatigue symptoms were examined in order to picture the cumulative fatigue for both investigated groups Rokuro Kosugo's questionnaire for cumulative fatigue symptoms was applied (Kosugo, 1982). The questionnaire consists of 59 items related to symptoms of chronic fatigue. The symptoms are grouped into six categories, obtained as a result of factor analysis. The categories of symptoms are as follows:

((Category A)), general fatigue, 15 items: (9) I feel drowsy at times every-day, (12) I often feel tired when getting up in the morning, (25) I sometimes feel all my strength is gone, (30) These days I feel tired in the whole body, (32) I feel unwell when getting up in the morning, (40) I have pain in my lower back, (41) I feel pain in my joints, (53) I sometimes have dim eyes, (58) My eyes feel strained, (59) I often feel stiff in my shoulders, (16) I sleep poorly, having dreams all the time, (38) My legs feel tired these days, (42) I can hardly recover from fatigue due to work, (51) I often want to lie down while working, (26) I am totally exhausted by the day's work.

((Category B)), weakened vitality, 15 items: (2) I lack patience, (8) I am too lazy to get about, (36) I spare myself in doing anything, (43) I find it troublesome to think about something, (45) I feel uneasy for some reasons or other, (56) I lose my energy in no time doing work, (47) I am out of spirits these days, (35) I do not feel like doing what I like, (20) My head feels in the blues, (46) I have no energy for some reasons or other, (10) I feel depressed.

((Category C)), irritability reflecting conditions of loading, 10 items: (3) I am apt to get angry about trifles, (7) My nerves are highly strung,

(13) I am not content with many things, (23) I am apt to cry loud or use violent language, (24) I feel nervous without knowing why, (31) I feel like to quarrel to the utmost, (44) I am quick at losing my temper, (52) I want to drive my cares away by doing something boldly, (54) I can not put up with sounds or voices, (19) I worry about trifles.

((Category D)), physical breakdown, 8 items: (1) I have a dull appetite these days, (11) My head feels heavy these days, (18) I can not get to sleep easily these days, (21) I have a poor digestion, (28) I often feel dozzy, (33) I find it hard to go to work everyday, (49) I am getting thinner than before, (27) I easily catch cold.

((Category E)), anxiety, 6 items: (14) I have something to worry about, (29) I have no self-confidence in doing anything, (50) I can not but feel myself onferior to anybody, (55) I am apt to be distracted, (5) I have worries to tell people, (17) I an anxious about my job even at home.

((Category F)), failing morale, 9 items: (6) My job is too monotonous, (22) I can not go about my work, (34) Atmosphere in my workplace is gloomy, (37) I can not get on well with my seniors, (39) My job relations go wrong, (48) I have no will to work on, (57) I have no interest in my job, (4) I do not like to continue my present job, (15) I just lead a dull life.

The number of each item in the questionnaire is given in parenthesis.

As Kosugo pointed out, the profile for these six categories is usually reproducible for the particular job and represents certain workload characteristics of the job. This questionnaire has proven useful for analysing workload in relation to chronic fatigue (Kosugo, 1982).

In our research, each examined operator assessed the intensity of each symptom applicable to her recent conditions using a ten centimeter unmarked scale. Each operator evaluated her own sensations in the course of her daily activities.

3. RESULTS AND ANALYSIS

Obtained results were analysed using profiles of categories of cumulative fatigue symptoms. The profiles obtained for dialogue and data entry operators are shown in Fig. 1. Both profiles were analysed using ANOVA.

As it follows from analysis the dialogue operators rate higher in intensity than the data entry operators in the following categories of symptoms: (B) weakened vitality and (C) irritability reflecting conditions of loading ($p < .05$). In categories (A) general fatigue, (E) anxiety and (F) failing morale the reverse occurs ($p < .05$). Both groups of operators rate the same in category (D), physical breakdown. Considering all of this, profile courses for dialogue and data entry operators differ significantly.

The profile for the dialogue operators is characterised by relatively higher values of categories (A), (B) and (C) and lower mean values of categories (D), (E) and (F). These differences are statisticly significant ($p<.05$). Contrary to the dialogue operators' profile, the profile for the data entry operators is characterised by high mean values of categories (A), (E) and (F) which differ

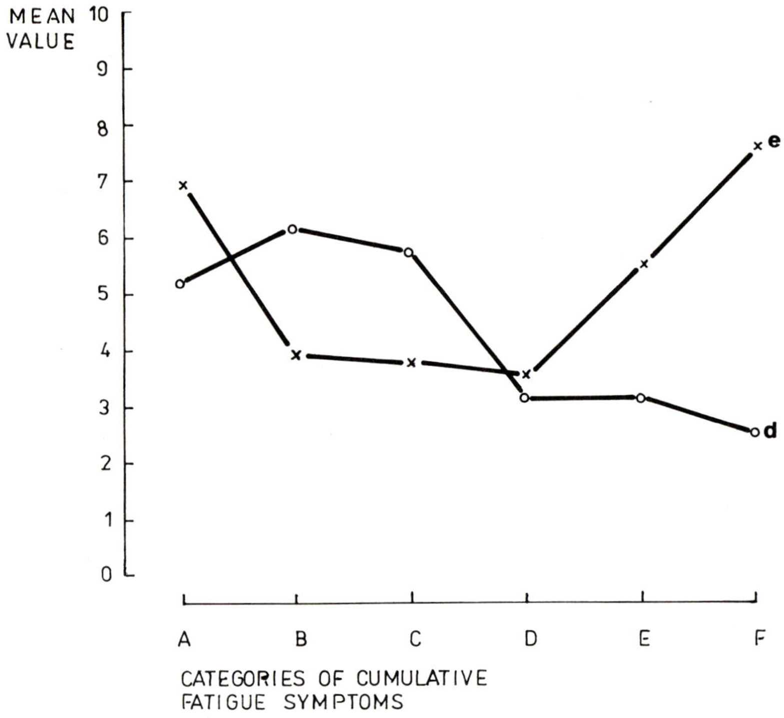

Fig. 1. The profiles for the six categories of cumulative fatigue symptoms for dialogue and data-entry operators (d - profile for dialogue operators (n = 39) and e - profile for data-entry operators (n = 44)). Mean values of six categories for dialogue operators: m.v. (A) = 5.1, m.v. (B) = 6.2, m.v. (C) = 5.8, m.v. (D) = 3.1, m.v. (E) = 3.1, m.v. (F) = 2.5. Mean values of six categories for data-entry operators: m.v. (A) = 6.9, m.v. (B) = 3.9, m.v. (C) = 3.8, m.v. (D) = 3.5, m.v. (E) = 3.5, m.v. (F) = 7.6. The values of standard deviation for each mean value are below 30 percent.

significantly from the lower mean values of categories (B), (C) and (D) ($p<.05$). The values for categories (A) and (F) are especially high. These values also differ significantly from the mean values for category (E).

Concerning the first question, the dialogue and data entry operators differ significantly in the five categories of cumulative fatigue symptoms ((A), (E),

(F) - higher values for data entry operators and (B), (C) - higher values for dialogue operators). This seems to confirm findings mentioned in the PROBLEM. According to this the data entry operators are characterised by a higher value of category (A) which is connected with visual and postural strain (items: 30, 40, 41, 53, 58, 59, 38), category (E) which is connected with mental or emotional load (Kosugo, 1982), and category (F), failing morale, which is connected with lower job satisfaction.

In regards to the second question relating to the problem of homogenity of both investigated groups of operators, cluster analysis was applied. Each group was seperately clustered into homogeneous subgroups. The nonhierarchical method of Overall and Klett (1972) was applied. As predictors all six categories of symptoms were used.

TABLE 1

The characteristics of clusters $C_{d,1}$ and $C_{d,2}$ obtained for dialogue operators (distance between clusters is 3.8 ($p<.05$)).

Characteristics of clusters	Clusters	
	$C_{d,1}$	$C_{d,2}$
Diameter of clusters	8.2	4.9
Number of operators in clusters	16	19
Range of operators' work period for clusters	from 6 months to 1 year and 5 months	from 2 years and 8 months to 5 years and 4 months

TABLE 2

The characteristics of clusters $C_{e,1}$ and $C_{e,2}$ obtained for data-entry operators (distance between clusters is 6.3 ($p<.05$)).

Characteristics of clusters	Clusters	
	$C_{e,1}$	$C_{e,2}$
Diameter of clusters	6.6	4.1
Number of operators in clusters	18	24
Range of operators' work period for clusters	from 2 months to 11 months	from 2 years and 3 months to 6 years

In each group of operators two clusters were found. These clusters are characterised in TABLES 1 and 2. All obtained clusters are homogeneous and differ significantly. This was proven using the maximum F-test (Bock, 1985) and the F-test after transformation of Mahalanobis distance. Four dialogue operators and two data entry operators did not join any of the clusters.

As it can be seen from both TABLES clusters $C_{d,1}$ and $C_{e,1}$ contain the operators with a relatively short work period and clusters $C_{d,2}$ and $C_{e,2}$ contain the operators with a long work period. It may be concluded that the work period is the variable which divides each investigated group of operators into two subgroups (clusters). Thus we have obtained four homogeneous subgroups of operators: dialogue operators with a short work period, dialogue operators with a long work period, data entry operators with a short work period and data entry operators with a long work period.

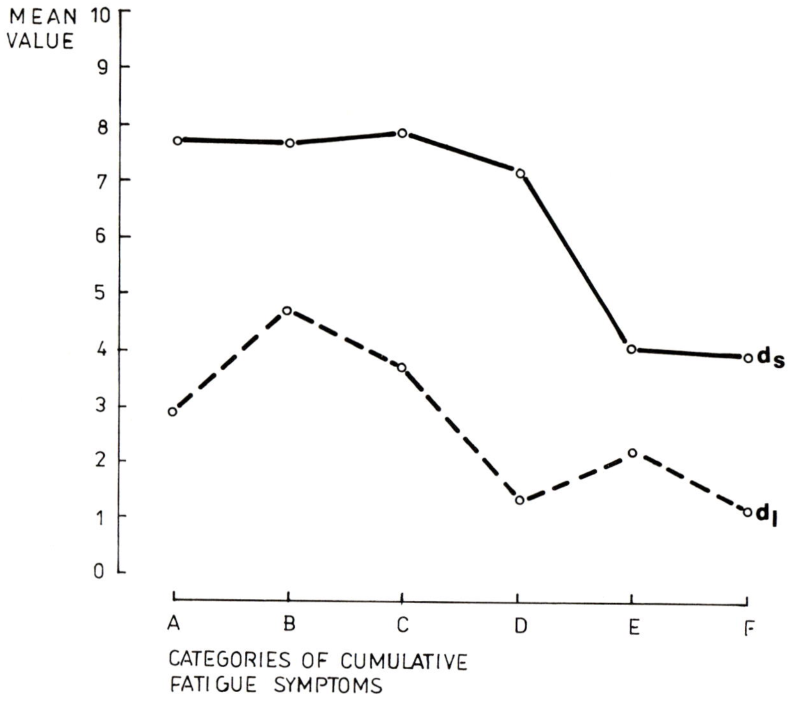

Fig. 2. The profiles for the six categories of cumulative fatigue symptoms for dialogue operators with a short and long work period (d_s - profile for operators with a short work period (n = 16) and d_l - profile for operators with a long work period (n = 19)). Mean values of six categories for dialogue operators with a short work period: m.v. (A) = 7.6, m.v. (B) = 7.6, m.v. (C) = 7.9, m.v. (D) = = 7.1, m.v. (E) = 4.0, m.v. (F) = 3.9. Mean values of six categories for dialogue operators with a long work period: m.v. (A) = 2.9, m.v. (B) = 4.6, m.v. (C) = = 3.6, m.v. (D) = 1.4, m.v. (E) = 2.7, m.v. (F) = 1.1. The values of standard deviation for each mean value are below 30 percent.

Mean values and standard deviations for six categories of symptoms were calculated for each subgroup. The obtained profiles for the two subgroups of dialogue operators and for the two subgroups of data entry operators are presented in Figs. 1 and 2, respectively.

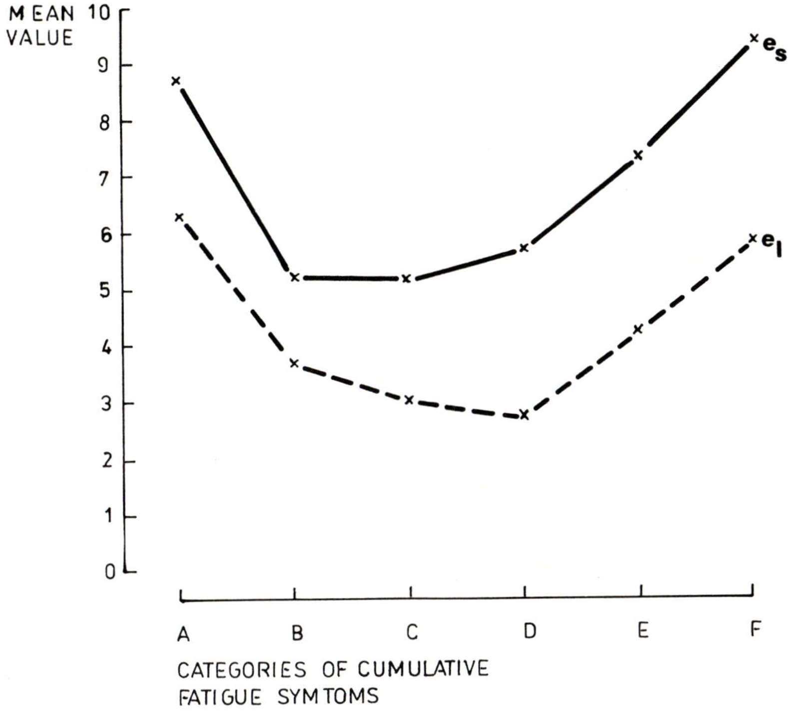

Fig. 3. The profiles for the six categories of cumulative fatigue symptoms for data-entry operators with a short and long work period (e_s - profile for operators with a short work period (n = 18) and e_l - profile for operators with a long work period (n = 24)). Mean values of six categories for data-entry operators with a short work period: m.v. (A) = 8.7, m.v. (B) = 5.2, m.v. (C) = 5.2, m.v. (D) = 5.6, m.v. (E) = 7.3, m.v. (F) = 9.4. Mean values of six categories for data-entry operators with a long work period: m.v. (A) = 6.4, m.v. (B) = = 3.6, m.v. (C) = 3.0, m.v. (D) = 2.7, m.v. (E) = 4.2, m.v. (F) = 5.9. The values of standard deviation for each mean value are below 30 percent.

ANOVA was applied to each group of operators (dialogue and data entry) in order to analysis the differences between operators with short and long work periods. In both cases it was proven that operators with a short work period are characterised by statisticly higher profiles (for all six categories) than operators with a long work period ($p < .05$).

4 CONCLUSIONS

It was found that cumulative fatigue causes significant differences between groups of dialogue and data entry operators. Dialogue operators rate higher than data entry operators in categories (B) and (C) which indicate that they have chronic fatigue due to general overload (Kosugo, 1982). Contrary to this, data entry operators rate higher than dialogue operators, in symptoms connected with (F) failing morale, (E) mental and emotional load and (A) general physical fatigue (visual, postural and general). This means that data entry operators have chronic fatigue due to visual, postural and mental load. As follows from our previous investigations (Marek, Noworol, 1987) the mental load of data entry operators is connected with a low level of arousal which follows understimulation. On the contrary, for VDU operators which do various tasks (dialogue operators) mental load is connected with a high level of arousal which follows overstimulation (overload). Extreme low job satisfaction which manifests itself by high ratings in category (F), observed in the data entry operators group, is probably connected with the mental load (monotonous work - undertimulation).

As has been shown in the study, operators with a short work period rate higher in all symptoms than operators with a long work period. This suggests that with time work adaptation begins to play a role. The operators with a long work period have the coping processes which minimize the level of chronic fatigue. Thus the symptoms of chronic fatigue manifest themselves with less strenght.

5 REFERENCES

Bock, H.H., 1985. On some significance tests in cluster analysis. Journal of Classification, 2: 77-109.

Elias, R., Tisserand, F. and Christmann, H., 1980. Investigations in operators working with CRT display: relations between task content and psychophysiological alterations. In: E. Grandjean and E. Vigiliani (Editors), Ergonomics Aspects of Visual Display Terminals. Taylor and Francis, London, pp. 211-217.

Johansson, G. and Aronsson, G., 1984. Stress reactions in computerized administrative work. Supplement 50, Department of Psychology, University of Stockholm, Stockholm.

Kosugo, R., 1982. Cumulative fatigue symptoms and conditions of work. In: K. Noro (Editor), Proceedings of the 8th Congress of IEA. Tokyo, pp. 144-145.

Marek, T. and Noworol, C., 1987. Mental fatigue of VDU operators induced by monotonous and various tasks. In: B. Knave and P.G. Wideback (Editors), Selected Papers Presented at the Conference on WORK WITH DISPLAY UNITS. Elsevier, Amsterdam, in press.

Overall, J.E. and Klett, C.J., 1972. Applied Multivariate Analysis. McGraw-Hill, New York.

Smith, M.J., Cohen, B.G.F., Stammerjohn, L.V. and Happ, A., 1981. An investigation of health complaints and job stress in video display operations. Human Factors, 23; 387-400.

Social, Ergonomic and Stress Aspects of Work with Computers,
edited by G. Salvendy, S.L. Sauter and J.J. Hurrell, Jr.
Elsevier Science Publishers B.V., Amsterdam, 1987 — Printed in The Netherlands

VIGILANCE AND ITS ROLE IN AI TECHNOLOGY: HOW SMART IS TOO SMART?

T. MOODY, M. JOOST, AND R. RODMAN

ABSTRACT

This paper raises questions and proposes potential guidelines for system designers and human factor specialists who are developing interfaces for automated systems that use Artificial Intelligence technology. The need to both evaluate and control for the vigilance decrement in these systems is the focal point.

Topics discussed include the role of Artificial Intelligence techniques in present day automation systems, and the potential problems associated with the use of these techniques in poorly designed systems. Shortcomings of current automated systems are examined and the role of vigilance in monitoring tasks is explored. In addition, techniques are presented that can be incorporated into the design of "intelligent" systems to limit the vigilance decrement effect that may be experienced by system users. Potential methodological procedures to determine the effectiveness of these techniques are also discussed.

1 INTRODUCTION

AI technology is challenging old definitions and concepts of automation in our society. No longer is the term automation limited to a system that replaces a human operator, nor performing tasks impossible for the human operator to do. The term now includes "smart" systems that help make the user's job less cumbersome, or reduce the mental workload involved in performing a task.

Wickens (1984) has defined three major areas where automation and AI technology are currently being employed. The first area involves those tasks which a human operator is unable to perform, because of either mental or physical limitations. Examples of these tasks include robotic arms that are designed to perform calculations and control functions in quided missle systems. This area of automation will be referred to as the extension area; the system extends the operator's performance beyond normal limits.

Another area in which automation has been used is with tasks that a human can do, but not well. Expert systems are playing a major role in this area. Examples of these include XCON (McDermott, 1980), PROSPECTOR (Duda et al., 1978) and MACSYMA (Martin and Fateman, 1971). XCON designs computer systems to meet the needs of the consumer purchasing the equipment. PROSPECTOR diagnoses geological data and identifies high probability locations of ore deposits. MACSYMA performs high level mathematical calculations symbolically. Humans can perform these tasks, but do so poorly (ie. slow and error prone). Therefore, this area will be referred to as the augmentation area.

The final area that Wickens identified is the use of automation in assisting operators in their tasks. Examples of these systems include the natural language query systems for database programs, and the self correcting typewriters that correct spelling mistakes as the typist types. This area will be referred to as the assistance area.

This paper will primarily focus on these last two application areas of automation, augmentation and assistance, though the reader should be able to intuitively see how all three of the above areas can be interrelated.

2 SHORTCOMINGS OF AUTOMATED SYSTEMS

Systems that are designed for the augmentation and assistance areas of automation, alter the task that the user must otherwise perform. The usual goal of these systems is to minimize user error and to correct the user's error whenever possible, thus reducing redundant input and the distraction of error correction. However, this minimization of user effort can produce some unexpected side effects that system designers may not anticipate. A list of these shortcomings is found in Table 1. The first two shortcomings listed will not be discussed in this paper, as several other papers have examined them in depth (Adams (1982) and Wilks (1985)).

The third entry in the list (user performing a monitoring task), is usually viewed as inherent in human computer interaction. These tasks range from non-critical, (eg. such as viewing a CRT and identifying a misspelled word in a typing task), to crucial (eg. identification of system malfunction based on changes in data displayed on a complicated control panel at a nuclear power plant).

Monitoring tasks of the latter type also necessitate that the user be able to distinguish "true" alarms from "false" alarms, the fourth shortcoming in Table 1. As the design of automated systems become increasingly more complex, there is a greater likelihood that one component may fail, resulting in erroneous data input and a "false" alarm. The user must be able to identify and differentiate between the alarms produced by the system effectively. The final shortcoming listed in Table 1, is that the user must also monitor personal mistakes. As "smart" systems are able to perform higher levels of error detection and correction, the user should have fewer remaining errors to detect, which may lead to unexpected results. Due to the vigilance decrement described in the following section, an attempt to reduce task complexity may have inadvertently increased the required mental effort.

3. VIGILANCE AND MONITORING TASKS

Monitoring tasks like those described above are often a major flaw of present day automated systems because people do not perform monitoring tasks well. The problems associated with performing sustained monitoring and signal detection tasks are described by the vigilance paradigm.

TABLE 1
Potential shortcomings in automated systems

Shortcoming	Consequence	Impact
1. System Failures Hardware and Software	1. User unable to use system 2. Possible loss of data	1. User frustration 2. Economic loss 3. Decreased acceptance of system
2. Control of Task Assigned to System	1. User develops a "false sense of security with regard to systems infallibility 2. User no longer acknowledges task responsibility	1. User no longer attends to potential problems with the task 2. Legal issues of responsibility for errors made, user or system designer
3. User must perform a monitoring task	1. Man performs monitoring tasks poorly	1. User error rate may increase 2. User may have too many tasks to perform 3. User may not have enough tasks to perform
4. User must be aware of "false alarms" produced by the system	1. User must be able to differentiate true alarms from false alarms	1. User may "lose faith" in the system 2. User may ignore true signals from the system
5. User error	1. User may not detect errors made	1. System may fail 2. System may provide erroneous results

The general characteristics of the vigilance paradigm are "that the operator is required to detect signals over a relatively long period of time and that the signals are intermittent, unpredictable, and infrequent" (Wickens, p. 33). The vigilance decrement, associated with the paradigm, can be defined as the deterioration of the person's ability to detect signals efficiently over time. The degree to which this vigilance decrement is observed is dependent on a complex interaction between the nature of the task, the type of signal to be detected, the individual, and the environment (Davies and Parasuraman, 1977). The vigilance decrement has been observed in several laboratory studies (Mackworth, 1948; Davies & Tune, 1969; and Parasuraman, (1979). These studies have used varying types of monitoring tasks in examining the vigilance effect. These include: the original "Clock Test" (Mackworth, 1948), the monitoring of auditory sonar outputs (Colquhoun, 1977), simulated radar watches (Baker, 1962), inspection tasks with multiple types of fault detection (Craig & Colquhoun, 1975), and electronic inspection tasks (Harris & Chaney, 1969).

Vigilance decrements have also been observed in a variety of field tasks including: long distance night driving (Riemersma et al., 1977), industrial inspection and assembly line tasks (Saito & Tanaka, 1977), and radar tasks (Mackworth, 1950).

4 THREE SIGNAL FEATURES THAT INCREASE THE VIGILANCE EFFECT

The vigilance decrement is intensified when the target to nontarget ratio is decreased (Baddeley & Colguhoun, 1969; Jerison et al., 1965). This means that a user is less likely to be aware of a signal that is produced infrequently, when compared to a signal that occurs often. The terms "infrequently" and "often" are not easily defined due to a number of complicating factors, the reader is referred to Broadbent (1971) for a complete discussion.

The vigilance decrement also increases when the signals presented are of short duration (Baker, 1963). Briefer signals produce "a more rapid decline in the percentage of signals correctly detected as well as resulting in a much lower overall level of efficiency" (Davies & Parasuraman, 1982, p.34).

The complexity of the signal has also been shown to enhance the vigilance effect (Childs, 1976). These studies have shown that increasing the complexity of the signal (eg. signals varying in two or more dimensions), decreases the subject's ability to detect the signal.

This decrement in signal detection has major implications for the use of "smart" systems. As the error detection and correction ability of the system increases, the error to non-error (or target to nontarget) ratio will decrease. The ability of the user to continue to identify errors will depend not only on the user, and their motivation level, but also the designer's ability to control for the vigilance decrement.

5 TECHNIQUES TO LIMIT THE LOSS OF PERFORMANCE FROM VIGILANCE

A list of techniques have been developed to decrease the vigilance decrement observed in sustained monitoring tasks (Table 2). The techniques have been divided into three sections, the user, the task, and the monitoring system, though overlap of these areas makes the division somewhat artificial.

5.1 The User

Fisk and Schneider (1981) have demonstrated that the vigilance decrement can be reduced through appropriate user training. Their technique develops what has been termed automatic processing (skilled behavior) (see LaBerge (1973) for more detail).

5.2 The Task

The tasks assigned to the user should be arranged so that periods of continual monitoring are interspersed with rest periods and/or other job related tasks. The optimal task time (continual monitoring without the vigilance decrement) varies for individuals (Davies & Parasuraman, 1982).

The vigilance decrement effect is also decreased when the user monitors only one target, and/or is continually allowed to have access to representations of the possible signals to be detected (Childs, 1976). When multiple signals are used, they should be easily differentiable. If the signals to be detected are numerous and cannot be displayed the standard or idealized product representation should always be available to the operator (Kelly, 1955).

5.3 The System

The rate that information is presented in displays should be set so as not to exceed the user's ability to process the information. Monty (1973) has suggested that input rates above one event per second will increase the vigilance decrement when the user has to monitor sequential events on a visual display. When possible, it is recommended that the operator be allowed to perform the task at his/her own rate (Grandjean, 1979).

When feasible, the system should produce false alarms to increase the signal to non-signal ratio (Baker, 1961). However, these false alarms must require some type of operator response, and the operator should not consider them to be false alarms.

Additional system design techniques include: operator performance feedback (Drury and Addison, 1973), visual displays located in central focus region, and the use of multisensory signals (Colquhoun, 1975).

In addition to the above recommendations, designers should examine user performance to identify the workload level produced by the task on a prototype system. Recent studies (Wiener et al., 1984) have supported the task as the task demand increases. Wiener found that operators, whose only task was to notice a change in the physical proximity of two signals (understimulation), detected these changes less frequently than operators who performed an additional tracking task requiring them to press a switch each time the cursor moved beyond predefined limits. However, when the operator had to monitor multiple channels (overstimulation), a decrease in signal detection was noted.

6 IMPLICATIONS

The intent of the information presented in this paper is not to suggest that Artificial Intelligence techniques will decrease the performance of operators of automated systems. However, the information reviewed does point to the conclusion that as Artificial Intelligence techniques play a greater role in automation, the system designer will need to be aware of the potential vigilance decrement that may occur in a poorly designed system.

System designers must test many prototype systems on a wide range of potential users to assess user response to the signals presented. This includes, but is not limited to, varying the modality in which signals are presented, the rate, the types of signals presented, the degree of error detection and correction done by the system, user training, and the length of time the user performs the task.

The development of user training procedures will play an important role

Table 2
Techniques to Minimize the Vigilance Effect

Technique	Consequence	Main Application Area
1. User training	1. Signal more easily detected by user	1. Any application area
2. Intersperse rest periods/other job tasks with monitoring task	1. Decreases user's feelings of boredom and fatigue 2. Improved job satisfaction	1. Industrial & military areas where monitoring is not only task or where staff rotation is possible
3. An image of the signal(s) to be detected or a target standard is displayed to the user	1. Decreases mental workload of user 2. Signal to be detected does not need to remain in memory	1. Industrial applications such as inspection tasks
4. Multiple signals easily differentiated	1. User can rapidly identify the signal and make the correct response	1. Military & industrial applications where signals can be modified and/or intensified
5. Event rate of presentation below 60 events per minute or self-paced monitoring task	1. Presentation rate does not exceed user's ability to process information presented	1. Applications where the rate of display can be controlled by user without adverse effects
6. Production of false alarms	1. Increase signal to non-signal ratio	1. Industrial inspection tasks
7. Operator feedback	1. increased level of motivation for operator 2. More accurate anticipation of signals	1. Industrial inspection tasks
8. Signals presented in central focus region	1. Search time for signal is decreased	1. All areas
9. Use of auditory, visual, or multisensory signals	1. User is provided with signal that is most easily detected for the task and environment	1. All areas
10. Optimal workload level	1. User is neither understimulated nor overstimulated	1. All areas

in the evaluation process of these prototype scenarios, though there is likely to be a tradeoff between time and resources allocatable to training and those needed for system development. However, system designers may find that these two tasks can be interweaved with the results of the training dictating the system requirements. Varying factors such as signal presentation modality and rate during training will allow the designer to investigate various types as well as degrees of error detection and develop algorithms that produce the optimal user response.

The designer must weigh the results of this type of analysis in terms of the effects that might occur if the operator misses a signal, or sees a signal that was not presented, and design the final system accordingly.

At this time, no available research can be found that gives guidelines on the optimal level of error correction and detection and the user's ability to locate or detect the remaining errors. However, as more "intelligent" systems are developed, there is an increasing need for research in this area.

7 REFERENCES

Adams, J.A., 1982. Issues in Human Reliability. Human Factors, 24: 1-10.

Baddeley, A.D. and Colquhoun, W.P., 1969. Signal Probability and Vigilance: A Reappraissal of the "Signal Rate" Effect. Brit.J. of Psych., 60: 169-178.

Baker, C.H., 1961. Maintaining the Level of Vigilance by Means of Knowledge of Results About a Secondary Vigilance Task. Ergonomics, 4: 311-316.

Baker, C.H., 1962. Man and Radar Displays. MacMillan, New York.

Baker, C.H., 1963. Signal Detection as a Factor in Vigilance Tasks, Science, 141: 1296-1297.

Broadbent, D.E., 1971. Decision and Stress. Academic Press, New York.

Childs, J.M., 1976. Signal Complexity, Response Complexity, and Signal Specification in Vigilance. Human Factors, 18: 149-160.

Colquhoun, W.P., 1975. Evaluation of Auditory, Visual, and Dual-Mode Displays for Prolonged Sonar Monitoring in Repeated Sessions. Human Factors, 17: 425-437.

Colquhoun, W.P., 1977. Simulataneous Monitoring of a Number of Auditory Sonar Outputs. In: R.R. Mackie(Ed.), Vigilance: Theory, Operational Performance and Physiological Correlates, Plenum, New York.

Craig, A. and Colquhoun, W.P., 1975. Vigilance: A Review. In: C.G. Drury and J.G. Fox(Eds.), Human Reliability in Quality Control, Taylor and Francis, London.

Davies, D.R. and Parasuraman, R., 1982. The Psychology of Vigilance, Academic Press: New York.

Davies, D.R. and Tune, G.S., 1969. Human Vigilance Performance. Elsevier, New York.

Drury, C.G. and Addison, J.C., 1973. An Industrial Study of the Effects of Feedback and Fault Devising on Inspection Performance. Ergonomics, 16: 159-169.

Duda, R.O., Hart, P.E., Barrett, P., Gaschnig, J., Konolige, K., Reboh, R. and Slocum, J., 1978. Development of the PROSPECTOR Consultation System for Mineral Exploration. Final Report SRI Projects 5821 and 6415, AI Center, SRI International, Menlo Park, CA.

Fisk, A.D. and Schneider, W., 1981. Controlled and Automatic Processing During Tasks Requiring Sustained Attention. Human Factors, 23: 737-750.

Grandjean, E., 1979. Fatigue in Industry. British Journal of Industrial Medicine, 36: 175-186.

Harris, D.H. and Chaney, F.B., 1969. Human Factors in Quality Assurance, Wiley: New York.

Jerison, M.L., Pickett, R.M. and Stenson, H.H., 1965. The Elicited Observing Rate and Decision Process in Vigilance. Human Factors, 7: 107-128.

Kelly, M.L., 1955. A Study of Industrial Inspection by the Method of Paired Comparisons. Psychological Monographs, 69(no. 394): 1-16.

LaBerge, D., 1973. Attention and the Measurement of Perceptual Learning. Memory and Cognition,1: 268-276.

McDermott, J., 1980. R1: A Rule-Based Configurer of Computer Systems. Tech. Rep. CMU-CS-80-119, Dept of Computer Science, Carnegie-Mellon, Pitts., PA.

Mackworth, N.H., 1948. The Breakdown of Vigilance During Prolonged Visual Search. Quarterly Journal of Experimental Psychology, 1: 5-61.

Mackworth, N.H., 1950. Researches on the Measurement of Human Performance Medical Research Council Special Report, No.268, H.M.S.O., London.

Martin, W.A. and Fateman, R.J., 1971. The MACSYMA System. In: Proceedings of the Second Sym. on Sym. and Algebraic Manipulation, Los Angeles, 59-75.

Monty, R.A., 1973. Keeping Track of Sequential Events: Implications for the Design of Displays. Ergonomics, 16: 443-454.

Parasuraman, R., 1979. Memory Load and Event Rate Control Sensitivity Decrements in Sustained Attention. Science, 205: 924-927.

Riemersma, J.B., Sanders, A.F., Wildervanck, C. and Gaillard, A.W., 1977. Performance Decrements During Prolonged Night Driving. In: R.R. Mackie (Ed.) Vigilance: Theory, Operational Performance and Physiological Correlates, Plenum, New York.

Saito, M. and Tanaka, T., 1977. Visual Bottle Inspection Performance in Highly Paced Belt-Conveyor Systems. Journal of Human Ergology, 6: 127-137.

Schmidtke, H., 1976. Vigilance. In: E. Simonson and P.C. Weisner (Eds.), Psychological Aspects and Physiological Correlates of Work and Fatigue. Thomas, Springfield, Ill.

Wickens, C.D., 1984. Engineering Psychology and Human Performance. Merrill Publishing, Columbus, Ohio.

Wiener, E.L., Curry, R.E. and Faustina, M.L., 1984. Vigilance and Task Load: In Search of the Inverted U. Human Factors, 26(2): 215-222.

Wilks, Y., 1985. Responsible Computers? In: Proceedings of the Ninth Intl. Joint Conference on Artificial Intelligence, Los Angeles, 1279-1280.

Social, Ergonomic and Stress Aspects of Work with Computers,
edited by G. Salvendy, S.L. Sauter and J.J. Hurrell, Jr.
Elsevier Science Publishers B.V., Amsterdam, 1987 — Printed in The Netherlands

MENTAL WORKLOAD IN SIMPLE AND COMPLICATED COMPUTER AIDED DESIGN

E. JÄRVENPÄÄ and V. TEIKARI
Helsinki University of Technology, Laboratory of Industrial Psychology,
Otakaari 1, SF-02150 Espoo (Finland)

ABSTRACT

In the study mental workload and strain were evaluated in four CAD-work groups, the mental demands of which were different. The design work groups studied were drafting, the design of the layout of printed circuit boards, parametrical design and 3-dimensional design. The methods used were questionnaires and interviews, job analysis and tests. CAD-designers' mental strain was in general moderate. Different mental strain profiles for different CAD-work groups were found.

1 INTRODUCTION

Technological development has changed the characteristics of work in many occupations both in factories and in offices during recent years. The computerization of the design work, CAD (Computer Aided Design), is one manifestation of these changes. Together with CAM (Computer Aided Manufacturing) it is one of the major technical factors causing drastic changes in production in this decade (Poitou 1981, p. 167). In this research we studied CAD-designers' short-term mental strain and its changes and compared it in different CAD-work groups.

2 MATERIAL

The mental workload of CAD-work was studied in two firms. In one firm we studied 13 CAD-designers whose work was to design the layout of printed circuit boards (material I). In the other firm we studied 42 CAD-designers, who designed different parts of power stations to the whole power station layout (material II). The results concerning material II are preliminary. The materials were divided into four groups on the basis of the job demands. The groups are :

1. very simple CAD-work: drafting "draftsmen", 2. rather simple: design of the layout of printed circuit boards: "layout design", 3. rather complicated: parametrical design: "parametrical design" and 4. very complicated: 3-dimensional design: "3 D design".

In the group "draftsmen" there were 20 persons whose work was to draw the sketches made by the designers. Their education was vocational school. Moreover there were 6 designers who designed and realized drawings by themselves. They were engineers and technicians.

In the group "layout design" (n=13) the work was to design the layout of printed circuit boards. The designers' education was vocational school.

In the group "parametrical design" (n=10) the work was to design different parts of power stations and e. g. concrete elements by given parametres. The designers were technicians and engineers.

In the group "3 D design" (n=6) the work was to design different parts of power stations 3-dimensionally.

3 METHODS

Mental workload factors were measured in materials I and II by job description and by questionnaires and interviews. Mental strain was measured by scales which measure workers' mental state at a certain moment (BMS-scale by Plath and Richter 1978, 1984, EZ-scale by Nitsch 1976, EWL-scale by Janke and Debus 1978). Moreover, D2-test (Brickenkamp, 1978) which measures concentration and attention, was used. In material II a questionnaire of long-term psychosomatic stress reactions was used. The measurements were made in the beginning and at the end of the working day.

4 RESULTS

4.1 All designers

The designers' mental state changed negatively during the working day in alertness (**, EZ), activity (*, EZ) and ability to act (*, EZ). Their mental state changed positively in relation to apathetic state (**, EWL), sensitiveness (**, EWL) and daydreaming (**, EWL).

4.2 Mental strain within different CAD-work groups

Drafting: very simple CAD-work

The draftsmen's mental state changed positively during the working day in relation to apathetic state (**, EWL), sensitiveness (**, EWL) and daydreaming (**. EWL).

Design of the layout of printed circuit boards: rather simple CAD-work

The designers' mental changed negatively during the working day. This came out in self-confidence (**, EWL), alertness (*, EZ) and concentration (*, EWL). (Järvenpää, 1986).

(*: p<.05, **: p<.01, ***: p<.001)

Parametrical design: rather complicated CAD-work

The designers' mental state changed negatively during the working day. This came out in the following dimensions in EZ-scale: alertness (***), motivation (**), ability to act (**), effort (**), social recognition (**), mental strain (*), activation (*), efficiency (*), tension (*), sociability (*), self-confidence (*) and sleepiness (*). In EWL-scale this came out in self-confidence (*) and in state of mind (*) and in BMS-scale in mental fatigue (*). In daydreaming the designers' mental state changed positively during the working day (*, EWL).

3-dimensional design: very complicated CAD-work

The designers' mental state changed positively during the working day. This came out in relation to apathetic state (*, EWL), sensitiveness (*, EWL) and monotony (*, BMS).

4.3 Comparison of mental strain between different CAD-work groups

In the beginning of the working day the mental state of the layout design group was worse than that of the draftsmen in introversion (*, EWL), but better in satiation (*, BMS). It was worse than that of the parametrical design group in social recognition (*, EZ), extroversion (*, EWL) and state of mind (*, EWL). Compared with the 3 D design group it was better in fatigue (*, EWL), apathetic state (*, EWL) and daydreaming (*, EWL).

In the beginning of the working day the mental state of the parametrical design group was better than that of the draftsmen. This came out in monotony (*, BMS), ability to act (*, EZ), alertness (*, EZ) and sleepiness (*, EZ). Their mental state was also better than that of 3 D design group in monotony (*, BMS), sleepiness (*, EZ), fatigue (*, EWL) and apathetic state (*, EWL).

At the end of the working day the mental state of the layout design group was worse than that of the draftsmen in sociability (*, EZ) and introversion (*, EWL). It was worse than that of the 3 D design group in activation (*, EWL) and monotony (*, BMS). Moreover, it was worse than that of the parametrical design group in monotony (*, BMS).

At the end of the working day the mental state of the 3 D design group was better than that of the draftsmen in satiation (*, BMS), but worse in daydreaming (*, EWL).

At the end of the working day the mental state of the parametrical design group was better than that of the 3 D design group in relation to apathetic state (*, EWL) and daydreaming (*, EWL).

5 CONCLUSIONS

Mental strain of the CAD-designers studied was not very high compared with other occupations which we have studied. Their mental state was at the end of the working day worse than in the beginning of it only in alertness, activity and ability to act. So the work was only moderately mentally loading. On the other hand their mental state changed positively during the working day in relation to apathetic state, sensitiveness and daydreaming.

The CAD-work groups studied seemed to have different "mental strain profiles". In layout design and parametrical design the work seemed to be mainly mentally loading and in drafting and 3 D design mainly activating. The layout designers' and the parametrical designers' mental state changed negatively during the working day. The mental state of the draftsmen and the 3 D designers was at the end of the working day better than in the beginning of it.

According to these results it seems that there is not only one CAD-work but many kinds of design tasks in which the job demands and mental strain are different.

Our results do not confirm earlier presented evaluations concerning the negative effects of new technology on the designers' mental strain. The crucial factors are job design and work organisation which do not so much depend on new technology as it has been thought.

Based on material I concerning the design of the layout of printed circuit boards we have made the following suggestions for the job design of CAD-work: CAD-designers should have lots of degrees of freedom concerning their work and working process because of the creative nature of their work. They should have flexible working times in general and especially in VDT-work. E. g. they should have possibilities to work longer periods at the VDT-unit when needed. Problems caused by CAD-equipment (delay and disturbancies) should be eliminated as much as possible. (Järvenpää, 1986).

6 REFERENCES

Brickenkamp, R., 1978. Test D 2. Aufmerksamkeits-Belastungs-Test. Handanweisung. 7. Auflage. Göttingen, Verlag fur Psychologie.

Janke, W. and Debus, G., 1978. Die Eigenschaftswörterliste. (EWL). Handanweisung. Göttingen, Verlag fur Psychologie.

Järvenpää, E.,1986. Mental workload in CAD-work: computer aided design of printed circuit boards as an example. In: L. Norros & M. Vartiainen (Editors), Psychological aspects of the technological and organizational change in work. Acta Psychologica Fennica XI, Proceedings of the symposium organized by the Finnish Psychological Society and the Psychological Society of GDR, Espoo 22.-24. October 1985, pp.107-122.

Mermet, J., 1981. (Editor), CAD in Medium Sized and Small Industries. North Holland Publishing Company, Micado.
Nitsch, J.R., 1976. Die Eigenzustandsskala (EZ-Skala) -Ein Verfahren zur hierarchisch-mehrdimensionalen Befindlichkeitsskalierung. In: J.R. Nitsch and I. Udris (Editors), Beanspruchung im Sport. Beiträge zur psychologischen Analyse sportlicher Leistungssituation, Wiesbaden, Schriftenreihe Training und Beanspruchung, pp. 81-102.
Nitsch, J.R. and Udris, I., 1976. Beanspruchung im Sport. Beiträge zur psychologischen Analyse sportlicher Leistungssituation. Wiesbaden, Schriftenreihe Training und Beanspruchung.
Poitou, J., 1981. CAD and working conditions. In J. Mermet (Editor), CAD in Medium Sized and Small Industries. North Holland Publishing Company, Micado, pp. 167-176.
Plath, H.-E. and Richter, P., 1978. Der BMS (I)-Erfassungsbogen - Ein Verfahren zur skalierten Erfassung erlebter Beanspruchungsfolgen. Probleme und Ergebnisse der Psychologie, Heft 65/78: 45-65.
Plath, H.E. and Richter, P., 1984. Ermudung-Monotonie-Sättigung-Stress (BMS). Verfahren zur skalierten Erfassung erlebter Beanspruchungsfolgen. Handanweisung. Psychodiagnostisches Zentrum. Sektion Psychologie der Humboldt-Universität zu Berlin.

Appendix 1.

Table 1 **The designers' mental strain in the beginning (B) and at the end (E) of the working day in different CAD-work groups.** In BMS- and EZ-scales the positive change in mental state is scored positively. In EWL-scale the positive change in mental state is scored positively (P) or negatively (N) in respective dimensions.

	Drafts-men (n=26)		Layout-design (n=13)		Parametri-cal design (n=10)		3 D-design (n=6)		
	B	E	B	E	B	E	B	E	
BMS-scale									
Mental fatigue	52.1	50.6	50.9	51.4	55.6	52.5*	50.7	53.5	
Monotony	50.6	52.2	51.0	50.7	55.9	55.8	49.3	58.0*	
Satiation	47.0	49.7	55.4	51.1	53.5	54.5	51.5	58.7	
EZ-scale									
Motivation	5.3	5.4	4.9	4.1	6.5	5.1**	5.3	5.8	
Activation	5.4	5.1	4.8	3.9	6.3	4.9*	5.2	5.7	
Efficiency	5.2	5.7	5.0	4.7	6.4	5.6*	6.0	6.0	
Effort	5.3	5.3	5.2	4.3	6.6	5.0**	5.2	6.2	
Sociability	4.7	4.5	3.9	3.1	5.3	4.1*	4.3	4.3	
Soc. recognition	5.2	5.3	4.5	4.4	6.4	5.2**	5.3	5.2	
Self-confidence	5.4	5.7	5.5	4.9	6.3	5.3*	6.0	6.3	
Mental strain	5.7	5.6	6.2	5.0	7.1	5.8*	5.5	6.0	
Tension	5.2	5.1	5.2	5.2	6.0	5.2*	5.3	5.3	
Ability to act	6.4	6.3	6.9	5.2	7.9	6.0**	6.0	6.5	
State of mind	4.8	5.0	4.9	4.8	5.6	5.0	5.0	4.5	
State of tension	5.6	5.3	5.5	5.4	5.9	5.7	5.3	5.5	
Alertness	6.5	6.4	7.0	5.1*	8.0	6.2***	6.5	6.7	
Sleepiness	5.8	5.8	6.5	5.0	7.8	6.0*	4.8	6.0	
EWL-scale									
Activation	11.3	10.5	13.5	8.0	13.9	11.8	11.5	15.8	P
Concentration	4.3	4.0	4.0	2.9*	4.3	4.0	3.7	4.5	P
Low. activation	3.5	3.2	3.2	4.2	2.6	2.3	4.5	2.2	N
Fatigue	1.7	1.5	1.1	1.7	0.8	0.9	3.0	0.7	N
Apathetic state	1.3	0.4**	0.7	0.9	0.8	0.2	2.5	0.7*	N
Extroversion	5.7	5.8	4.0	4.5	6.9	5.9	5.2	4.5	P
Introversion	1.0	0.9	2.1	2.1	1.2	1.3	1.5	2.2	N
Self-confidence	4.2	4.0	4.3	3.5**	5.0	3.5*	4.3	5.3	P
State of mind	5.2	5.4	4.3	4.2	7.3	4.5*	5.2	6.0	P
Irritation	2.5	2.5	3.5	2.5	1.7	1.1	2.3	2.0	N
Sensitiveness	1.5	0.9**	1.5	1.2	0.9	0.9	1.5	0.7*	N
Hate	0.7	0.7	1.5	0.9	0.7	0.2	0.8	0.2	N
Depression	1.7	1.7	1.1	1.2	0.7	0.2	1.5	0.7	N
Daydreaming	4.3	3.1**	3.5	3.8	4.6	3.1*	6.7	6.2	N

(*: $p<.05$, **: $p<.01$, ***: $p<.001$)

CONTENTS

Social, Ergonomic and Stress Aspects of Work with Computers,
edited by G. Salvendy, S.L. Sauter and J.J. Hurrell, Jr.
Elsevier Science Publishers B.V., Amsterdam, 1987 — Printed in The Netherlands

HUMAN-INTERFACE ARCHITECTURE
-ITS SIGNIFICANCE FOR OFFICE AUTOMATION SYSTEM-

HITOSHI WATANABE
NEC Corporation 33-1, Shiba 5-Chome, Minato-ku, Tokyo 108, Japan

ABSTRACT

Basic concept and structure of C&C (Computers and Communications) Office System is presented here, where the OSA (Office System Architecture) plays a most important role. HIA (Human Interface Architecture) is emphasized as the essential core of OSA. Also outlined is the HIA response time guideline and multi-media communications.

1 INTRODUCTION

With the remarkable development of information technology, various types of information handling equipment have been introduced into many offices. Work methods have also been radically altered. These are just some of the changes resulting from OA or Office Automation. These information systems will play an important role in supporting human intellectual activities in offices of the future. In OA systems, various kinds of equipment are integrated to support intellectual activities.

In this paper, the system of office system architectures are described as a basic and essential concept for smooth running office system. Among office system architectures, the Human Interface Architecture is considered the most important, but has not been fully studied. This article proposes a basic framework for this concept with special emphasis on response time guideline and multi-media representations.

2 C&C OFFICE SYSTEM

2.1 Basic concept of the C&C office system

Work performed in an office is varied and diverse. However, the essential feature of an office, no matter how different in appearance, is that it is a place where people are engaged in intellectual activities (not mere physical labor).

The following three items can be considered as the preferred support environments for such an office:

1 Support of each individual's information generation work
   An environment permitting people to give full play to their intelligence in the office for information generation.

---

*NEC CORPORATION 33-1, SHIBA 5-CHOME, MINATO-KU, TOKYO, 108 JAPAN

2 Support of information transfer/circulation.
An environment permitting the accurate and timely transfer of information as a result of human intellectual activities.

3 Support of information storage and retrieval activity.
An environment permitting required information to obtained instantly.

Human intellectual activities in an office can be considered essentially as information creation activities, and at the same time, these activities are problem-solving activities.

Information creation (generation/processing), transmission (distribution/networking), and storage (memory/filing) are the basic information handling functions of "modern communications". These environments are nothing more than the utilization of "modern communications", which integrate and systematize the functions of computers and communications ("C&C"), in the office. OA systems based on modern communications are called C&C Office Systems.

2.2 Office systems must be truly useful for people

An office should be a place where people can give full play to their capabilities to accomplish meaningful tasks. The goal of OA is to "improve the effectiveness of people working in offices" and to "humanize the work environment for people working in offices". In such office systems, human activities should concentrate on the basic and core portions of office work, that can never be replaced by office robots, no matter how much science and technology progress. To accomplish this, the major goal of OA is to accelerate the creation of valuable information (wisdom) by realizing office environments such as:

1 Information gathering , storage and retrieval systems which enable related information to be obtained quickly.

2 A system which enables the retrieved related information to be analyzed from various aspects.

3 A system which enables interactive information exchange between an office worker and the office systems to produce solutions by repeating heuristic interaction based on the above analytical results.

Information of a high value, which is produced as a result of human creative activities, produces no effects unless it is understood and utilized by other people. It can be said that "offices are places where information is created (produced) and utilized (consumed)". In offices, information is created, transferred, and utilized, and information with a higher added value is reproduced and result. Information at this stage should be arranged and transferred in a format which is easily understood and communicated.

In this context, the key activities, of the office are to mutually couple human creative activities, that is, to produce communication among people. Supporting creative activities has been the strong point of so-called information processing systems (computer systems), and one of the roles of communications systems is to support the distribution/transfer of information. But these systems need to be integrated to function as one system so as to accelerate correct decision-making and quick action to help improve the effectiveness of human activities. This is the aim of "C&C" (Computers and Communications) Office Systems using integrated systems of computers and communications.

Having reached this conclusion, it can be said that the basic human activity in the office is "information creation". "Creation" in the office is "finding solutions to problems", and the general procedure for this process can be described as follows:

1 To collect as much information as possible regarding the task.
2 To combine and analyze the collected information.
3 To find a proposal that permits a solution through human wisdom and experience.
4 To evaluate to what extent a task can be completed by several solution proposals, as well as evaluate their advantages and disadvantages.
5 To select the best solution within the conceivable range by repeating this procedure.

People who skillfully follow this procedure are considered "capable". Is such capability inherent? It does not seem so. It should be considered that people have the potential for creative ability. People are frequently occupied by routine work, which is not essential, and this prevents them from giving full play to their abilities. If people are able to work in the type of office environment described here, they can demonstrate their creative powers to the fullest extent of their abilities.

2.3 Basic physical structure of the C&C office system

In general, the office system can contain various subsystems, which can be divided functionally into three composite elements and understood as follows:

(a) Workstations (WS's)
(b) Local area networks (LANs)
(C) Office processors (OPs)

WS's are various desk-top information handling equipment, characterized by the existence of at least one type of human-interface. Ideally, WS's should be compact and easy-to-use. WS's should be connected to a LAN or Wide Area Network (WAN), because of the necessity of business communications among people, or smooth information transmission among people (particularly people engaged in intellectual production activities) at work in the office. Furthermore, office workers will not be satisfied with the limited functions available through stand-alone type WS's. Sooner or later, their requirements (desire) escalate. Therefore, a back-up arrangement by means of Office-Processor should be established for WS's, for common use of document processing, information storage, or other functions which are common in the work-group. LAN and WAN are regarded as the connection means for those Office Processors and WS's with human-interface.

2.4 Functional structure of C&C office system

The essential function of an office system is to effectively support office work in terms of intellectual communication among human beings. This support is carried out by acting as the alternative of direct and mutual communication among office people by the following two subfunction stages:

1 Communications between humans and systems
2 Communications between systems

There exists functional layer structures behind the human interface of the Office System. The four layers inside the work stations and office processors and the additional network layers are as follows:

(1) Application layer

This is a software layer to realize concrete office work support functions (e.g., office job processing, document processing etc.), and is subdivided into two sublayers:

i) Office Application Sublayer, and
ii) Office Common software Sublayer

(2) Basic software layer

Various basic functions, such as basic file management, data communications control, input/output control, arithmetic calculation, language processing, etc., are shared by this layer.

(3) Hardware layer

This layer includes various types of equipment comprising the office system (CPU, terminal, personal computer, facsimile terminal, etc.) and their functional modules.

(4) Electronic device layer

Basic devices and components comprising equipment and functional modules (general-purpose or custom LSI circuit, components, etc.)

The "thickness" of these layers depends on the system. Basically, human activities can be supported by the logical coupling of these four layers.

In the integrated office systems shown, various office information generated as a result of human activities is smoothly processed and transferred through each hierarchical level to be absorbed and utilized at the final destination, i.e. human activities on the other side.

3 ARCHITECTURE

3.1 Models of office systems

As mentioned earlier, the basic configuration of the office system comprises WS's (Workstations), Networking System's such as LANs (Local Area Networks) and WAN (Wide Area Network), and OPs (Office Processors). Real integrated office systems are a complex hierarchical configuration combining WS's, LANs (WANs), and OPs to suit an enterprise's scale and system purposes. Actual combinations of such equipment is the so-called (static) Physical configuration of the system, and the combination alone cannot work as a system. Dynamic characteristics must be specified for the (static) physical configuration to permit the combination to work as a system. The dynamic characteristics of an office system are office information handling characteristics, which show the dynamic properties of office information being generated, processed, transferred, stored, retrieved, and utilized through and over the (static) physical configuration of the office system. Thus, the actual model of the office system can be understood through the physical configuration and (dynamic) office information handling characteristics.

As mentioned in the previous section, and office system functional structure was extracted independent of the physical configuration of the office system while focusing on its functions. Office information must flow smoothly throughout each functional layer of the functional configuration to allow an office system to operate as a "modern communication" system. The standard information handling system for office information that flows throughout the office system functional layer interfaces can be called Office System Architecture. Unless office system architecture is established, office information cannot flow freely, and the integrated office system cannot exist. Table 1 shows the overall relationship of the actual and functional models of the office system, as well as static configuration and dynamic characteristics.

TABLE 1 Structure of Office System

	Static Configuration	Dynamic Characteristic
Real Model of Office System	<u>Physical configuration</u> Complex of various OA equipment (WS's, LANs, WANs and OPs)	<u>Office Information Handling Characteristics</u> Actual situation of office information being generated, processed, transferred, stored, retrieved, and utilized through the office system physical configuration.
Functional Model of Office System	<u>Functional Configuration</u> Application Layer: OA Application Layer / OA Common Software Layer Basic Software Layer [Network Layer] Hardware Layer Electronic Device Layer	<u>Office System Architecture</u> System of standards that enable office information to flow smoothly through functional configuration of office system. HIA : Human Interface Architecture OIA : Office Information Architecture DINA: Distributed Information Network Architecture CA : Computer Architecture (CMA; Communication Architecture)

3.2 Architectures that comprise office system

The concept of architecture in systems was first introduced in the 1960s as Computer Architecture (CA). This architecture was a system to regulate information exchange between a basic software layer and a hardware layer. On the other hand, in the area of conventional communications, compatibility had been its essential nature, and Communication Architecture (CMA) as a traditional system of standards has been established based on many years of experience. In the 1970s, distributed data processing through communication channels became important, and NEC has been proposing the Distributed Information Network Architecture (DINA).

The architecture mentioned above is still insufficient and new Office Information Architecture (OIA) has to be established to allow the smooth flow of a wider variety of office information (numerical data, text, graphics, voice, pictures, etc.) that is handled in offices. This should be a standard system for smooth information circulation between the OA common software layer and basic software layer. It comprises Information Interchange Architecture (IIA), a protocol to smoothly transfer office information, and Information Content Architecture (ICA) to formulate the construction of office information to be transferred. Office information, which is a complex combination of multi-media such as numerical data, text, graphics, pictures, and voice can be handed over to the Distributed Information Network Architecture (DINA), and flow in the system smoothly, only after it conforms to this OIA (IIA/ICA).

Considering that the purpose of the integrated office system is to strengthen the tie between the intellectual activities of office workers and the goals of the organization, the last but most important item is nothing but information circulation among people (minds, or intellectual activities) and the office system. The system of standards to regulate this can be called the Human Interface Architecture (HIA). This architecture concerns the relationship between the total human activities present within the system and the system itself, and contains environmental and operational standards, mechanical or psychological operation standards, code systems, and other items relevant to the keyboard, character pattern, icons, images, voice, etc.

4 HUMAN INTERFACE ARCHITECTURE (HIA)

4.1 Defining HIA

In the above office system architecture, the Network Architecture and Office Information Architecture determine the mechanism of information exchange among the "machines" making up an office system and the representational format of the information to be exchanged. The Human Interface (HI), however, is defined as an interface for information exchange between human beings and machines or among human beings. Accordingly, the

Human Interface Architecture (HIA) is the systematization of a method of performing this information exchange. For simplicity, this section discusses interface between human beings and machines only.

Under the above conditions, the HIA is defined as "The systematization of the way of transferring information between an office system and human beings that required for problem-solving or business activities." Figure 1 shows the scope of HIA.

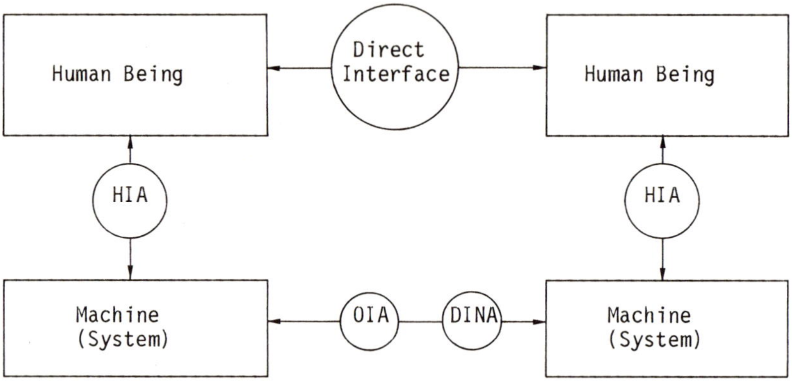

Fig. 1 Scope of HIA

4.2 HI features

The degree to which an interface is appropriate for human beings may generally be assessed according to the following criteria:

(1) User-friendliness (mainly for information input) and understandability (mainly for output)
(2) Efficiency
(3) Operability
(4) Learnability (ease of learning, effectiveness of training, etc.)

In practice, the problem becomes more complex when emotional (personal likes/dislikes, etc.) factors are taken into consideration. Even when restricted to these four points, an interface must consider individual differences. This is particularly true in the case of points (1) and (2) which are mutually conflicting. A veteran user would request an interface stressing efficiency at the expense of user-friendliness, while a novice would require a friendly and understandable interface which sometimes sacrifices efficiency. It is also necessary to permit a gradual shift toward a more efficient interface as the user gains experience. Figure 2 shows some examples of interfaces for different target users.

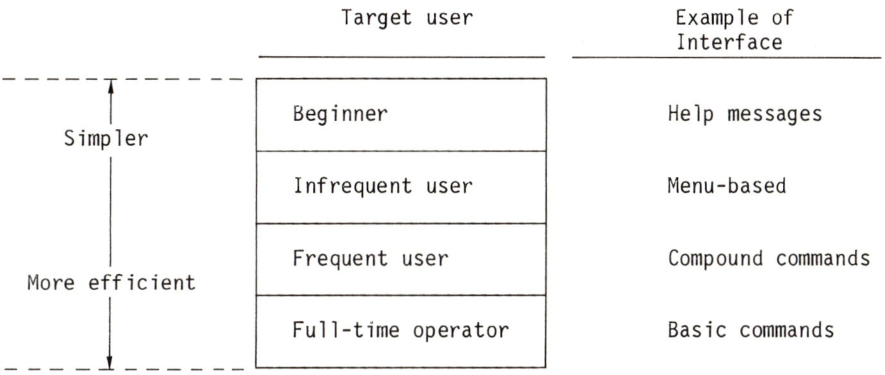

Fig. 2 Examples of Interfaces for Different users

4.3 HI function levels

The function levels required for data transfer in a format appropriate for the users are summarized in the Figure 3. This transfer of information is performed after conversion of the necessary raw information, taking point (1) through (4) in the preceding section into consideration.

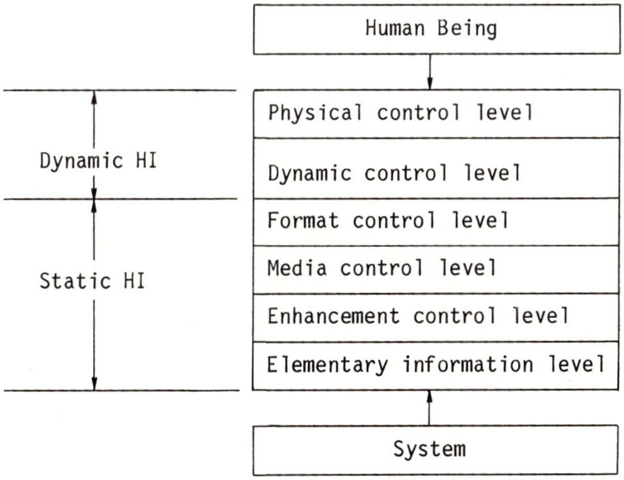

Fig. 3 Function Levels of a Human Interface in an Office System

The function of each level is summarized as follows:

(i) Elementary information level. This is the level of information which is the required minimum to be transferred between machines.

(ii) Enhancement control level. Elementary information per se is usually difficult to understand for human beings. Therefore explanatory statements, supplementary information and information that facilitates operation must be added to the elementary information for a viable interface with human beings. The extent to which these augmentations of elementary information are made differs for the various types of users. The enhancement control level adds this supplementary information to elementary information for easy operation and understanding, or bilaterally removes such augmented information. This level mainly controls the different levels of efficiency needed for different users.

(iii) Media Control Level. The media control level implements the information handled by the enhancement control level through the optimum combination of multi-media in terms of points (1) through (4) in Section 4.2. It also converts media as required. As multi-media, table formats, sentences, text, graphs, images, and voice are used. The significance of multi-media in Human Interface will be discussed later. (See Section 6.)
Figure 4 indicates characteristics of various media along with expressiveness, understandability, efficiency, and distributability.

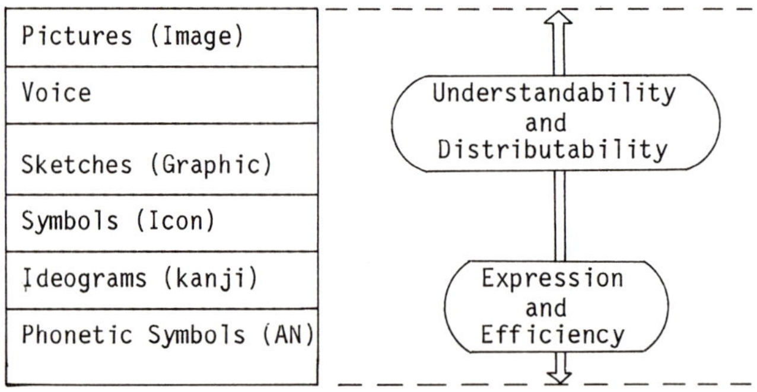

Fig. 4. An Illustration of Media Characteristics.

(iv) Format control level. The format control level controls the representational format in each medium. It handles, for example, character size, color selection, layout, image accuracy, and expansion and contraction.

(v) Dynamic control level. The levels listed above control the static representational format of information. In contrast, the dynamic control level controls the informational transfer procedure between human beings and machines. This level controls time-related factors, such as conversation, display and voice output speed and response time whose importance is discussed in the next section.

(vi) Physical control level. The physical control level controls the hardware that carries out information transfer. Specifically, it controls the physical interfaces, including keyboard, mouse, and CRT display control.

5 RESPONSE TIME GUIDELINE IN HIA

Time is money. It is of great value. When unexpected delays impede the progress of a task, many people become frustrated, irritated, and finally angry. A too long response time leads to more frequent errors and lower satisfaction. On the other hand, a too quick response time makes people less comprehending, and wrong decisions are made.

The response time in computers and communications (C&C) systems is the number of seconds which it takes from the moment a user initiates an activity until the C&C system begins to present results on the screen or printer. The user initiates, waits, for the system to respond, watches the results appear, thinks for a while, and initiates again.

How long will users wait for the C&C system to respond before they become frustrated? There are related design situations that may clarify the question of acceptable response time.

Therefore, we propose a response time guideline as a part of HIA. We introduce two types of response time. T_h and T_c.

T_h: Human beings response time.
T_c: C&C System response time.

T_h is the time for human beings to analyze the display and understand the situation to initiate an activity. T_c is the time for the C&C system to process given information task. The real system has the following three types:

(a) $T_h >> T_c$
As the C&C system response time is far beyond the human thinking or cognition time, human beings, comprehension and satisfaction deteriorate.

(b) $T_h > T_c$
If the information is processed more quickly than or the same as expected, people will be happy.

(c) $T_h < T_c$
If the information is processed much more slowly than expected, users will become frustrated.

If a system satisfies case (2), it is said to be a good system. Otherwise, many efforts have to be made to reduce the computer response time or find an appropriate time.

Although the human beings response time T_h will be changed on past experiences, abilities or adaptiveness of individual users, it is classified into four categories.

Class XA: Extremely simple case of class A

Class A : User can predict the result completely and proceed to the next step without confirming the result.

Class B : User can predict the greatest part of the result, but not completely, and has to spare time to confirm details.

Class C : User cannot predict the main part of the computer result, and has to spend time to understand the result.

Class A is the so-called "rapid reaction task". People always prefer a rapid interaction, if the task is simple and easy. The response time of "0.2 seconds" is set, since the blinking of eyes takes this time. Class XA is the extremely simple case of class A such as Echo Back, and its response time is set to "0.1 second".

Class B is the so-called "smooth reaction task". Many tasks belong to this class. The response time of "2 seconds" is set, since the human brain's short memory can temporarily hold chunks of information for 2 seconds. In many experiments a "2 second response time" is shown to be an acceptable time for people to wait on the display terminal.

Class C is the so-called "slow reaction task". Lengthy response times are generally detrimental to productivity, increasing error rates and decreasing satisfaction. The response time of "20 seconds" is set, since the error rates increase a rapidly from a 20 seconds response time in experiments.

The system engineers who are going to build a specific system with human machine interaction have to consider to which class their tasks belong. Once they decide the task class, they have to design the system to satisfy the system response time. The C&C office systems are a complex combination of workstations, networks and various types of processors, and tasks to be processed are also mixtures of each class type. Therefore, in designing a C&C system, information handling and Information storage capabilities of each systems component such as Workstation, Network and Processor should be allocated so that most of the tasks may satisfy the response time guideline of corresponding class types.

Although, the real system must be designed to take into account practical constraints such as technical feasibility, costs, speed of task performance or error rates, these classes and respective response time values may provide the appropriate guiding principle to C&C system designers. Table 2 shows detail descriptions of each class.

TABLE 2 Response Time Guideline of Each Class

Class	Response Time	User Expectation	Data Handling Complexities	Examples
XA	0.1	Extremely Easy		Echo BAck
A	0.2	Easy to predict	Less (1)	Switch on/off Character editing
B	2.0	Hard to predict	Little (10)	Graphic Editing (Draw; Redraw, Move, Rotate) Information Retrieval
C	20.0	Nor predictable	Huge (100)	Numerical Computation Simulation

6 SIGNIFICANCE OF MULTI-MEDIA SPECTRUM IN HIA

Future Human Interface Architecture should provide measures for presentation, processing and understanding of more intellectual information such as a thought or idea, which is created by the human brain, through a multi-media environment. To meet such requirements, further study on the property of information and multi-media that carry such information is necessary. In this section, a model of the multi-media communication transferring intellectual information among human beings through the C&C system is proposed, which may be a useful foundation for multi-media human interface.

When one's idea is created, it may be assumed that this idea has its own native Idea Expression Spectrum (IES), amount and range of information necessary to express the idea.

The Idea Expression Spectrum of an idea is difficult to decompose into a set of small pieces of information having different attribute values, and to observe directly from the outside. However, there are several ways of expressing a part of the Idea Expression Spectrum, in what we call multi-media, such as data, text, voice, graph, image, gesture, etc.

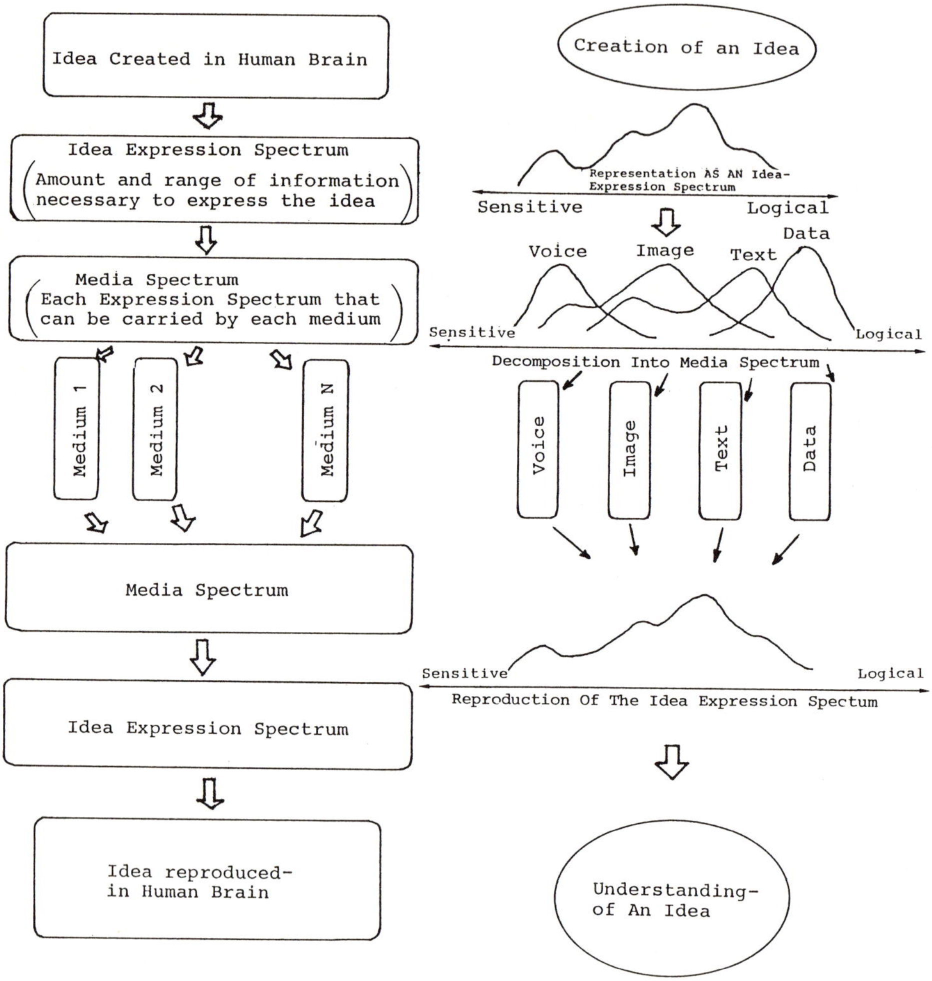

Fig. 5 Multi-Media Communication Model, through which a created idea is transferred

This situation is analogous to the color of an object, where reflecting light from the object is composed of certain spectrum of light-wave-lengths. One can recognize this object through such light-spectrum. However, it is not necessary to recognize the object using the full range of light-spectrum. Instead the object can be recognized almost exactly through three primary waves, i.e. red, green and blue.

Unlike light-spectra, there are no primary waves representing the original Idea Expression Spectrum, though one can define a part of the Idea Expression Spectrum expressed by a medium as a Media spectrum. It is clear that every medium has its own native limitation of expression capability and that no single medium can express all kinds of information. So in many cases the original Idea Expression Spectrum can be approximately represented approximately by the proper combination of several media spectra. Therefore, it is quite natural to use multi-media when one wants to communicate his idea more exactly.

Each media spectrum is transferred, processed and manipulated through the C&C system, then those media spectra are recombined into an expression spectrum, which approximates the original Idea Expression Spectrum.

In the recombination phase, coherence of those media spectra is very important. If those spectra are out of phase, the entire figure obtained from them becomes hard to grasp. They should be integrated so that the original idea is clearly presented. Thus the decomposition and the integration technology of multi-media spectrum with coherence is the key to good use of multi-media.

In Figure 5 an idea in expressed as a spectrum, which is decomposed into several media spectra, and transmitted to the destination. Received media spectra are combined and they reproduce a spectrum, which approximates the original idea.

7 CONCLUSION

An Office Automation system is "Modern Communications (computers and communications system)" realized in an office, that supports the intellectual and creative activities of the people working there. We call it a C&C office system. To construct a useful C&C office system, it is essential to establish Office System Architecture (OSA), which consists of Communication Architecture (CMA), Computer Architecture (CA), Distributed Information Network Architecture (DINA), Office Information Architecture (OIA) and HUman Interface Architecture (HIA). Of those architectures, the importance of human Interface Architecture (HIA)) was emphasized and the structure of HIA was presented. For the dynamic characteristics of human Interface Architecture, the response time between human beings and the C&C system is one of the fundamentals and the guiding principle. The response time guideline for each class of task was proposed and its significance for C&C system design was also discussed. In present and future human-C&C system interface, Multi-media interface and communication are the most important and essential areas. In this connection, a multi-media communication model and the Multi-Media Spectrum concept were proposed. Using the spectrum concept, a thought (idea) is expressed as an Idea Expression Spectrum and is decomposed into a set of several Media Spectra. Each Media Spectrum is manipulated through a proper C&C system, then those media-spectra are recombined into compound spectrum, which represents an approximation of the original idea Expression Spectrum. Although the proposed model of the multi-media interface concept is still premature, the concept of multi-media spectrum may be useful for human interface study in a multi-media environment of the C&C system.

Should the considerations about human interface stated in this article be a little useful for further study of the interface between human beings and C&C system, the author would be greatly delighted.

Finally, the author would like to express his thanks to Mr. A. Itoh, Mr. K. Niwa and Dr. S. Gotoh of NEC Corporation for their valuable discussions and support.

Social, Ergonomic and Stress Aspects of Work with Computers,
edited by G. Salvendy, S.L. Sauter and J.J. Hurrell, Jr.
Elsevier Science Publishers B.V., Amsterdam, 1987 — Printed in The Netherlands

CHANGING ROLES IN AN ELECTRONIC INDUSTRY - ENGINEERS USING CAD SYSTEM AND SECRETARIES USING WORD-PROCESSING SYSTEM

GUNILLA BRADLEY

Department of Sociology, Stockholm University, Stockholm, Sweden

1 INTRODUCTION

Research on computerization and the psychosocial work environment has been performed at the University of Stockholm during 1974-1986. The research group comprised researchers in psychology, sociology, economy, and computer science. The research was financed by the Swedish Work Environment Fund. It comprised two phases, RAM:1 and RAM:2.

This article describes part of RAM:2. It started 1982 and ended 1986. It was an extension of the RAM:1 project (summarized in Bradley 1986a). The aim of RAM:2 was to describe and analyse the effects of micro-computerization on the working environment. Two categories of personnel were studied at the firm of Philips Electronic Industries Ltd: Female secretaries using a word-processing system and male and female engineers using CAD-CAM systems.

We analysed the work, its characteristics, and how it is organized. We studied three related problem areas:

1) Data-processing and communication structure at work sites
2) Data-processing, leadership, and collaboration
3) Interaction between:
   - Data-processing - Working conditions - Health
   - Data-processing - Working conditions - Life styles outside place of work.

2 SUBJECTS AND METHODS

All subjects were employed at Philips Electronic Industries Ltd, situated in the Stockholm area. Secretaries: All the secretaries at a major technical department. Engineers: 1) Graduate engineers using CAD systems for designing integrated circuits (chips). 2) Circuit technicians using CAD systems to prepare each of the drawings needed for the various stages of making chips.

Data were collected using semi-structured interviews, direct observations and questionnaires.

We cooperated with local reference groups that included representatives from management, local trade unions, the personnel department and the company health service. The results were discussed among the secretaries and the engineers respectively, with people from the product marketing organization and with other key company staff.

3 RESULTS

For the complete results of the two studies presented under 3.1 and 3.2, see list of references.

3.1 The secretarial role and word-processing

Changes in the objective job content. 40% of the women stated that they had to write more. For example, massive documents were not being sent to external typewriting bureaus as much as before. 60% stated that the editing work had increased. Other tasks had also increased e.g. booking travel and hotels, arranging meetings and conferences, and preparing tables and pictures. The informal network task of "co-ordinating and unifying" had increased for some secretaries, and decreased for others.

Most of the secretaries would not like to work without word-processing (compare viewpoint of engineers).

That the writing job has increased may be commented on as follows. The computer boosts efficiency. It is good for the employees and the company that human resources are wisely used. Unfortunately, there is often no time left for personal contacts with colleagues or customers. Nor is there time for further training to cope with the new technique.

Instead, the workload of the individual increases, e.g. the number of items to be processed per unit of time. Our own ambitions and our demands on ourselves also increase. Better planning of the organization can help to reduce these problems.

The work tasks must be distributed among various professions at various levels at the same time in the history of the company. Then it becomes possible to reshape job functions. Today, the latter are often characterized by one-sidedness. This holds true for chiefs and other top administrators as well as for persons with more routine tasks. Up to now, computerization has mainly been introduced at lower levels of personnel.

There is confusion about job description. Behind the scenes, people talk about changes in the job descriptions for employees using word processors - Either they are "up-graded to terminal workers" or "down-graded to clerks". Such words also reveal problems about the speaker's own professional identity. This problem is often rooted in the history of office automation.

The confusion in the nomenclature of positions is also related to another habit. We tend to look upon the organization as static, not as a process of continuous development.

Here are six major ways of redesigning work roles during computerization (hence avoiding the danger of routine work tasks):

1) Adding other tasks to the "production chain", tasks now existing at the *same* level
2) Adding tasks of planning and controlling, now found at *another* level
3) Analyzing *whether* computerization is suitable (ethics, quality, etc)
4) Analyzing *which* tasks ought to be done by machine or by man
5) Revising the management role
6) Identifying and clarifying the *goal* of the work.

Changes in the subjective work content. Referring to psychosocial aspects of the *job content*, most secretaries report that this situation was unchanged. About half of them (52%) stated that the stimulation and excitement in the work had increased; 44% that the variety of the tasks had also increased. 40% said they could use more of their own ideas and knowledge.

Secretaries now had more *influence* on their own work as to working pace, planning and choice of method. This demonstrates the advantage of using micro-computers in word-processing, thus avoiding a central computer. Micro-computers also mean that employees are not dependent upon telecommunication services at each instant. This dependence may be a short chapter in the history of computer technology. Furthermore, there exist other strategies for introducing computers.

Working pace and dependence on machines. 64% stated that demands for attention and accuracy had risen after the use of word-processing. 60% stated that working pace now depended on the computer and its equipment.

In actual practice, computerization is being used as a tool for rationalization. Intensification of working pace and increased dependence on machines result. Greater external demands for performance are made possible by the new technology. This easily leads to internal demands which we impose on ourselves in our working as well as private lives. Our *relation to time* is one of the key questions for the future.

The women became anxious about the *development* of their own *professional role*, and the risk of becoming replaceable. On the other hand, there was a belief in the future. They had notions about the positive possibilities of developing their own occupational group. Women talked about how to regain their previous versatility in the profession, whether called operator, assistant, or clerk. Most of our subjects emphasized their interest for working in areas such as budgeting, planning, economic and personnel administration. They were very positive towards further training during paid hours of work.

Obstacles to personal contact. 44% mentioned psychosocial obstacles:

* Different professional groups have different notions of speech and status
* Workload and stress
* Personal factors linked to the work situation.

Physical confinement to the word processor was considered the main obstacle to personal contacts.

Discussions among the employees had become more frequent and more technical. The language gap that exists between secretaries and technicians or managers was tending to change (partly due to their educational backgrounds). Secretaries were now meeting to talk about their problems. The use of word-processing systems had become a direct reason for keeping in touch with one another. *Informal contact networks* were expanding as microcomputerization extended. There are many forms of collective action for women and men. These forms vary from country to country according to the degree of emancipation among women and men.

The professional role. Most secretaries were working with a group of engineers. 52% of the secretaries thought that their own professional role had changed with word-processing. No change had occurrred in the role of management (72%) - nor in the teamwork and cooperation between secretaries and management (76%).

Private life and leisure. Most of the usual leisure activities had changed in connection with the new technology. These effects were not found in the companies studied in RAM:1.

RAM:2 showed increased collective action, expressed in the form of secreterial meetings not related to union activities. Female employees tend to have a low interest in union organization, due to many causes. The interest for courses of study had clearly increased - an important condition for a positive development of the professional role in connection with the new technology. Questioned about what they wanted to achieve during this time away from paid work, 60% of the secretaries listed studies, trips, and cultural activities.

3.2 The engineer's role and CAD

We studied two types of personnel - engineers and technicians. Engineers work out the design of new integrated circuits. They have at least four or five years of academic studies and are graduate engineers. In contrast, the actual work of detailed draughtmanship is done by circuit design technicians. They have attended secondary technical school followed by several months of specialized training.

Technicians felt they had to have CAD in order to cope with the complexity of the new circuits and the demands for accuracy and faster draughtmanship. Engineers felt that CAD gave them better possibilities of simulating new designs and making creative analyses.

Company supervisors noticed that there was less fluctuation in pace of work. Company demands had increased as to accuracy, attentiveness and efficiency.

We were able to study the views of staff that had been using CAD various length of time (years). As time went on, they found that the work became less stimulating and more monotonous. Their work tended to fall in prestige, and they became more and more dependent on the computer and the CAD system.

None of the engineers wished to go back to the old days without CAD. 75% of them said that they could use their previous experience when working with CAD. Most of them also felt they could develop themselves to keep up with the advance of technology. The general attitude was still one of optimism.

Advantages of CAD. There was a wide range of answers. Here are the most frequent:

* Variety of tasks
* Tasks performed faster
* Easier to survey whole situation (better overview)
* Exalted tasks, "on a higher level"
* Higher product quality
* Less need to check details
* Tidier, more attractive design
* More accurate

Subjective experience of work with CAD - positive changes. Through our questions, we tried to distinguish between the results due to CAD and the results of other changes in company organization. Our subjects quoted the following improvements in their work.

* More influence on own work content
* More influence on pace of own work
* More influence over scheduling of work breaks
* More stimulation and challenge in tasks
* More responsibility

Disadvantages of CAD

* Problems of physical environment at work
* Great dependence on computer
* Disadvantages inherent in CAD system
* Tendency to isolation from colleagues

Subjective experience of work with CAD - negative changes

* Work pace dependent on computer and system
* Uneven pace of work
* Demand for attention and accuracy
* Company demand for efficiency

Engineers often replied that CAD was not suitable for creative thinking and problem-solving. They often maintained: "It is I who must solve the problem." Circuit technicians, however, wanted automation of parts of their duties. They also emphasize personal contacts with colleagues when planning the work, not merely contacts via computers.

Stress. The great dependence on computers was often mentioned as a major stress factor. The syndrome "taking the job home with you" was considered as frustrating and exhausting.

Much of the stress among the engineers was linked to technical problems, response time (too much waiting) and faulty operation and breakdowns.

We evaluated over-stimulation and under-stimulation. A special set of questions was used. Subjects replied:

* Too large amount of work
* Too little daily face-to-face contacts with co-workers
* Too conflicting demands as to professional role
* Too tied to place of work

Health. Surprisingly many subjects mentioned:

* Eye troubles
* Backache
* Headache
* General tiredness

Our study showed that male engineers and circuits technicians have started to suffer from the same symptoms as female terminal operators. And they expressed their discomfort in our interviews. However, they were not visiting health care services to the same extent as women to talk about their complaints, This agreed with the traditional behaviour connected to the male sex role.

A parallel can be drawn with the insurance company we studied in 1977. During the summer of 1985, we did follow-up interviews. We asked key staff members with a good grasp of the situation. They worked at different levels in the company, and held a range of different tasks. They told us the following.

The staff of the Systems Department had become highly dependent on their computer terminals. Improvements in company organization had been made elsewhere in the company, but not in the Systems Department. Redesign of work and finding parallel work tasks without the use of display terminals seemed to be more difficult among the computer people. Interactive programming had become a major work task.

The Systems Department was now suffering from the problems that had previously afflicted other departments, e.g. Customer Service, which had longer experience of intensive work with display terminals.

Private life and leisure. The engineers using the new technology tended to emphasize the instrumental role of working life. Leisure should contribute the emotional component. Being together with the family played an important role. This is contradictory to older research results that showed engineers had a very clear intrinsic motivation. A review of the theories about the relationship between work and leisure, together with our empirical results, led us to formulate some hypotheses:

* Life styles are adapting to the changing content of working life.
* Computerization in working life tends to diminish computerization tied to private life.
* Computerization in working life tends to strengthen the importance given to the emotional and reproductive functions of private life.

* Computerization in working life calls for continuous education and training. This leads to a demand for an active leisure with possibilities for growth and self-realization.

4 CONCLUSIONS

We derived various action strategies from the results of the two studies at Philips. They refer to the "Swedish action tree" (Bradley 1986b).

* Organization of tasks (6 strategies)
* Technology itself (e.g. response time)
* Training to prevent redundancy of circuit design technicians and other staff with relatively short training
* Training secretaries in economic and personal administration
* Training in problem-solving and creative thinking
* Training in self-confidence
* Dealing with stress: prevention and cure

As a result of our research at Philips, a decision was taken by company management in autumn 1986. There will now be a continuous investigation of the impact of computerization. Discussions on actions of prevention and cure will also be held.

Time spent at the terminal must be reduced. One important solution lies in reorganizing jobs and job functions. The major strategies of designing/re-designing work roles during computerization are mentioned in section 3.1. These strategies are also applicable for engineers and circuit technicians.

The high pace is striking. A second seems to shrink to a picosecond when one enters the Silicon Valley of Sweden. The *perception of time* is a key question. We must find strategies at three levels (society, company, and individual) to cope with this phenomenon in a computerized society.

REFERENCES

Bradley, G., Bergström, C., Sundberg, L. (1984), *Sekreterarrollen och ord- och textbehandling - kontorsyrke i förändring* (The Secretarial Role and Word-Processing - Office Profession in Transition). 221 pp. in Swedish.

Bradley, G. (1986a), *"Psykosocial arbetsmiljö och datorer"* (Psychosocial Work Environment and Computers), Akademilitteratur, Stockholm 1985. 286 pp. in Swedish + 19 pp. summary in English. To be published in English in 1988 by Taylor & Francis.

Bradley, G. (1986b). *Visual Display Terminals and Psychosocial Work Environment* (in B. Knave & P.-G. Widebäck, *Work With Display Units 86*. Selected Papers from the International Conference, Stockholm, Sweden, 12-15 May 1986.

Bradley, G., Bergström, C., Lindeberg, S. (1987). *Ingenjörsrollen och CAD-system* (The Engineer's Role and CAD System). In manuscript. 246 pp. in Swedish.

Social, Ergonomic and Stress Aspects of Work with Computers,
edited by G. Salvendy, S.L. Sauter and J.J. Hurrell, Jr.
Elsevier Science Publishers B.V., Amsterdam, 1987 — Printed in The Netherlands

THE INTERACTIVE COMPUTER AS A SOCIAL STIMULUS IN HUMAN-COMPUTER INTERACTIONS*

LEO R. QUINTANAR, PH.D.
System Architecture Department, AT&T Information Systems,
11900 N. Pecos Street, Denver, Colorado 80234 USA

CHARLES R. CROWELL, PH.D.
Department of Psychology, University of Notre Dame,
Notre Dame, Indiana 46556 USA

PATRICK J. MOSKAL, PH.D.
Human Factors Division, Naval Training Systems Center,
Orlando, Florida 32813 USA

The theory that interactive computers can be social stimuli in human-computer interactions is developed. A series of our experiments are reviewed that identify (i) the ingredients of user/computer dialogs that induce users' perception of a dialog as human-like, (ii) the differences between experienced and naive computer users in their perceptions and preferences towards different dialog styles, and (iii) the differences in social psychological processes under human- and machine-like (nonhuman) user/computer dialogs. Results from these experiments suggest that user/computer dialogs are judged along consistent social dimensions, and that perceived causation and social facilitation processes do vary with dialog style. Implications of these findings for social-psychological views of human-computer interaction are discussed.

1. INTRODUCTION

A largely unexplored area of human-computer interaction research has been the theoretical consideration of the interactive computer as a social stimulus (Kiesler, Siegel, & McGuire, 1984), i.e., a stimulus that elicits social responses from the user by simulating or implying the presence of another person (Allport, 1968). As such a stimulus, an interactive computer may influence the user's attitude and behavior in ways predictable from social psychological theory. By this view, the degree to which social processes in the user are elicited by the computer would likely vary as a function of the perceived "humanness" of the user/computer dialog. To better understand this possibility, we examined the implications for human-computer interaction of causal attribution and social facilitation theories.

Attribution theory is normally defined as "...the study of perceived causality....attribution referring to the perception or inference of cause" (Kelley & Michela, 1980). An application of this theory is suggested by Heider's (1944) view that persons are more likely to be perceived as causal agents than nonpersons, and the observation that human-like behavior by interactive computers seems associated with a loss in users' sense of control, self-responsibility, and self-motivation (Shneiderman, 1980; Smith, 1980).

*Portions of this research served as the first author's doctoral dissertation (Quintanar, 1982) in partial completion of requirements for the Ph.D. degree at the University of Notre Dame, August, 1982. Experiment 2 was actually performed last, but is presented earlier here for clarity. This paper was supported in part by P. J. Moskal's appointment to the research participation program at the Naval Training Systems Center (NTSC), administered by Oak Ridge Associated Universities through an agreement between the U.S. Dept of Energy and the NTSC. Requests for reprints should be sent to Charles R. Crowell.

Social facilitation theory (Geen & Gange, 1977) states that the presence of other persons produces an increment in general arousal, which in turn serves as a drive that energizes dominant responses and inhibits subordinate responses (usually measured in terms of response latency). Task difficulty is often used to categorize these responses under the assumption that performance on simple tasks engenders dominant responses whereas performance on difficult tasks elicits subordinate responses The application of this theory to human-computer interactions suggests that users may be more "socially facilitated" when interacting with a computer that uses a human-like than a machine-like dialog style.

We evaluated these social theories of human-computer interaction in three experiments.

2. EXPERIMENT 1

Our intent in this experiment was to identify those components of a human-computer dialog that users perceive as the most and least human-like.

2.1 Method

Sixteen undergraduate male volunteers served as subjects in a 2 (Affect) x 2 (Conversation) x 2 (Diversity) x 2 (Self-references) within-subjects design. The presentation order of experimental conditions was counterbalanced across subjects using a Latin-square technique. Subjects were randomly assigned to the levels of the Latin-square.

The four independent variables were selected on apriori grounds as generic components of a human-like dialog style and were manipulated such that the various characteristics were either present or absent in the computer's dialog with the user. These manipulations were operationalized as follows: (a) AFFECT: evaluative-emotional responses suggesting that the computer "understands" emotions; (b) CONVERSATION: well-linked flowing sentences with complete thoughts, informality, use of the subject's first name, acceptance of "yes/no" responses to questions rather than "1=yes/2=no"; (c) DIVERSITY: variation in the wording of prompts, responses, and pauses, and use of upper- and lowercase text; (d) SELF-REFERENCES: use of the pronouns "I" and "me" by the computer, and no reference to the computer as a machine.

The stimulus materials for each cell of the experimental design consisted of three video screens of dialog where the computer introduced itself as a teaching assistant. Each set of screens was designed to present the same conceptual information while varying only in the human-like attributes of the computer's dialog. IBM 3278 video display terminals (hardwired to the IBM 370/168 host computer) were used as interactive stations.

Subjects were informed that the experiment involved people's judgements about computers. Each subject participated in 16 different human-computer dialog styles. After each style, subjects rated characteristics of the computer on 24 bipolar adjective scales, and dialog understandability.

2.2 Results

In our analysis of the results, we sought to assess the underlying judgmental dimensions that subjects used while rating the different dialog styles, and the effects of our manipulations on subjects' perception of the computer dialog along these dimensions. A major goal was to discover which combination of the manipulations produced the highest and lowest "human-like" rating of dialog style.

To determine judgmental dimensions, a factor analysis was performed on subjects' ratings of the characteristics of the computer. The 24 bipolar adjective scales completed by the 16 subjects in each of the

TABLE 1

Experiments 1, 2, 3

Factor Analysis Upon the Twenty-Four Bipolar Scales of the Interactive Computer's Characteristics Combined over Manipulations of Dialog-Style

Bipolar Scales		Humanness			Incongruity		
		Exp 1 39.8%	Exp 2 36.8%	Exp 3 22.6%	Exp 1 21.5%	Exp 2 25.8%	Exp 3 15.6%
IMPERSONAL	- PERSONAL	-.86*	-.88*	-.76*	.03	.11	.06
HUMAN-LIKE	- NONHUMAN-LIKE	.89*	.89*	.77*	-.06	-.06	.08
UNINTELLIGENT	- INTELLIGENT	-.09	.02	-.55*	.65*	.73*	.23
MECHANISTIC	- NON-MECHANISTIC	-.83*	-.80*	-.42	-.12	.01	-.10
COLD	- WARM	-.91*	-.88*	-.80*	.11	.20	.23
EMOTIONAL	- UNEMOTIONAL	.84*	.87*	.67*	.08	.07	.07
UNCOMPREHENDING	- COMPREHENDING	-.26	-.26	-.34	.64*	.63*	.48
ALIVE	- DEAD	.86*	.87*	.56*	-.08	-.16	-.43
CONSCIOUS	- UNCONSCIOUS	.77*	.78*	.54*	-.29	-.23	-.39
CONSTRAINED	- FREE	-.82*	-.75*	-.47	.02	.18	.04
ILLOGICAL	- LOGICAL	.00	.14	.28	.81*	.80*	.40
NONSEXUAL	- SEXUAL	-.09	-.51*	-.54*	-.33	-.06	-.06
UNAWARE	- AWARE	-.59*	-.54*	-.54*	.48	.37	.28
SINCERE	- INSINCERE	.67*	.51*	.50*	-.37	-.56*	-.21
BAD	- GOOD	-.67*	-.48	-.46	.51*	.68*	.44
SELFISH	- UNSELFISH	-.55*	-.49	-.13	.26	.37	.13
IMPOLITE	- POLITE	-.57*	-.49	-.30	.46	.57*	.41
RATIONAL	- IRRATIONAL	-.03	-.02	-.05	-.71*	-.81*	-.46
SOCIABLE	- UNSOCIABLE	.89*	.86*	.33	-.06	-.19	.07
INOFFENSIVE	- OFFENSIVE	.45	.17	.09	-.59*	-.72*	-.69*
HONEST	- DISHONEST	.42	.22	-.18	-.60*	-.74*	-.62*
CREATIVE	- UNCREATIVE	.81*	.86*	.55*	-.08	-.09	.23
RELIABLE	- UNRELIABLE	.15	.05	-.14	-.81*	-.81*	-.79*
INACCURATE	- ACCURATE	.02	-.01	-.09	.80*	.81*	.78*

Note. A large positive factor loading indicates a high score towards the1st pole of the corresponding bipolar scale whereas a large negative loading implies a high score toward the 2nd pole. Loadings greater than or equal to absolute .50 are marked by asterisks. Beneath the label for each factor is the percentage of explained variance after varimax rotation.

16 cells of the experimental design were combined to form a 256 (Cases) x 24 (Ratings) matrix. This matrix was then subjected to a principal components analysis and followed by a varimax rotation (Kim & Mueller, 1978). As shown in Table 1, two factors appeared as the primary judgemental dimensions. Based on the scale loadings, these factors were labeled Humanness and Incongruity, respectively.

To assess the effects of our manipulations on subjects perceptions of the computer, separate 2 x 2 x 2 x 2 analyses of variance (ANOVAs) were performed on factor scores (Tucker, 1971) calculated from the Humanness and Incongruity factors. Results for the analysis of the Humanness factor score indicated an anomalous higher-order interaction but more importantly main effects for the affect, conversation, and self-reference manipulations such that the presence of each ingredient resulted in an increase in users' perceived humanness of the computer ($\underline{F}(1,15)=59.30$, $\underline{p} <.001$, $\underline{F}(1,15)=48.38$, $p <.001$, and $\underline{F}(1,15)=9.73$, $\underline{p} <.01$). Results for the Incongruity factor score showed that while there was some effect of our manipulations (e.g., affect and self-references increased perceived incongruity), nevertheless subjects did not regard any of the computer's dialog styles as excessively incongruous since the grand mean was -.67 and all cell values were negative.

To assess which combination of the manipulations produced the highest and lowest perceived humanness of the computer, a Newman-Keuls analysis was performed on the means of the affect x conversation x self-reference cells (collapsing over diversity) using the Humanness factor score. As shown in Table 2, when the ingredients of affect, conversation, and self-reference were all present, subjects perceived the computer to be more human than any other condition. When these ingredients were absent, subjects' perceived the computer to be less human than in any other condition.

TABLE 2

Experiment 1

Mean Humanness Factor Scores as a Function of Human-like Attributes of the Computer's Dialog

	Self-References	No Self-References
Affect		
Conversation	1.82^{a}	1.22^{b}
No Conversation	$.60^{cd}$	1.04^{cb}
No Affect		
Conversation	$.85^{bc}$	$.29^{d}$
No Conversation	$-.48^{e}$	-1.44^{f}

Note. Cell means with no common superscript differ from one another ($p < .05$) by a Newman-Keuls range test. As recommended by Winer (1971) for within-subjects experimental designs, separate error terms associated only with the relevant cells were used to evaluate each comparison of means.

An analysis showed that dialog understandability did not covary with perceived humanness of dialog styles, and that understandability was very high across all styles.

3. EXPERIMENT 2

This experiment investigated perception and preference differences between experienced and naive computer users (within a similar context as Experiment 1), and also sought to determine whether an improved Diversity manipulation increased perceived humanness of dialog.

3.1 Method

Thirty-two volunteers served as subjects in a 2 (Computer Experience) x 2 (Affect) x 2 (Conversation) x 2 (Diversity) x 2 (Self-references) mixed model design. The single between-subjects factor was users' level of experience (experienced/naive) with interactive computers. The within-subjects manipulation, stimulus materials, questionnaires, and procedure were identical to Experiment 1 with the following exceptions: the Diversity manipulation was enhanced for more variation of user input to the computer (e.g., "yeah", "ok", and "sure" could be used as synonyms for "yes"), and a user preference rating on each dialog style was added as a dependent measure.

3.2 Results

In our analysis of the results, we again examined the underlying judgmental dimensions that subjects used while rating the different dialog styles, and the effects of our manipulations on subjects' perception of the computer dialog along these dimensions. Major goals were to discover differences due to user's level of computer experience, and to establish whether our new Diversity manipulation increased perceived humanness of dialog.

A factor analysis was performed on subjects' ratings of the characteristics of the computer. The results replicated our findings for Experiment 1, i.e., the two primary judgemental dimensions that emerged were the Humanness and Incongruity factors. Interestingly, these judgemental dimensions appeared identical for both high and low computer experience groups. Table 1 shows these factors combined for the two groups.

ANOVAs on the factor scores also replicated our previous findings for Affect, Conversation, and Self-References. In addition, our new Diversity manipulation did impact perceived humanness ($\underline{F}(1,30)=7.46$, $\underline{p} < .01$). User's level of experience had no effect on perceived humanness of the dialog styles, and dialog understandability was again very high across all styles.

To assess differences in users' preference of dialog styles, a 2 x 2 x 2 x 2 x 2 ANOVA was performed. Results revealed a two-way interaction between Level of Experience and Self-References ($\underline{F}(1,30)=4.18$, $\underline{p} < .05$). A simple effects analysis indicated that experienced users preferred user/system dialogs <u>without</u> self-references over those with self-references whereas naive users generally liked both types equally. A two-way interaction was also found between Affect and Conversation ($\underline{F}(1,30)=4.34$, $\underline{p} < .05$) such that all users tended to prefer a conversational style over a non-conversational style under conditions without affect, but liked both types equally when affect was present. There was also a two-way interaction between Diversity and Self-References ($\underline{F}(1,30)=5.49$, $\underline{p} < .05$) such that users preferred no self-references over self-references when diversity was absent, but liked both types equally when diversity was present.

4. EXPERIMENT 3

This experiment investigated differences in users' social processes induced under human- and machine-like user/computer dialogs. Our central hypotheses were: (i) CAUSAL ATTRIBUTION - users who unexpectedly fail a quiz taken with a computer should blame the computer more (and less their own "lack of effort") under a human-like than a machine-like dialog style, and (ii) SOCIAL FACILITATION - users will be more aroused when interacting with a computer under a human-like than a machine-like dialog style, and thereby exhibit a "social facilitated" performance pattern characterized by relatively faster responses on low difficulty items and slower responses on high difficulty items.

4.1 <u>Method</u>

Forty-eight undergraduate volunteers served as subjects in a 2 x 2 x 2 x 3 mixed model factorial design. The between-subjects factors were Dialog Style (human-like/machine-like) and a control factor of Retake Knowledge (which had little effect and will no longer be discussed). The within-subjects factors were Quiz (initial/retake), and item difficulty (low/medium/high). Subjects were assigned randomly to the between-subjects conditions, and had not participated in our previous experiments.

The manipulation consisted of user/system dialogs designed to be either the most human-like or most machine-like based on the results of our earlier experiments (the initial screens were identical to those used before). Study materials were drawn from a general psychology textbook. A total of 60 quiz questions, distributed over two quizzes, were selected from a question bank pretested to establish item difficulty. Each quiz was constructed to tap the same concepts while maintaining the same overall level of difficulty. One-third of the items of each quiz were of low, medium, and high difficulty. Also, a procedure was used that induced subjects to freely choose not to study the material in a fashion similar to "forced noncompliance" (Darley & Cooper, 1972) in order to maximize the realism of their upcoming quiz failure.

The human-computer session consisted of (1) an introduction to the computer, (2) an initial quiz that inevitably led to failure, (3) a brief study period, (4) a retake quiz, and (5) a farewell message that concluded the session. Each quiz question was presented one at a time and subjects were given immediate feedback (by the computer) on the correctness of their answer. Dependent measures consisted of questionnaire responses during the session and behavioral measures collected by the computer.

4.2 Results

In our analysis, we first examined the effectiveness of our dialog style, failure, and item difficulty manipulations. Analyses of subjects' ratings on perceived humanness of dialog and quiz failure, and an analysis of scores on quiz items indicated that these manipulations were successful. Next, we again examined underlying judgmental dimensions of various ratings of dialog style characteristics by performing a factor analysis. As shown in Table 1, results again replicated Humanness and Incongruity factors.

To assess changes in failure attributions toward the "causes" of the computer and personal effort, ANOVAs were performed on subjects' attribution ratings. Results revealed a triple-order interaction betwee dialog style, attribution target, and quiz ($F(1,40)=4.86$, $p<.05$) as shown in Figure 1. Simple effects analyse and two-tailed t-tests were performed. Within the machine-like style, effort and computer attributions were significantly different from zero after the initial quiz ($t(24)=-3.49$, $p=<.001$ and $t(24)=2.20$, $p<.05$) but not after the retake quiz. Within the human-like style, only the computer attributions after the initial ($t(24)=1.98$, $p=.05$) and retake quiz ($t(24)=2.60$, $p<.05$) were different from zero. Apparently, subjects experiencing failure under a machine-like style blamed effort more and quickly lost their sense of the computer as a positive causal influence over time, as compared to subjects under a human-like style who blamed their lack of effort less and viewed the computer as a consistent positive influence over time. An analysis of subjects' thought-listings (a separate free response measure) also indicated that more internal and less external causes were listed after initial failure under the machine-like as compared to the human-like style ($F(1,40)=4.92$, $p<.05$ and $F(1,40)=5.96$, $p<.05$). Analyses on other measures indicated that the presumed consequences of those attributions for emotion, expectancy of future success, and motivation were also evident. Together these findings support our attribution hypothesis, although there was no evidence that subjects actually blamed the computer for their failure.

Our assessment of social arousal differences under different dialog styles began with ANOVAs performed on various self-report measures. As predicted, results indicated that subjects were more uncertain of the computer's next response and more distracted by the computer's behavior under the human-like as compared to the machine-like dialog style ($F(1,40)=4.27$, $p<.05$ and $F(1,40)=4.60$, $p<.05$). Next, we sought to assess the predicted social facilitation effect by examining the "think time" response latency measure collected during each quiz (the time spent reading and thinking about each question before answering). Separate ANOVAs were performed on logarithmically transformed think time measures (cf. Winer, 1971) for all items pooled together as well as for items scored as correct or incorrect. Although all

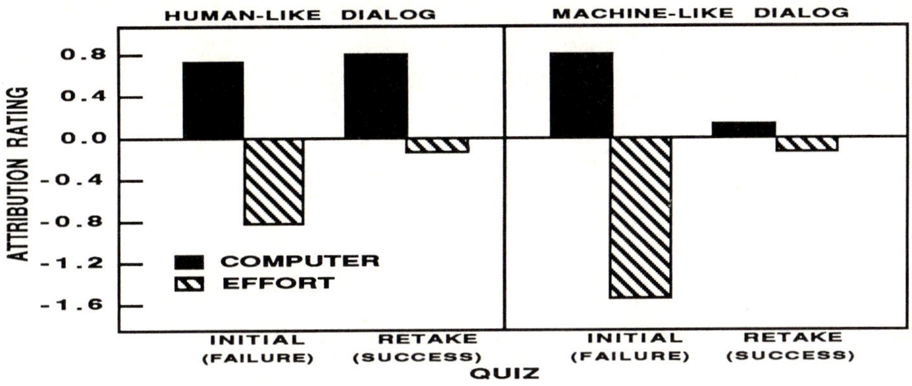

Fig. 1. Attribution ratings for initial and retake quiz as a function of dialog condition.

analyses were generally consistent, the best evidence for social facilitation effects was revealed in data for incorrect items. Results indicated a three-way interaction between dialog style, item difficulty, and quizzes ($F(2,56)=3.19$, $p < .05$). Since social facilitation processes were expected to be most isolated during the initial quiz (before attribution processes were instigated by failure), an analysis of simple effects was done separately for each quiz. Results revealed a two-way interaction of style and difficulty for the initial quiz ($F(2,70)=6.04$, $p < .01$), as shown in Figure 2, but not for the retake quiz. Further analyses on the two-way interaction indicated that the simple main effect of dialog style was significant for the low ($t(41)=1.68$, $p < .05$) and medium ($t(46)=1.77$, $p < .05$) but not high levels of difficulty. Together these findings provide evidence of social facilitation in human-computer interaction: Improved performance under the human dialog style for easy (low difficulty) items but degraded performance under this style for harder (medium difficulty) items.

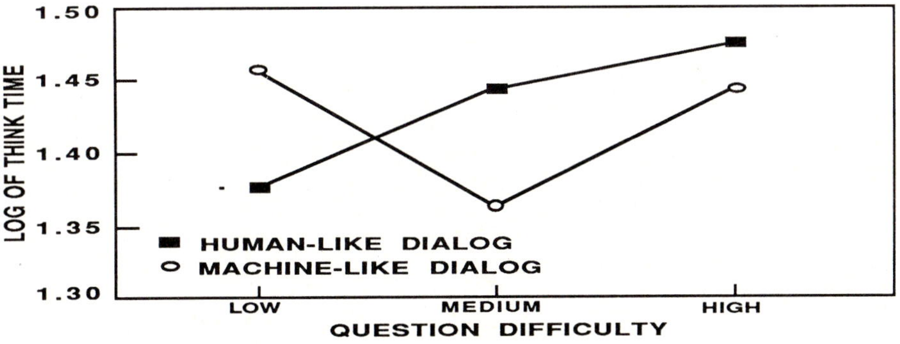

Fig. 2. Latency to respond as a function of dialog condition and item difficulty.

5. DISCUSSION

The conception that users react toward an interactive computer as a social stimulus to a degree that varies as a function of the humanness of the computer's dialog appears to be supported by our findings. It follows that the discipline of social psychology should be useful for the prediction and interpretation of psychological processes elicited in the user by interaction with a computer. In this respect, our results are consistent with Fulton's (1985) conclusion that "...the social psychological level of analysis....[and] the socialization of humans holds a key to at least part of the human-computer interaction problem."

Our findings also suggest that attribution and social facilitation theoretical domains can now be extended to cover relationships between humans and human-like entities. By applying social psychology to human-computer interaction, our research addresses Ramsey & Atwood's (1979) concern that more substantive theoretical perspectives are needed for research on human factors and ergonomic design issues. Our research also sheds light on Bork's (1980) concern about the unknown effects of adding human-like characteristics to the student-computer dialog in computer-aided education. The present findings also suggest directions for future research in the areas of dimensionality of users' perception of computers, active human-computer social processes, and factors influencing perceived humanness. As computers become ever-present in our society, the study of the social issues surrounding human-computer interaction will become paramount to the healthy and effective functioning of people in a rapidly robotic, software-driven, and technologically complex environment.

6. REFERENCES

Allport, G. W. The historical background of modern social psychology. In G. Lindsey & E. Aronson (Eds.), The handbook of social psychology (2nd ed., vol. 1). Reading, Massachusetts: Addison-Wesley, 1968.

Bork, A. Preparing student-computer dialogs: advice to teachers. In R. P. Taylor (Ed.), The computer in the school: tutor, tool, tutee. Teacher's College Press, 1980.

Darley, S. A., & Cooper, J. Cognitive consequences of forced noncompliance. Journal of Personality and Social Psychology, 1972, 24, 321-326.

Fulton, M. A. A research model for studying gender/power aspects of human-computer communication. International Journal of Man-Machine Studies, 1985, 23, 369-382.

Geen, R. G., & Gange, J. J. Drive theory of social facilitation: twelve years of theory and research. Psychological Bulletin, 1977, 84, 1267-1288.

Heider, F. Social perception and phenomenal causality. Psychological Review, 1944, 51, 358-374.

Kelley, H. H., & Michela, J. L. Attribution theory and research. Annual Review of Psychology, 1980, 31, 457-501.

Kiesler, S., Siegel, J., & McGuire, T. W. Social Psychological Aspects of Computer-Mediated Communication. American Psychologist, 1984, 39, 1123-1134.

Kim, J., & Mueller, C. W. Factor analysis: statistical methods and practical issues. (Sage University Paper Series on Quantitative Applications in the Social Sciences, 07-014). Beverly Hills and London: Sage Publications, 1978.

Quintanar, L. R. The interactive computer as a social stimulus in computer-managed instruction: a theoretical and empirical analysis of the social psychological processes evoked during human-computer interaction. Ph.D Doctoral Dissertation, University of Notre Dame, August, 1982.

Ramsey, H. R., & Atwood, M. E. Human factors in computer systems: a review of the literature. Englewood, Colorado: Science Applications Inc., 1979.

Shneiderman, B. Software psychology: human factors in computer and information systems. Cambridge, Massachusetts: Winthrop, 1980.

Smith, H. T. Human-computer communication. In H. T. Smith and T. R. G. Green (Eds.) Human interaction with computers. New York: Academic Press, 1980.

Tucker, L. R. Implications of factor analysis of three-way matrices for measurement of change. In C. W. Harris (Ed.), Problems in measuring change. Madison, Wisconsin: University of Wisconsin Press, 1963.

Winer, B. J. Statistical principals in experimental design (2nd ed). New York: McGraw-Hill, 1971.

Social, Ergonomic and Stress Aspects of Work with Computers, edited by G. Salvendy, S.L. Sauter and J.J. Hurrell, Jr.
Elsevier Science Publishers B.V., Amsterdam, 1987 — Printed in The Netherlands

EXAMINING THE SILENT MAJORITY: A CURRENT PERSPECTIVE ON PREDICTIVE VARIABLES ASSOCIATED WITH MANAGERIAL AND CLERICAL ADP USERS

STEPHEN R. RUTH and ELLA P. GARDNER
Department of Decision Sciences, George Mason University, Fairfax, Virginia U.S.A.

ABSTRACT

The authors report on a new center's ongoing research in the effects of computer-based automation on managers and administrative and clerical personnel in large organizations. Their focus is not on programmer and ADP personnel but on the tens of millions of persons who must use the systems on a daily basis. A surprising number of persons show phobic reactions to their work environment because of the introduction of a computer. Specific recommendations are made to take advantage of findings to improve productivity and reduce stress.

1 INTRODUCTION

Most of the literature on human-computer interaction has concentrated on careful studies of the behavior of programmers and analysts in their work environments. Scales have been developed to evaluate the satisfaction, frustration, stress, level of aspiration and many other variables of interest to this population. Yet in terms of total users, programmers and analysts represent only about 3 percent of the population. At George Mason University's Center for Human Factors Studies, we concentrate on the other 97 percent--managers, clerks, data entry personnel, accountants, auditors, engineers, nurses, and many more. So far we have completed nearly a dozen studies of segments of this population. The results have been surprising in that they suggest that some of the most important variables that could be of interest to managers trying to improve productivity are being missed altogether. More important, there are few indications that studies of this type are being done before major decisions affecting workstation changes. In this paper, we will discuss briefly some of the major studies we have completed and some of the actions that seem to be required by managers.

2 SUMMARY OF STUDIES

2.1 Computer Phobia Study

In several informal studies of managers and administrative and clerical workers in computer training sessions at George Mason University's Center for Human Factors Studies, it had been observed that some of the workers were noticeable anxious about using a computer. This led to a group of studies into what some have called "technostress" (Brod, 1982). Brod defined technostress as a condition resulting from the inability of an individual or organization to adapt to the introduction and operation of new technology. When the new technology is computers, technostress may be termed computer anxiety or computer phobia.

Booz, Allen and Hamilton, Inc. conducted a study which indicated that about a third of all knowledge workers would be wary of the computer terminal rather than receptive to it (How to conquer fear of computers, 1982). The term knowledge workers includes professionals and executives. A similar study was conducted by Weinberg (1984). Of 400 students and 180 business people, he found that over 30 percent had some fear of computers. About 5 percent had physical reactions which would indicate an actual phobia. The other 25 percent felt a milder form of anxiety when confronted with computers.

The George Mason University Center for Human Factors Studies, in one of its first attempts to examine this phenomenon, confronted the issue of computer fear explicitly instead of taking the indirect approach of previous studies. In cooperation with a psychologist who specialized in phobias, the GMU team formulated a questionnaire to measure computer fear and anxiety directly. Surveys were returned from 372 persons, 86 percent of whom were managers or other professionals (Gardner et al, 1985). In the sample, we found 14 percent anxious or phobic about computers. Some differences between phobics and those normal with respect to computers were identified. While 57 percent of the sample were managers, only 36 percent of the phobics were managers. Twelve percent of the total sample was over 50 years old, but 46 percent of the phobics were over 50. Also, while only 37 percent of the sample were women, 64 percent of the phobics were women.

More phobics than others agreed that it takes a long time to learn to use a computer productively, were anxious and concerned about the pace of automation in the world, sensed that computers increased the level of frustration of the job, were afraid they would damage a computer if they used it, were frustrated by computerized bills, and were sometimes frustrated by increasing automation in the home. They had a greater fear than others that the computer might someday displace some of their co-workers and believed they could do their jobs very well without computers. The results of the study confirmed that computer anxiety was a problem for a significant portion of managers and professionals.

2.2 Use of microcomputers in government finance organizations

The interesting results from our phobia study encouraged us to extend the approach to other types of populations. We were approached by the Government

Finance Officers Association (GFOA) to survey the use of microcomputers at the city, county, and state levels. This would be of interest for two reasons. It would tell us the extent of computer use by one group of end users, and it would give us a chance to replicate segments of our phobia study.

Because of the client's concern that some of the explicit references to fear could cause reduced response rate, the questionnaire was changed in the direction of asking whether other workers seemed to have anxiety symptoms. A typical question in this context would be: Do you see evidence of the following attitudes in people using microcomputers in your organization? (Select all appropriate answers)

a. Refusal to use a microcomputer

b. Anxiety over using the microcomputer

c. Fear of learning a new microcomputer application

d. Continually putting off computer tasks

e. Frequent misuse or abuse of microcomputer equipment

Over half the respondents (51.7 percent) reported at least one of these attitudes, with anxiety being reported most frequently, followed by fear of learning a new application (Ostrowski et al, 1986). Less frequent behavior involved continually putting off microcomputer tasks and refusal to use a microcomputer. Few reported actual misuse or abuse of the equipment. Over 250 out of 513 respondents indicated a willingness to participate in a survey on computer anxiety. This response confirmed that many still believed computer anxiety to be a problem.

The results of the study also confirmed that microcomputer use is growing dramatically. For example, in 1982, the International City Management Association (ICMA) conducted a national survey of microcomputer use in local governments (Griesemer, 1983). The results of that survey indicated that 13 percent of responding local governments were using microcomputers. The GMU GFOA study in 1986 found that 88 percent of the responding organizations used microcomputers. A direct relationship existed between organizational size and budget and microcomputer use; as size and budgets grew, so did the likelihood of microcomputer use.

Basic word processing and file management were functions largely the purview of secretarial-clerical level employees. The professional, analytical, and managerial classes of employee were most likely to use spreadsheet, database management, and project management software. Senior managers indicated relatively little microcomputer use; when they did use them, they were likely to be using spreadsheet software. The impacts of microcomputers on management and productivity were generally positive. They had a positive influence on employee satisfaction, quality of decisions, and ability to maintain schedules.

To sum up, this study gave further strong indications that, while use of microcomputers is rising dramatically, there is a significant population of users whose fear or anxiety about computers will have an adverse effect on productivity.

2.3 Longitudinal view of computer anxiety

Doing a factor analysis of the data from the phobia study enabled us to do a longitudinal comparison of the attitude of people toward computers over the past decade and a half. Lee administered a 20-item questionnaire on computer attitudes to 3000 North Americans over 18 years of age (Lee, 1970). He applied factor analysis and found that the variance within the data could be attributed to two high-order variables: the "beneficial tool of man" and the "awesome thinking machine" perspective. An example of a question in the beneficial tool of man factor was, "They will help bring about a better way of life for the average man." One component of the awesome thinking machine perspective was, "There is no limit to what these machines can do."

When Morrison replicated the study in Australia, he found a different situation (Morrison, 1983). Using the same questionnaire, he found the largest factor defined a negative attitude toward computers. The major components of this factor were:

- With these machines, the individual person will not count for very much anymore.
- Sometime in the future, these machines may be running our lives for us.
- These machines help to create unemployment.

We ran a factor analysis on 35 questions on our computer attitude survey. We found five major factors which we called enjoyment, quality of life, fear, anxiety, and service (Gardner et al, 1987a). The factors called enjoyment and quality of life were clearly the major contributors to the total variation in the sample. This result was consistent with the population sample, 86 percent of whom were classified as relatively healthy concerning computers (neither anxious nor phobic). The findings appear to substantiate the generally held feeling that people were getting more accustomed to computers. The negative factors were present--fear and anxiety--but they did not have nearly the influence as in the two earlier studies.

2.4 Bottom-line approach to reducing computer anxiety

In order to help those who do have computer phobia or computer anxiety, the authors have developed and tested a three-step approach (Ruth et al, 1987) in organizations where computers are being introduced on a major scale. The three steps are: Evaluation, Assessment, and Action. In the evaluation stage, participants are asked to complete the computer anxiety instrument. They are classified into three groups: phobics, computer anxious, and normal with respect to computers.

During the assessment stage, those who were classified as phobic have the opportunity to be treated by a psychologist or psychiatrist specializing in phobias. Computer phobia is similar to other phobias in that the individual affected has an unnatural fear. Some example of other phobias are fear of height (acrophobia), open spaces (agoraphobia), and water (hydrophobia). A study by the National Institute of Health revealed that 11 percent of all U.S. adults manifest some type of phobia (A fight to conquer fear, 1984).

The action phase consists of therapy for the phobics, training for the anxious, and listening to comments made by all employees. The treatment for computer phobia is similar to that for other phobias. The individual is gradually introduced to the computer. Treatment may take two months and cost $2000 per worker.

As mentioned earlier, GMU has trained several thousand end users on microcomputers. Those who were anxious responded well to a one-week training program in word processing, file management, and spreadsheet applications. The cost is approximately $400 to $600 per worker. The cost of these treatments is believed to be much less than the productivity that would be lost otherwise. In a year, an expected increase in productivity of 10 percent for non-phobics would more than pay for the treatment.

A lot can be learned about the current system by assessing the comments made by those who are "normal"--as well as those made by the phobics and computer anxious. These actions, therapy for the phobics, training for the anxious, and listening to comments from all the participants can be expected to release workers from their concerns about machines and gain substantial productivity improvements.

2.5 Health effects of computer use

Health complaints sometimes arise after the introduction of computers, especially in clerical applications. These problems can be classified under various types of job stress. Caplan et al (1975) of the University of Michigan developed a model theorizing that job stress leads to both psychological and physiological strains. These strains lead to health complaints. Examples of job stress are uncertainty about what is required of the person on the job, the presence of two or more conflicting demands from supervisors, peers or subordinates, or uncertainty about job career security. Psychological strains like boredom or workload dissatisfaction may develop. Health complaints such as backaches and headaches may follow.

Smith et al (1981) at NIOSH (National Institute of Occupational Safety and

Health) applied this theory to VDT operators. They found that clerical VDT Operators showed much higher levels of visual, musculo-skeletal, and emotional health complaints, as well as higher job stress levels, than did control subjects and professionals using VDTs.

Sauter et al (1983) at the University of Wisconsin continued the research with VDT operators. Not only did they survey users and non-users, but they also took physical measurements at the workstation. They found that VDT users showed more eye strain than did non-users. Also, chair and workstation configuration were particularly important predictors of musculo-skeletal disturbances. Finally, corrective eyewear use and ambient lighting were important predictors of visuo-ocular disturbances.

At George Mason University's Center for Human Factors Studies, we surveyed a sample of administrative and clerical workers from the U.S. Department of the Navy (Gardner et al, 1987b). Eight health complaints were found to increase with VDT use: tearing and itching of eyes, periods of depression, difficulty sleeping, ringing or buzzing in ears, neck pain (that radiates into shoulder, arm or hand), periods of extreme anxiety, cramps in hands and fingers, and alarming pain in chest. In addition, more VDT users experienced job boredom than non-users. Those who used a computer or terminal more than five hours a day experienced more work pressure and more control by management than those who used a terminal or computer one to four hours per day.

2.6 Revised procedure to test effects of various interventions on users

The combination of the findings from the phobia study and from the VDT study at George Mason University has led us to amend our three-step plan of evaluation, assessment, and action. We now evaluate employees using the computer anxiety instrument and also an amended version of the instrument developed at NIOSH to measure computer anxiety, job stress, psychological strains, and health complaints. In addition, blood pressure and heart rate are measured.

For those diagnosed as suffering from computer anxiety or computer phobia, the treatment is as described above. To eliminate physical problems associated with workstation design, we go into the workplace and adjust desk and chair heights, height and angle of the CRT screen and keyboard, ambient lighting, and lighting of the screen. Workers are trained to make their own workplace adjustments.

To alleviate symptoms associated with keeping the body in the same position for long periods of time, we prescribe exercises designed to relax the eye, neck, shoulder, arm, hand, wrist, finger, ankle, torso, and leg muscles through specific exercises and, more generally to relieve tension through the entire body through exercises such as deep breathing.

These new hypotheses we are now testing are:

- Workstation redesign will reduce the physical complaints associated with VDT use.
- Self-reported physiological and psychological stress will be reduced through operator involvement in stress reduction relaxation exercises and VDT orientation training.

After the interventions described, the instruments are administered again and blood pressure and heart rate measured to see whether job stress, psychological and physiological strains, health complaints and computer anxiety have been reduced. That remaining would be attributable to other causes.

2.7 User-centered criteria for selecting software

One major area to examine next would be the user interface to the system software. There is a great need to specify metrics of standards of performance for software as well as hardware. A specific set of criteria was recommended by Fried (1982). These principles have been expanded by the GMU group (Knittle et al, 1987) resulting in a series of user-centered metrics for the evaluation of interactive systems. These metrics are being consolidated into a series of weighted measurements to make it possible eventually to evaluate user software with something like the precision now possible for hardware.

3 AGENDA FOR ACTION

The purpose of these studies at George Mason University's Center for Human Factors Studies has been to characterize the concerns and fears of VDT users about their work environment. Our view is that the primary users of the results of these studies should be managers, not academics. For example, if an organization is considering the introduction of a new technology, it is relatively inexpensive to examine the prospective users beforehand. If there are genuine problems of greater stress for VDT operators than for the rest of the work force--the type clearly indicated in our samples--remedial action can be taken before major changes are made. If there are computer phobics and computer anxious employees--perhaps thousands of them in a large organization--should management deliver a new computer-based workstation before trying to help the employees with appropriate intervention?

Our studies do not show that every population of supervisors and administrative and clerical employees has the same percentage of those with problems and requires the identical response. Instead they suggest that the prudent manager should examine this population carefully from time to time and make decisions on technology introduction based on the results. To do otherwise would seem to ignore a simple, effective analysis capability now available to any organization.

4 REFERENCES

A fight to conquer fear, 1984. Newsweek, April 23, pp. 66-72.

Brod, C., 1982. Managing Technostress: optimizing the use of computer technology. Personnel Journal, 61: 753-757.

Caplan, R.D., Cobb, S., French, J.R.P., Van Harrison, R., and Pinneda, S.R., Jr., 1975. Job demands and worker health (U.S. DHEW (NIOSH) Publication No. 75-160), National Institute for Occupational Safety and Health, Cincinnati, Ohio.

Fried, L., 1982. Nine principles for ergonomic software, Datamation, 28: 163-166.

Gardner, E.P., Render, B., Ruth, S.R., Ross, J., 1985. Human-oriented implementation cures 'cyberphobia'," Data Management, November, p. 29.

Gardner, E.P., Young, P., Ruth, S.R., 1987a. Productivity, fear and acceptance of computers: A factor analysis approach. Submitted for publication.

Gardner, E.P., Ruth, S.R., Render, B., 1987b. The stress of computer use. Submitted for publication.

Griesemer, J.R., 1983. Microcomputers in local government. International City Management Association.

How to conquer fear of computers: Executive guide, 1982. Business Week, March 29, pp. 176-178.

Knittle, D., Ruth, S.R., Gardner, E.P., 1987. Establishing user-centered criteria for information systems: A software ergonomics perspective, Information and Management, in press.

Lee, R., 1970. Social attitudes and the computer revolution, Public Opinion Quarterly, 34: 53-59.

Morrison, P.R., 1983. A survey of attitudes toward computers. Communications of the ACM, 26: 1051-1057.

Ostrowski, J.W., Gardner, E.P., and Motawi, M.H., 1986. Microcomputers in public finance organizations: A survey of uses and trends. Government Finance Review, 2: 23-29.

Ruth, S.R., Gardner, E.P., Render, B., 1987. Examining the problems of computer-based anxiety: A systematic approach. In: Proceedings of the ACM Computer Science Conferences, St. Louis, Missouri.

Sauter, S.L., Gottlieb, M.S., Jones, K.C., Dodson, V.N., Rohrer, K.M., 1983. Job and health implications of VDT use: Initial results of the Wisconsin-NIOSH study. Communications of the ACM, 26: 284-294.

Smith, M.J., Cohen, B.G.F., Stammerjohn, L.W., Jr., 1981. An investigation of health complaints and job stress in video display operations. Human Factors, 23: 387-400.

Weinberg, S. and Fuerst, M., 1984. Computer phobia: How to slay the dragon of computer fear. Banbury Books, Wayne, Penn.

Social, Ergonomic and Stress Aspects of Work with Computers,
edited by G. Salvendy, S.L. Sauter and J.J. Hurrell, Jr.
Elsevier Science Publishers B.V., Amsterdam, 1987 — Printed in The Netherlands

SKILLS ANALYSIS IN HUMAN-COMPUTER INTERFACE - A HOLISTIC APPROACH

LILLIAN G MALT
Lillian Malt (SkAn) Training Services,
Quirral House, Thursley, Surrey GU8 6QW, England.

Abstract

Skills analysis, born out of task analysis draws on research of ergonomists, neurologists, physiologists, psychologists and learning theorists. This enables a holistic approach to providing work satisfaction through the use of task methods, training and work organisation that fulfil worker and company needs. SkAn as applied to the different aspects of human computer input is dealt with: opinions of workers and statistical results related to human attitudes to the task are given.

Introduction

Skills Analysis (SkAn) may be said to have been born out of the work of Bryan and Harter (1897) advanced by the Gilbreths, in the early 1900's; and spurred by Analytical Training for war time needs both in Britain and America. With changing attitudes to the needs of workers as human beings, the Seymour brothers under the guidance of Mace and Myers of the Nat. Inst. of Industrial Psychology added a psychological dimension which was cross-fertilised by the work of Elton Mayo, the Hawthorne Experiment (Ref.) and by the London Tavistock Inst. of Human Relations, to give it full socio-technical status. Because of its accent on the needs and feelings of people, it includes aspects that are now referred to as cognitive ergonomics and macroergonomics. The work of such disciplines has been part of the SkAn philosophy since the 50's and its proving ground has always been the shop floor rather than research laboratories.

What Skills Analysis (SkAn) is

Skills analysis aims to provide work satisfaction through developing task methods and training that will allow experienced worker standards (EWS) to be attained easily and in an optimal time span. The holistic approach of SkAn allows for changes in human requirements as they occur, in the work place and in society. It embraces element, task and SkAn techniques. Its work methods are designed to fulfill psychological and physiological needs. SkAn training systems are based on educational psychology and learning theory.

Measurement of effectiveness of SkAn methods

The success of skills analysed task methods and its complementary training are measured both against work output and human satisfactions. Quantity and quality of output provide objective measures of results;

peoples' opinions are subjective measures and are not only expressed verbally, they are reflected in productivity, absenteeism and, terminally, in labour turnover. Objective statistics can be obtained for these.

Human factors to be considered

In any enterprise there are four parties whose participation is essential and who will look for need fulfillment:

a the person who performs the task
b management who control the task and run the company
c financial investors
d customers whose continued satisfaction enables the company to make a profit and remain in business.

To fulfill all human needs, 'work should be meaningful and provide rewards, not only financial rewards but rewards in terms of human satisfaction.' (Ref. Seymour D W)

Definitions of: a skill, job analysis and skills analysis

A skill may be defined as the developed capacity to perform to a high standard, easily, regularly and with minimum fatigue. This definition embraces the twin focus of skills analysts to enable high production standards to be achieved - with ease. Minimum fatigue predicates that human physical and psychological needs are met.

A job analysis lists **what** is done. A skills analysis defines **how** the job is best performed. 'How an experienced and skilled worker performs, including sensory information and neuro-muscular responses required.' (Ref. Salvendy and Seymour) In undertaking the analysis of any task, these aspects are considered:

1 the physical requirements for performing the task at a skilled level - the equipment, tools etc.
2 the human requirements for skilled performance, both motor and sensory elements and how the skill level may be improved upon.

Work is done with equipment, within a human group, by feeling, sentient individuals who require multifaceted satisfactions from the task, not least that of success. Success is a prime motivator and attainment of a skilled level of performance in an optimal time span is one important measure of success.

This is a list of aspects considered when a task is skills analysed. (SkAnd)

SKILLS ANALYSIS (SkAn) TASK ASPECT LIST

Physical Aspects					Cognitive Aspects	
Human Specific				Task Specific		Socio-Technical
1 Sight	2 Hearing	4 Breathing	5 Body Physical	6 Environment	7 Input Format	8 Human Factors
Ambient light Task light Glare - reflection Screen - contrast colour flicker External environment	Audio signals Ambient noise Music Speech 3 Touch Kinaesthesis (reach & distance)	Thermal conditions Air conditioning - humidity toxic agents passive smoking Ionisation	Autonomic tone Posture - all limbs Gravity Physiology	Office layout - human density traffic Interior design & decor Work station - keyboards screens footrests copy holders	Standards - quality quantity faults Devices - overlays Software design Screen layout Language	Co. Organisation Communications - lateral vertical Group relations Task value - intellectual aesthetic ethical Work Flow

To emphasize that work is performed by sentient beings, the chart has factors affecting task productivity under two broad headings. Physical aspects in Sections 1 - 5 and cognitive or sensory in Sections 6 - 8.

It is generally accepted that physical stress should be avoided for human beings, there are only differing views on how this can best be done. It is perhaps less frequently acknowledged that mental stress induced by a task does have physical effects, and that this too should be an area of management concern. Certainly the inability to perform to production standards within a reasonable time span causes emotional stress and is a prime cause of labour turnover in the first three to six months of employment.

<u>SkAn in use</u>

Skills analysis has been used successfully in many industrial tasks. This paper concentrates on keyboard input, a vital component of the human-computer interface. Section 6 in Fig.1 lists the physical aspects and Section 7 has cognitive aspects of keyboard input.

The research of Roger W Sperry and his colleagues on the split brain of animals and accident victims proved that the left hemisphere (LH) controls linguistic reactions, while musculo-spatial reactions are controlled in the right hemisphere (RH). (Ref. Myers & Sperry). Sperry's concepts were confirmed for touch keying through observations of skilled copy typists, many of whom are able to continue keying while conducting a conversation or listening to a radio play or to a cricket or baseball match. 'An experienced copy typist may be typing through a cortical reflex from the visual information of the manuscript to the execution of the appropriate sequences of finger motions. The more logical and conscious the keying is, the more the left hemisphere may be required.' (Ref. Yamada). Experimental confirmation has come from other Japanese researchers including Okadome (Ref.)

It is important to recognise that computer and word processor input are similar but not identical to typewriter input. Both involve tapping keys to produce output, it is the cognitive aspects that are different. The number and range of functions of the word processor or the computer make a qualitative difference to the input process. While copy typing a typist concentrates on the layout of the letter or document or form that is being typed. The use of the few functions available on a typewriter does not occupy much of the LH thinking process during keying. However when functions multiply as they do on some electronictronic typrwriters, Word Processors and computers: function keys and command sequences are sporadically, but frequently used within the keying process. If operators can learn to use function keys and commands in the RH, i.e. if it is possible to develop the musculo-spatial reflex actions for function keys and commands, as for text keying, then skilled input - fast, accurate, easy, without fatigue becomes possible.

The degree to which this development can take place will be influenced by task related factors such as: 1. keyboard design (shape) 2. keyboard layout (letter and function key arrangement) 3. selection of characters for commands. Development of a RH cortical reflex for computer input is retarded by factors such as:

a the unergonomic design and letter layout of the standard Sholes (Qwerty) typewriter that has almost universally been accepted as the input to computer device. It makes optimal placement of function keys difficult.

b the design of input systems where little thought is given to the accessibility of characters chosen for functions and commands.

c the code language selected for command sequences. If codes use simple mnemonics, 'if they are closer to the natural sound, commands are more easily learned and accessed.' (Ref. Watanabe)

With such constraints on RH reflex development, it is not surprising that this facility is not easily developed. SkAn training has concentrated much attention on developing RH cortical reflex action for all keyboard input. This involves the development of SkAn training programmes - frequently self-paced audio-visual or audio graphic - for software and firmware systems. Human understanding of the capacity of computers and programmes is an essential element for the optimum use of all Information Technology (IT) equipment.

Categories of input/output

Humans interact with their computers to produce specific output. Different cognitive skills are required for different outputs. Broadly speaking there are four categories of input/output that may be described as:

1. Copy keying - numeric and/or text input, composing

Keyboard operators who work a 7-hour day read in the text to be keyed while feeding in from their own internal memory store the requisite commands for layout and text processing. The extent to which the command structure is stored as a musculo-spatial reflex determines the ease, accuracy and speed of input. If the command structure, the layout of the keyboard and the mnemonics allow for Right Hemisphere storing, input will not be hard work and will be faster and more accurate.

2. 'Think-keying' by executives, journalists, authors and others

'Think-keying' is a term applied to authoring directly on to a keyboard. Thinking in words is a function of the linguistic, the left cerebral hemisphere (Left Hemisphere). The words are then transferred to the Right Hemisphere (RH) for musculo-spatial translation into keystrokes. It is logical that if the Left Hemisphere (LH) is involved in originating the copy to be keyed, it will operate more efficiently if all finger actions are controlled by the RH. 'Think-keying' is as effortless as 'think-writing' where input is by touch keying.

2.1. Executive input

For executives using PC's the 'think-keying' process is similar to that of the journalist, although not usually so sustained. That so many executive PC's remain unused or underused may well be due to the strain felt by those who are unable to touch key. If they cannot find characters on a keyboard without involving the left brain their thinking process is handicapped.

2.2. Journalists and authors

Until recently, most journalists were not touch key operators. The fastest two fingered operators in the world must have been, and may still be, journalists who 'bash' away at typewriters sometimes at very great rates and with a fair degree of accuracy. This was however accomplished at a cost in terms of human strain. The stressful states of journalists so often depicted in novels and on films were not due entirely to 'going-to-press' deadlines, but were exacerbated by the overuse of the eyes, darting from the keyboard to copy and back, and the interference in the Left Hemisphere caused by thinking where to locate the correct keys.

3. Editing

Keyboard editing is similar to the input tasks described above, and yet is different. There are two kinds of editing: proof reading for literal and other keying errors, and text editing concerned with syntax and content.

Traditionally, coloured-pencil editors marked up errors on galley proofs leaving the actual correction of the typescript to keyboard operators. As desk-top editing systems proliferate, editing corrections are increasingly being made directly onto screens. For those who cannot 'think-key' with the fluency with which they can 'think-write' the left thinking hemisphere is stressed by the addition of the keying task.

4. Programming

Programmers may be said to be involved in a three-dimensional thinking process. One dimension holds the concept of the systems analysis or the programme specification and its desired end result. A second dimension holds the programming language into which the concept is to be translated, while a third dimension is occupied in finding keys. The left brain is wrapped in logic translating a critical process, entirely given to thinking through the effect of the programme commands issued. Something of a dual musculo-spatial input is required. One that will allow for translating the programme specification into the command structure and another that allows for kinaesthetic access to keys.

Easy touch keying makes programming input faster, but more importantly there are fewer errors. An error in programming can create days of work: there is no spell-check software of any kind for detecting programming errors. On the Standard Sholes (qwerty) keyboard where special characters are used mainly in shift, programmers have great difficulty in touch keying.

Although the four categories have different cognitive requirements, the musculo-spatial requirements remain the same. A pre-requisite for accurate, fast and almost effortless input is that the keys should be accessed by touch both for text keying and for WP and computer commands. The development of a kinaesthetic sense of reach and distance not only allows operators to use their eyes for continuous copy intake but also frees the left brain hemisphere for the thinking or cognitive aspects of keying.

Development of a SkAn training programme

To attain such skill levels, it is necessary to develop motor-sensory pathways that will process information through to fingertips. The ability to touch key will to a large extent determine the degree of eye strain experienced. If the text copy and the screen are at the same height, and the eyes are free to concentrate on reading with only minimal glances on the same horizontal plane to the screen, the fixing of the central fovea is not stressful. However if the eyes have constantly to look down to the keyboard and refocus from that vertical movement to the horizontal planes, fatigue of the central fovea soon develops. Except for glare, eye movements are probably the greatest cause of eyestrain in screen use. Once touch keying is established, the eyes are free to concentrate on intake. This reduces screen reading and again minimises eyestrain.

The aim of SkAn training programmes is to develop training methods that will in the shortest possible time develop motor-sensory pathways that make high skilled performance possible. Movements of high speed keyboard operators with speeds ranging from 80 to 120 words per minute (wpm), that is 24,000 to 36,000 keystrokes per hour (kph) were studied minutely. (Performance on production rather than in test conditions was observed). This provided information about:

a) mini-finger movements
b) stages in the development from slow rhythmic to fast a-rhythmic keying, massively confirmed in a Commission for European Community (CEC) research project. (Ref.)
c) reading ahead of keying. Confirmed by Shaffer (Ref.) and many others since
d) the correlation between reading and keying speed
e) very importantly, posture at the keyboard.

Malt SkAn findings and training were probably responsible for the adoption of what came to be know as the 'four right angled' posture.

SkAn error (fault) analysis

The MacNeilage (Ref.) classification of typing errors was adopted and developed into a SkAn grid on which errors are easily analysed. This provides a guide for the diagnosis of error causes and for corrective actions.

As the analysis proceeded, it was observed that posture had a dramatic effect on accuracy as well as on speed. While, of course, each individual operator had been the arbiter of his/her own comfort, when SkAn revealed that faulty posture caused one-third of all errors, an objective and immediate measure of correct posture for each individual became available. In addition, the necessary motivation to maintain good posture was established when operators found that not only were they more comfortable, but the most annoying of all errors, those that they were not aware they had made were eliminated. Ong of Singapore (Ref.) confirms similar productivity improvements from attention to posture by using only footrests and copyholders.

To assist development of musculo-spatial reflex actions in the RH, SkAn practise material works at first, on letters keyed by three fingers of the left hand only, then on three fingers of the right hand. Careful introduction of each of the fingers prevents all errors usually caused by neural confusions. To aid rapid development of mini-movements keyboard table mats are used. Arm rests that rob fingers of gravitational aid are not used. Kinaesthetic reach and distance movements are accurately developed within finger guides that allow the fingers to move in the vertically diagonal directions of the standard keyboards. Through eliminating causes of errors mental interruptions and tension are reduced allowing naturally for increases in overall speed and for success in the job.

Results of SkAn training for keyboard operators

SkAn training methods were first developed for typist training in the late 50's. Previously only 2% of trainees reached 24,000 kph (keystrokes per hour), this rose to 10%. In 1963 SkAn, training was developed for keyboard operators in the printing industry. The success of the SkAn method has been objectively proven by almost invariably reduced labour turnover, less absenteeism and the fact that not one of the many thousands of keyboard operators trained in this way including operators in United States, England, Australia and other countries have developed any kind of back, wrist or other Repetitive Strain Injury (RSI).

Results

When experienced operators undertake SkAn retraining, both quantity and quality improvements are recorded. Results of three to four weeks retraining for 100 experienced keyboard operators in the printing industry are given:

Speed improvement	**from 21% to 53% with a mean average of 38%**
Accuracy improvement	from 40% to 88% with a mean average of 69%

Correction costs are frequently reduced by 40%.

Typical of SkAn results in non keyboarding applications are these for new entrants in an electronics company:

Reduction in off-the-job training time	from 8 weeks to 4 weeks
Reduction in annual novice labour turnover	from 70% to 32%
Reduction in time to reach experienced worker standard	from 12 to 4 months.

Operator Comments

These comments from operators give a variety of reasons for the success of the training:

"I have suffered with backache for about 20 years .. rarely had a complete night's sleep .. treatment at the hospital .. since taking the Alpha Course and being taught ... how to sit properly I have not had any backache and am having good nights."

"I frankly never thought that I personally could go much faster than I was at the beginning of the course but I have improved 25%." (Actual improvement was 35.5% over 24,000 kph)

"I can operate faster, make less errors and will not be as tired."

"It has helped me to realise that bad habits of many years duration can be broken .."

"I find less strain and less monotony by concentrating .. and most of all, it is an invaluable aid to accuracy."

"Most of all, the course has shown me that I have a higher potential than I thought I had."

"I thought I was setting at my maximum speed ... but you proved that I was capable of much more. This is really the most valuable thing, I can also apply it to my everyday life."

"I do feel much more confident and less strain .. "

References

Commission for the European Community (CEC) Keyboard Research Programme, 1983. Data Tapes from J de Smith (Systems) Ltd, Tunbridge Wells TN1 1DJ, U.K.

Dvorak, A., Merrick, N. L., Dealey, W. L., Ford, G. C.,, 1936. Typewriting Behavior: Am. Book Co, NY, pp504-505.

MacNeilage, P. P., 1966. Typing errors as clues to serial ordering mechanisms in language behaviour. Haskins Laboratories, New York.

Mayo, E., 1949. The social problems of an industrial civilisation. Routledge & Kegan Paul Ltd., London.

McGregor, D., 1960. The human side of enterprise. McGraw-Hill Book Company, New York, Toronto, London. pp33-57.

Myers, R. E., Sperry, R. W., 1953. Interocular transfer of a visual form discrimination habit in cats .. Anatomical Record Vol. 115 pp.351-352.

Okadome, T., Yamada, H., 1983. A brain wave study of cerebral hemisphericity in English typists. *Changing Natural Language Input and Output Appropriately (59101004). Tokyo University March 1986. pp195-220.

Ong, C. N., 1984. VDT Work Place Design and Physical Fatigue: a Case Study in Singapore. **Ergonomics and Health in Modern Offices. Taylor & Francis, London and Philadelphia. pp484-494.

Salvendy, G., Seymour, W.D., 1973. Prediction and Development of Industrial Work Performance. John Wiley & Sons, New York, London, Sydney, Toronto. Seymour, Dr. A. H. (Unpublished papers)

Seymour, W. D., 1966. Industrial Skills. Sir Isaac Pitman & Sons Ltd, London.

Shaffer, L.H., 1975. Intention and performance. Psych.Review, 83. pp375-393.

Watanabe, H., Yamada, H., et al. 1983. Cognitive Aspects of Reflex-Code Typing for Japanese Text.* ibid. p73.

Weber, A., Sancin, E, Grandjean, E., 1984. The effects of various keyboard heights on EMG and physical discomfort.** ibid. p482.

Yamada, H., 1983. Certain Problems Associated with the Design of Input Keyboards for Japanese Writing. Cognitive Aspects of Skilled Typewriting. Springer-Verlag, New York, Heidelberg, Berlin, pp348-359

Social, Ergonomic and Stress Aspects of Work with Computers,
edited by G. Salvendy, S.L. Sauter and J.J. Hurrell, Jr.
Elsevier Science Publishers B.V., Amsterdam, 1987 — Printed in The Netherlands

MANAGEMENT OF COMMUNICATIONS IN CUSTOMER/SYSTEM INTERFACE TRAINING

W. S. LEICHT

Associate, L. G. Malt Training Services, Box 152 Peck Slip Sta., NY, NY 10272 USA

ABSTRACT

Communication between individuals, departments, semi-autonomous R&D, and an outside vendor affected customer training of a typographic system manufacturer. Complexity called for shared purposes and aims, which had not developed; coordination difficulty ensued. Training, lacking notification from systems development and reports from sales, marketing, and implementation, could not keep current. Consultants hypothesized that 1. clear purpose and aim statements, 2. better communication, and 3. enhanced training methods, would lead to 4. improved customer training design and delivery; this intervention succeeded. They also hypothesized that top management would soon adopt and propagate the communications changes; it did not.

1 INTRODUCTION

The human-computer interface is a system embedded in a system. Understanding human-computer interaction means understanding the social system within which humans and computers function. Specifying an interface in terms of human users' cognitive and physiological needs may not be enough. Marketing and organizational requirements may need consideration and can include cognitive and physical ergonomics.

In 1985 a company that manufactured and sold computer composition systems commissioned a training study. Customer training needs had become a priority as the company expanded, developed new markets, and reorganized. The work was to include: 1. Assessment, 2. Customer Training, 3. Policy Development, 4. Training Course Specification, 5. Training Course Production and Implementation. A major aim was to enable it to create its own courses.

2 PRESENTING PROBLEMS

Customer 'training failures' and resulting computer system returns persuaded the company to seek advice. An earlier study had recommended replacing "small classroom" with "self-paced" training for more effectiveness at lower cost.

Low morale prompted management also to request training staff evaluation.

3 RE-ASSESSMENT

The company in 1985 engaged consultants to re-assess and determine whether implementing previous recommendations would solve present problems. The consultants reported that much more was required.

3.1 Process

The managing director granted access to all customer service training personnel and agreed to confidential interviews. Later, he approved including an outside vendor that served as an original equipment supplier.

Assessment was to take a month. The consultant began interviewing training department management and staff. Concurrently he was participant-observer in several new-client training courses. He thus evaluated the system first-hand, experienced current methods of training 'naive' learners, and observed their reactions to learning command procedures and operations. Then he interviewed managers of site-implementations, sales, and marketing and observed on-site training of new operators.

3.2 Results

The evidence showed that communications problems (which had figured in the first assessment) had intensified as the product complexity grew and as staff attrition (during unprofitable years) had broken previous informal linkages between operating departments. The primary training mission of getting customers 'off to a running start' called for new linkages and training methods.

3.2.1. Purpose Good training should improve sales; but the presenting problems (training morale and system return after sale) cast doubt on whether training staff understood it. In fact they did, but were not convinced that the company valued training. They referred to themselves as 'orphans' (citing instances like learning of major system enhancements inadvertently when customers informed them or receiving assignments to train new clients with little warning and meager information from sales).

3.2.2. Training Customer Operators Several departments, training, sales support, and engineering, had their own training operations, often overlapping responsibilities. Job definitions were out-of-date or missing. Ad hoc coordination usually got them through.

The training department's three full time trainers were knowledgeable specialists and concerned about their learners, however their training skills needed development and updating.

The sales support department's four systems specialists were supposed to customize procedures to actual production during installation. However, they often conducted full training courses on site.

3.2.3. Training Design Training design problems occurred early in product initiation. Development teams typically disbanded before completing a handbook, let alone a training design or a course. Training design coupled with trainer training might filled the gap, but the training department had lost its design group. Trainer briefing and course creattion occurred ad hoc, with incomplete information, and often facing an implementation.

3.2.4. Communications and Management Inadequate liaison between training and departments affecting its performance was the single most serious cause of problems (even 'failures'). Information on enhancements, product introductions, and marketing did not reliably reach training. For example, a new software enhancement was about to be introduced. No trainer understood what it was nor why it existed. They perceived it variously as being 'already in the field' or 'barely beyond the development stage.' Therefore rumor, contact out of channels, and speculation substituted for 'good' information.

3.2.5. Staffing Trainers were former 'hot-metal' typographers, had served as apprentices, worked in a shop, and believed in the apprenticeship system. All enjoyed teaching. All were intelligently interested in technology but isolated from both 'shop floor' and 'hi-tech' R&D. Consequently they 1. had limited understanding of learning, 2. knew little of learning methods, and 3. had fallen out of touch with production realities. The company had no plan to keep them abreast of high technology and training.

3.3 Management Audit

"Where were top management in all of this?" might be legitimate at this point. In fact with intense product development and marketing they had just returned the company to profitability and a position in world markets. Now they were turning to organizational development.

Management had undertaken self assessment and had engaged a management development firm. They were doing task analysis and job definitions from the top down. It believed that company survival would depend upon rapidly evolving and marketing technologies for human-computer interaction in electronic publishing and document processing.

But the executive committee of senior management was ill-prepared for the pace. It needed to assimilate and promote a new corporate ethos (symbolized by "pro-active thinking") before remolding the organization to react flexibly to changing circumstance. The "training failures", for example, pointed up resistance to 'lateral' communication.

4 RECOMMENDATIONS

Formal recommendations covered training and communications. A verbal audit report to the managing director covered organizational development.

4.1 TRAINING RECOMMENDATIONS

4.1.1. Manuals

1. Create operator handbook from systems manual

2. Write training manual to consultant specification
3. Write exercises based on typical client production documents
4. Write ready reference to commands

4.1.2. Training Design

1. Establish group to produce courses, updates, and custom training
2. Write self-paced training script to consultant specification
3. Instruct trainers in learning theory and training methods

4.1.3. Morale

1. Provide generous notice of product enhancements and releases
2. Schedule courses or duplicate equipment to eliminate conflicts

4.2 COMMUNICATIONS RECOMMENDATIONS

1. Select and train a representative group to analyze communications
2. Appoint communications officer as staff (not line) to report at a high level

4.3 SENIOR MANAGEMENT

1. Conduct executive seminar on change management
2. Charter project to analyze communications

5 EVALUATION AND DISCUSSION OF RECOMMENDATONS

Top management agreed with the thrust of the analysis and approved two seminars, one week each for the trainers and for a "communications project team" (to be nominated by senior management).

They further agreed to invite an outside vendor to the communications seminars, because it represented a "top priority for action." They also agreed to "new handbooks, manuals and training methods to be developed" for that vendor system first.

6 INTERVENTION

Consultants prepared and delivered two seminars. The author specified the self-paced customer training outline and example.

6.1 COMMUNICATIONS GROUP TRAINING

Communications Project and seminar nominees were second level managers (8).

6.1.1. COMMUNICATIONS PURPOSE The purpose of communications training was to ensure that sales, training, and R&D shared relevant information.

6.1.2. COMMUNICATIONS AIMS The aim of communications training was to develop a group capable of facilitating lateral communications between departments. An implied aim was to stimulate top management to think "pro-actively" about communications between divisions and to use the Communications Project.

6.1.3. COMMUNICATIONS GOALS

1. To link training and product creation and enhancement more closely

COMMUNICATION PURPOSES/AIMS	ATTITUDE/SKILL	SKILL (continued)	WORKSHOPS	WORKSHOPS COMPANY FUTURE
Monday-Wk 1	**Tuesday-Wk 1**	**Wednesday-Wk 1**	**Thursday-Wk 2**	**Friday-Wk 2**
Introductions Course Purposes Course Aims	Session Reports Attitudes: Effects Development Interventions	Rvw Highlights Purpose & Aim (2) Getting Information	Group Needs Purpose & Aim (3) Workshop (5): New Product Introductions	Workshop (8): 'Turbo,' A Failed Communication
BREAK	BREAK	BREAK	BREAK	BREAK
Communications Promote/Inhibit Tell & Persuade Checklist	Purpose & Aim (1) Communication Skill Questioning	Reporting (A) Discussion Leading Decision Making	Reporting to Other Levels (B) Workshop (6): W/P Interface	Workshop (9): Failure - Sales, System, Training?
LUNCH	LUNCH	LUNCH	LUNCH	LUNCH
Role Play: Company Crises Project Purposes & Aims	Workshop (1): Command Structure	Workshop (3): A New Product	Workshop (7): Graphic System	Workshop (10): Pagination
BREAK	BREAK	BREAK	BREAK	BREAK
Understanding & Being Understood	Workshop (2): Customer Profile	Problem Solving Brain Storming Workshop (4): Company Media	Aggression/Conflict Correcting Errors	Project Future Company Future Course Evaluation
HOMEWORK Session Highlights	HOMEWORK Purpose/Aim Topics	HOMEWORK Practical Topics	HOMEWORK Prepare Workshop	HOMEWORK [none]

Figure 6.1: Course Structure for the Communications Project

2. To link training and sales starting with the customer profile procedure
3. To train a group to identify similar needs to management

6.1.4. COMMUNICATIONS STRATEGY The strategy was to give second level managers an intensive one week course on company purposes, aims, goals and communications during which they could become the core of a Communications Project. This project would identify interdepartmental communications needs and propose solutions.

6.1.5. PROCEDURE The course delivered appears as Figure 6.1, page 5. Communications Project training occupied 8 hours a day in a hotel away from the company. Most participants spent the full period, but the vendor was absent from the confidential final day and a half. The course spent roughly 3 days on communication basics and 2 days of workshops on specific problems.

The participants were intelligent and wanted to help the company, though unused to extensive analysis. Almost immediately they identified serious communications problems caused by departments that acted as "six different companies" within the company and further difficulties in lateral communication caused by stiff hierarchy. Within five days they had formed strong bonds and had received the blessings of top management for two project proposals.

Developing bases for change was harder. Participants had trouble understanding 'purpose': ***why to act*** ('objective' was our synonym), 'aim': ***what direction to pursue***, 'goal': ***measurable target***, and 'strategy': ***how to proceed.***

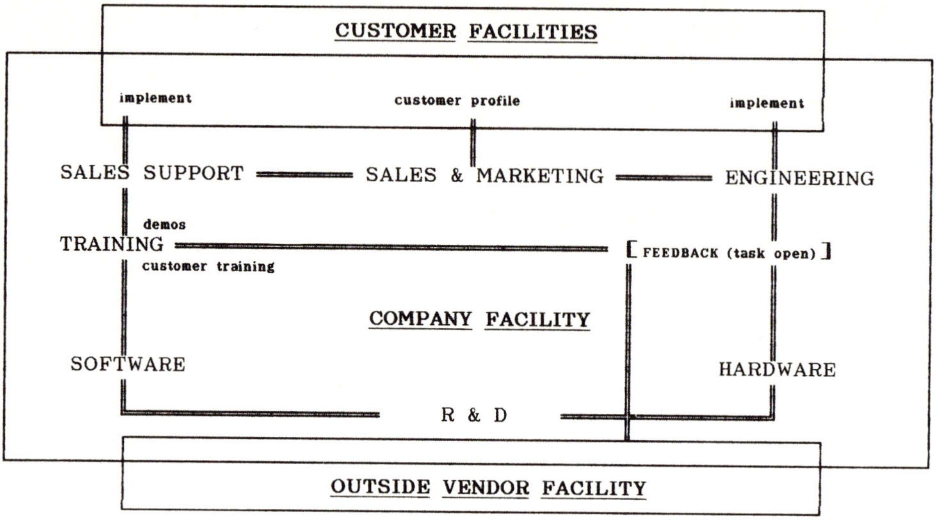

Figure 6.2: Interrelationships of Departments, Vendor, and Customers

A particularly useful tool to elucidate the problem was a chart (Figure 6.2, page 6) that showed interrelationships of departments, outside vendor, and customers and thereby clarified their interests.

6.2 TRAINER TRAINING

Trainer nominees came from training (4) and from customer support (4).

6.2.1. TRAINING PURPOSE The purpose of improving training was to ensure that systems "stayed sold," and to uphold a reputation for customer support.

6.2.2. TRAINING AIMS Re-assessment had revealed that trainers were individually capable and well-disposed. It also revealed their need for 1. timely information on sales and product release, 2. higher morale and mutual support, 3. better understanding of learning, 4. coaching on teaching technique, 5. methodology to produce and use self-paced instruction.

The aims were to help them become more collegial, more confident, and more competent. It implied moving away from the apprenticeship model of instruction, but preserving concern for craftsmanship. Most had no formal training in either instructional methods or theory.

6.2.3. TRAINER TRAINING GOALS

1. Grasp basic learning theory
2. Learn new training methods and gain confidence through practice
3. Experience collegial support and instructional criticism

LEARNING Thursday	SKILL TRAINING Friday	STRUCTURE (1) PRACTICE (2) Monday	PRACTICE Tuesday	COURSE DESIGN (1) TRAINER ROLE (2) Wednesday
Introductions Course Purpose Industry Training Learning Theory Why People Work	Group Process Environment Interpersonal 1860-1960 Hot Metal 1960-2000 Computer	Rvw Highlights Understanding & Being Understood Critiquing Present: Units$_{(1)}$	Present: Module Outlines	Rvw Motivation Questioning Section, Module & Course Design Present: Units$_{(1)}$
BREAK	BREAK	BREAK	BREAK	BREAK
S-R Bonds (A) Neurophysiology Kinesthesis Left/Right Brain Near Conscious	S-R Bonds (B) Whole/Part Learning Association Learning Environment Training Techniques	Present: Units	Present: Sessions	S-R Bonds (C) Link & Develop Reinforcement Sessions & Units Standards Learning Retention
LUNCH	LUNCH	LUNCH	LUNCH	LUNCH
Experience Self-Paced Training Flexible Thinking	Standards Reinforcement Learning Retention Meeting Needs	Mnemonics Applied Learning Command Language	Session Critique Sales & Training Training Failures Operator Manual	Trainee Queries$_{(2)}$ Group Project Collegial Approach Managing Upwards Self-Paced Training Reinforcement
BREAK	BREAK	BREAK	BREAK	BREAK
Attitudes Understanding & Being Understood	Course Structure: Sec/Module/Unit Role Play: Units	Course Sequence$_{(2)}$	Course Practice: Sec/Module/Unit System Problems	The Instructor: Learning Manager Managing Upwards Course Evaluation
HOMEWORK Prepare Units	HOMEWORK Prepare Units	HOMEWORK Outline Sec/Module	HOMEWORK Prepare Units	HOMEWORK [none]

Figure 6.3: Course Structure for Trainer Training

4. Establish a training design team
5. Specify a course for the most critical system
6. Provide follow-up support for the self-paced training design

6.2.4. TRAINING STRATEGY The strategy was to gather the trainers for a one week course on training purposes, aims, goals and methodology and to provide sufficient time and support for group sharing to lead to more collegial behavior.

6.2.5. PROCEDURE The course delivered appears as Figure 6.3, page 7.

6.3 SELF-PACED TRAINING SPECIFICATION

The aim of the self-paced training specification was to provide a structure to guide the training design team in producing a course. The consultant specified modules (in outline) that instructors could assemble, as appropriate, into a course for each trainee category. The consultant also specified an approach to writing a standard course unit and provided a sample of a unit. Early modules formed a generic core, later ones dealt with application and customization.

6.4 TOP MANAGEMENT ROLE

No intervention at the top management level took place.

7 RESULTS

The "purpose of the project was to provide training and consultation on two aspects: managing change and the upgrading of instructional techniques. . . ." At the outset "management considered the new assessment and chose to address the long term opportunities presented in the recommendations accepting that the results would appear more slowly." (Extracts from internal Company Project Summary, 1986)

7.1 COMMUNICATIONS RESULTS

The Communications Group after wrestling with the concepts of 'purpose,' 'aim,' 'goal,' and 'strategy,' grasped them, then used the terms well to define their tasks to the executive committee and get two projects chartered.

The participants learned that their purposes, aims and goals were in harmony, then formed bonds and a group identity. After three of the five training days they were so elated over their sense of mission that consultants had to lower their expectations and to counsel patience: their actual power in the company consisted not in rank but enthusiasm, group cohesiveness, and ability.

Thinking in terms of purposes and aims led them to the need for interdepartmental connections. This applied to their observation that the organization operated like a "set of six separate companies," rather than as a single entity with six divisions. Realizing their common interests, the members assured one another that they would attempt to set up 'early warning' when for example a part was running out, a customer had a peculiar training requirement, or a new software release was pending. During the course they had learned to think 'flexibly,' 'laterally,' 'vertically,' "pro-actively" - more creatively.

This group completed its first task successfully two months after training.

The training department appeared on the customer profile distribution shortly after the course. Information flowed directly from outside vendor to training and back. The vendor stated that, although not sure whether the the result was worth three full days, they now understood the company organization much better and could provide and receive critical information reliably.

The training department manager still serves as a clearinghouse for distributing information to and from the vendor. He has also begun working on sales administration. Improved sales information should help well beyond training.

It appears that communication between R&D and training has not improved. Management turnover may be more responsible than procedures or receptivity.

7.1.1. EVALUATION "This group received a charter to undertake several important projects . . . . It completed or made substantial progress on all . . . . it had both proven its value and demonstrated the usefulness of its methods in adapting to change." (Extracts from internal Company Project Summary, 1986)

7.2 TRAINING RESULTS

7.2.1. TERMS OF THE TASK Like the Communications Group, the trainers first grappled and then grasped the concepts of 'purpose,' 'aim,' 'goal,' and 'strategy.' Analyzing the purposes of training for a customer was unfamiliar. Having done it, they reported successfully applying it to teaching. Much time also went to transmitting standard instructional ideas. The underlying structure of a training unit, as differentiated from a machine operation, was new. They received devices such as Allow Practice Check and Correct ('APCC') enthusiastically. Teaching interpersonal techniques (see Figure 6.3, page 7) likewise was both necessary and gratefully accepted.

7.2.2. GROUP FORMATION The trainer group already had close relationships and years of working together in many cases. However, the idea of collegial support and criticism of curriculum and training was new. Establishing common language and concepts for trainers based at the company and those in the field was particularly important. A collegial spirit did evolve.

7.2.3. EVALUATION, TRAINING GROUP In the trainees words, they "had gained confidence and also learned not to limit thinking to a particular line of thought, to keep an open mind and to think positively rather than negatively. The group felt that they had a greater insight into the learning process and the learning techniques which would enable them to carry out training more efficiently."

7.2.4. TRAINING DEVELOPMENT Consultants produced a full self-paced training specification that did not (as a result of changes in the market) become a course, but did lead to revised 'small classroom' training.

8 DISCUSSION

Poor lateral communication rooted in company structure and ethos and revealed by re-assessment was the problem presented as "training failure" and interpreted as a training need. General improvement required changing leadership style. Despite the managing director's support, top management resisted.

Trainer training successfully altered the training style because all those responsible for action participated, because they began to get current information, and because they felt the company cared.

Our hypothesis was that the Communications Project might catalyze change in corporate ethos by attacking problems through second level management.

Many ('modernist') theorists believe that corporate ethos changes stepwise from the top down: "A manager can manipulate worker behavior by manipulating . . . externals. . . . almost any kind of behavior can be elicited." (Sullivan 1985, p. 543). Our intervention does not contradict this view, but the Communications Project succeeded where those responsible for action were in the group. They had power and authority to set up meaningful lateral relationships and did so, thus improving information flow in and out of training.

Thus the Communications Project did after all demonstrate the value of changes in thinking style by creating lateral connections. This should help in reverse, eg. to keep R&D in touch with customer needs. F. A. M. J. Faas (1985) points out "While it was originally believed that technological feasibility is the more important factor in technological innovation, the research efforts report that 3 out of 4 successful innovations originated in demand recognition."

Sun Tzu (1963) says these are "(III.19) ways a ruler can bring misfortune upon his army: 20. When ignorant that the army should not advance, to order an advance or ignorant that it should not retire, to order a retirement. . . . This engenders doubts in the minds of the officers. . . . (VIII.8) There are occasions when the commands of the sovereign need not be obeyed." In the same sense when a task group receives its purpose, aims, and a charter from a chief executive, it legitimately may 'violate boundaries in the territory' to which it has been ordered. Of course its actions must remain in keeping with its charter. Therefore clear mission statements are essential.

Organizations under pressure must develop, crystallize, and persuasively publicize appropriate purposes and aims. Accurate, timely information ensures that leadership will not be ignorant of the affairs of their divisions. In this company promoting such communication was at the heart of the intervention. Values and roles were too fixed; "a shift from the exchange of value to the exchange of meaning" was needed (Sullivan 1985, p 545). Success in evolving appropriate human-computer interaction was proportionate to the quality of communication.

9 CONCLUSIONS

The Communications Project improved information flow and in turn, training.

Apparently the Project threatened departmental boundaries and did not catalyze modification of corporate ethos. That probably would have required ***initially*** changing senior management style. Training aims alone were insufficient to ensure project survival, even with the managing director's support.

Trainer Training successfully implanted a fresh idea of the sales purpose of customer training and transmitted to the trainers appropriate aims, goals, and methods. Collegiality rapidly evolved during training and probably permanently will support development of customer training.

10 REFERENCES

Faas, F. A. M. J., "How to Solve Communications problems on the R&D Interface," **J. of Management Studies, 22** 1985, p 83.

Sullivan, Jeremiah J., "Human Nature, Organizations, and Management Theory," **Acad. of Management Rvw. 11**, 1985 (534-551)

Sun Tzu, 1963, **The Art of War**, tr Saml B Griffith, Oxford U Pr, Oxford

Social, Ergonomic and Stress Aspects of Work with Computers,
edited by G. Salvendy, S.L. Sauter and J.J. Hurrell, Jr.
Elsevier Science Publishers B.V., Amsterdam, 1987 — Printed in The Netherlands

QUESTIONABLE ASSUMPTIONS AND MISCONCEPTIONS IN ERGONOMIC RESEARCH ON HUMAN-COMPUTER INTERACTION: A DISCUSSION

JACK LITEWKA
President, Applied Ergonomics, P.O. Box 9978, Berkeley, CA 94709, U.S.A.
(after July 10, 1987, c/o Professor Jane W. Ellis, School of Law, University of Washington, Seattle, Washington 98105, U.S.A.)

ABSTRACT

Noting differences between the conclusions of ergonomic researchers and his own hands-on (workplace) experience, as well as contradictions among researchers, questions are raised about the validity of certain assumptions in ergonomic research focused on the human-computer interface. A few of the more common fallacious assumptions in this area of research are high-lighted. A recommendation is made calling for closer cooperation between research and hands-on ergonomists in the design of ergonomic studies.

1 INTRODUCTION

In reviewing the literature of the past few years relating to the ergonomics of computer workstations, I was aware that some of the conclusions reached in studies undertaken by a variety of researchers not only contradicted each other but also often had little correlation to many workplace observations and successful interventions that I, and other workplace ergonomists that I know, have made over the years.

I have identified a number of research assumptions which I consider highly questionable. A few of these are dicsussed here.

2 COMMON QUESTIONABLE ASSUMPTIONS IN WORKSTATION RESEARCH

2.1 Assumption #1: That 10-minute trials are valid tests.

If one wanted to identify potential marathon runners, would the screening process consist of a 100-meter-dash test? For those unfamiliar with the world of runners, the answer is an emphatic "No!"

I think a similar type of mis-analogizing takes place in many computer-keying studies: 10 minutes equals 1/42 of a work day, 1/210 of a work week, and approximately 1/10,500 of a working year. It is risky to predicate behavior, formulate conclusions, and issue recommendations on experiments of relatively short duration.

How can one know a priori the length of a test or trial necessary to produce valid results and conclusions? The answer is: one cannot know. Only field experience and trial-and-error create proper guidelines, and it of course depends on what is being assessed. Here are some guidelines based on my experience and observations in the field.

If we are talking about measuring keying speed and accuracy and then being able to extrapolate to day-long/week-long/month-long productivity, a 30-minute test can provide a high degree of correlation.

If we are talking about evaluating the effects of various postures on productivity and on pain causation and alleviation, then longer -- sometimes much longer -- observation periods are required. It can take hours (or even days) for a posture's defect to show up as symptomatic discomfort or pain. In some cases, discomfort arises within minutes, but one still needs to evaluate if the discomfort is due to "bad" posture or to the sudden neuro-muscular readjustment that the new posture requires: i.e., there are occasions when discomfort is a danger signal and other times when it is a transition stage toward improvement. In still other cases, it may take months or years for symptoms to emerge that are caused by "bad" posture or equipment design.

If we are talking about alleviating discomfort/pain by means of postural re-alignment and the neuro-muscular retraining period that must follow, then the answer is: it depends. Sometimes relief happens instantly, sometimes gradually, sometimes not at all. Usually, if the changes being made are major, then as much as 2 days is required for the new posture "to take" -- i.e., for the the initial stage of comfort and kinaesthetic readjustment to take place. But it can take many weeks in instances of postures that have been held for more than 10 years and/or that call for dramatic postural readjustment. The assistance of an experienced ergonomist or trainer can greatly ease the transition.

Is it meaningful, then to have a keyboard operator test a work-station design, posture, chair, or keyboard for 10 minutes? Based on my experience, it is not. And, when varying positions of equipment and human postures are being assessed, even hour-long tests can be misleading.

2.2 Assumption #2: Asking operators their preferred positions and set-ups leads to the healthiest, most ergonomically sound postures.

A fallacious assumption closely related to 2.1 (above) concerns subjective reports of operator "comfort."

Years ago, when I went through a keyboard-ergonomics apprenticeship with Lillian G. Malt to become a trainer for Malt Keyboard Dynamics, she told me a story about a linotype operator that seemed far-fetched... until I had similar experiences at The New York Times and the Philadelphia Inquirer. At both newspapers, I was retraining operators from linotype machines to typesetters driven by QWERTY/Sholes keyboards. As I watched them work at the linotype machine, I noticed -- especially with the "old timers" -- considerable and even severe body contortion. One shoulder was higher than the other; one shoulder was more advanced than the other; the lower spine was in extreme kyphosis; the upper back and neck were bent far forward; and the left and right wrists were bent in opposite ways. When they stood up, the contortions did not entirely disappear: their bodies had been transformed by years of conforming to the peculiar demands of the linotype machine´s design.

In the new keyboard room, I attempted to get them into a proper posture for working at a QWERTY/Sholes keyboard, and they complained bitterly. They were not "comfortable". Sitting at the linotype machine was "comfortable", but sitting at this modern keyboard was "torture".

So what does "comfort" mean? Some might think that a silly question because the answer is apparently obvious. After all, each of us knows when we´re comfortable and when we´re not. End of subject.

Not really. It has often been observed that extra-soft sofas are good sellers because people sit and instantly feel "comfortable". It has also been noted that after an hour or so (and often much sooner) on mushy sofas, the same people will complain about being "un-comfortable". So what does "comfort" mean. An accurate definition of comfort would have to specify, at the very least, the <u>duration</u> of the proposed comfort and the <u>function</u> being performed during the state of comfort (e.g., reading, drafting, typing, etc.).

During the 5-week courses with linotype operators, I learned what they liked to do in their off-the-job hours -- and it was oddly limited. Did these men like to go to movies? No. Did they like social dinners? No. To make a long story short, anything that required them to remain in "normal" sitting postures was anathema to many of the linotype operators because it was "un-comfortable".

While this is an extreme example, I have found that the lessons it contains hold true for less dramatic situations: (a) reports of "comfort" and "discomfort" are not only highly subjective but also can change as operators adjust or maladjust to varying situations over time; (b) operators´ preferences cannot be assumed to be optimal, healthy, and ergonomically sound; and (c) a carefully considered training program of appropriate duration is required to undo existing neuro-muscular "habits" and then to develop new neuro-muscular patterns. We should constantly remind ourselves that people, when placed in new situations, tend to recreate/revert to their "normal" behaviors.

I am not denying the importance of feedback from operators at every stage of a study or training program: such feedback is essential and is one of a number of factors to consider. As ergonomists, we have to "translate" what operators mean when they say they are not comfortable. This is very difficult because often they themselves do not know what they mean. If placed in a new posture, they might say, "I´m not comfortable." They may be trying to say, "I can´t sit this way," or they may be trying to say, "I can´t perform in this position." What a workplace ergonomist does to help operators depends on the operator´s real meaning -- and remedial strategies ought to vary accordingly.

2.3 Assumption #3: There is a "correct height" for the tabletop that ensures proper operator posture, comfort, and productivity.

Human bodies differ dramatically not only in height, weight, and shape, but also in limb lengths and in proportions of limb sections (e.g., upper arm to forearm). What is important is the relation of a particular body to the workstation. A "correct height" for a table, in and of itself without reference to the user´s body, is a meaningless abstraction.

As a rule-of-thumb, it is better -- if variable-height tables are not an option -- for a table height to err on the side of being too high rather than being too low because it is possible to "build up" beneath a person with adjustable footrests and chairs. Lowering the floor because the tabletop is too low for an individual usually is not an option. Obviously, a wider range of adjustability for tables, chairs, and footrests is preferable.

I have read that some researchers are surprised that tall people often prefer lower tabletops. It should not be surprising. Standing height is different from sitting height. The ratios of body segments and parts vary greatly. Some people, for example, are long in the torso and short in the legs; others are the reverse. Arm lengths also differ significantly. In the case of, say, a tall person with long legs, a medium-long torso, and very long upper arms, a low tabletop will usually be required. (Other "ideal" measures are also meaningless abstractions, for example, the "ideal" distance from stomach to keyboard: clearly, forearm length and hand-and-finger length have more than a little to do with what that ideal distance is for any given individual.)

2.4 Assumption #4: An ergonomist does not need to have a first-hand, intimate knowledge of the task being performed in order to study it.

If a similar statement were made about a dance teacher or a basketball coach, eyebrows would immediately be raised. I am aware that "doing" is not the same as "teaching" or "studying". On the other hand, I am not aware of any dance teacher or basketball coach who has not danced or played basketball. One need not have been a great performer to be a good teacher or coach, but one does have to have had the experience of that type of activity -- mentally, physically, and emotionally -- under differing conditions and periods of time to develop a "feel" for what´s going on the subtlest of levels.

The importance of "feel" crops up repeatedly -- and its absence looms large. One of its most disturbing appearances is in those cases in which researchers describe a posture being assessed and refer the reader to a photograph that purportedly demonstrates the posture -- when it does not. The description and photograph are similar enough so that one can be certain that it is not a mistake. Rather, it becomes clear that the researchers do not recognize or do not consider it significant that the posture supposedly being tested is not the one actually being tested. The researchers either do not have an "eye" for noticing the non-achieved posture or they seem to believe that a few degrees here or there do not matter. This discrepancy between the described and the actual posture lessens or in some cases completely undermines the validity of the postural assessment.

This lack of "feel" also appears in the form of "the loose analogy". For example, comparisons have been made between the sitting positions of a cab driver and a keyboard operator. Yet, there is little meaningful relationship between these two tasks and what they require of the human being. Even if a cab driver is comfortable in the driver´s position, one cannot assume that a keyboard operator, pianist, or potter would be comfortable in that position.

The appropriate postural adjustment to alleviate a specific pain can be minute -- it is sometimes startling what a big difference a small change can make -- and require a trained eye and feel for the body. Is it likely that one will notice subtleties, analyze complaints, and feel/intuit what another person´s body is experiencing, doing, or needs to do if one does not have keyboarding experience? My own observations and readings tell me that it is not.

3 GENERAL REMARKS ON AN APPROACH TO ACHIEVING A DESIRED POSTURE

This is not the place that I will enter the ideological debates about what the ideal posture is for operators at a computer workstation. Instead, I want to discuss what I believe to be the correct approach for arriving at the optimal posture for any keyboard operator -- always with health and productivity foremost in mind.

When posturing any person for any task, one should begin with the part (or parts) of the body actually doing the preponderance of (or the most difficult part of) the work. Working backwards from the desired result is a classic problem-solving technique, and it works wonderfully when employed to achieve a desired posture. In the case of a human-computer interface, the fingers and the eyes are the primary "workers."

For the record, I will state that the posture that E. Grandjean has termed "wishful thinking" is, in my experience, easily achieved, easily maintained, and without question the posture that generally is the most effective and therefore the most preferred. (The "wishful thinking" posture, to simplify, consists of right angles at ankles, knees, waist, and elbows; lower legs, upper arms, and torso-neck-head oriented vertically; and thighs and forearms oriented horizontally.) And, while my leanings are toward a constant posture and while I am concerned with the problems caused by constant alteration of position, I am a strong believer in frequent stretch breaks and mini-alterations in one´s sitting position.

But even if you subscribe to a different "ideal" posture than I do, the approach should be the same: begin by achieving what you believe to be the correct (optimal) positioning of the fingers. (Because it is usually fairly easy to position the monitor in a way that is acceptable for the eyes, the needs of the eyes can be accommodated later on.) After you have the fingers right, position the hands correctly, then the forearms, and then the upper arms. Having accomplished that, you can now position the torso, and then the upper legs, lower legs, and feet. Now, position the monitor so that the neck is relaxed and the eyes are able to read the screen. (This can be more difficult than it seems, but an experienced hands-on ergonomist can accomplish this in a few minutes.)

Warning: If you start with a concern for an abstract, ideal table height or seat height or other rigid measure, it often results in the ergonomic equivalent of painting oneself into a corner.

A final note on posture... Just as people learn in different ways and at different rates, the ergonomic trouble-shooter must recognize that different bodies "learn" in different ways and at different rates: intervention strategies must be implemented with this in mind.

4 CLOSING REMARKS

In the computer workstation area, the number of cases of repetition strain injury (RSI) and other ailments is rising. RSI can develop in as short a period as days or weeks, although more frequently it takes many months for serious cases to develop. It varies from being an annoyance to being crippling, as occurs in severe cases of carpal tunnel syndrome.

There have been a few attention-getting court cases in which the damages awarded to employees were considerable: employers and/or equipment manufacturers have been held liable for physical injuries incurred while keyboarding. Numerous workstation-related grievances have been filed by employees, and other court cases brought by employees are still pending.

Of the thousands of professional keyboard operators trained by Malt Training Services (London), Malt Keyboard Dynamics (formerly the U.S. franchise of Malt Training Services)), and Applied Learning Corporation (King of Prussia, Penn.), no cases of RSI have developed. The keyboard professionals trained included data-entry operators, typesetters, classified-ad takers, and so forth. In many cases, trainers arrived at a client company and were told about finger and wrist pains: in almost every case, a combination of postural adjustment and neuro-muscular retraining eliminated the symptoms reported.

The costs in damage to human beings and in dollars paid by businesses are increasing. These injuries and payments are unnecessary: all that is required are improvements in equipment design and in the neuro-muscular retraining of professionals using computer workstations. To realize these improvements, collaboration between ergonomic research scientists and hands-on ergonomists is necessary.

Social, Ergonomic and Stress Aspects of Work with Computers,
edited by G. Salvendy, S.L. Sauter and J.J. Hurrell, Jr.
Elsevier Science Publishers B.V., Amsterdam, 1987 — Printed in The Netherlands

"80-20 OA" METHODOLOGY TO COST JUSTIFY OFFICE AUTOMATION

DAVID L. SHAY
White Collar Quality and Productivity, Peat Marwick Main & Co, 345 Park Avenue, New York, New York

ABSTRACT

The "80-20 OA" White Collar measurement is a proprietary methodology developed at Peat Marwick to quantify benefits provided to knowledge workers from the installation of new technology. "80-20 OA" contains a series of manual and microcomputer supported diagnostics which Peat Marwick has developed to report on productivity, quality, and effectiveness changes for knowledge workers and other white collar workers. The experience to date is that office technology can produce sizable enough productivity and quality gains so that management can convert these to tactical, competitive, or strategic advantages.

PRESENTATION SUMMARY

The paper will recap the "80-20 OA" methodology, principal findings, and share some of the findings from studies at Du Pont, GE, GTE, Kodak, Martin Marietta, Peat Marwick, University of Chicago, and other locations.

PRIMARY ELEMENTS OF "80-20 OA" METHODOLOGY

"80-20 OA" Methodology

The "80-20 OA" methodology is based on The Pareto Law which states that 80% of the value added contribution of an organization is probably based upon 20% of the key work efforts performed. Perhaps 20% of employees are providing the majority (80%) of this value added.

Key management team members supply the knowledge of the critical issues facing the organization. The employees who perform these activities describe the processes used to deliver the key products or services. The consultants zero in on key issues and work efforts only, rather than studying, surveying, or measuring all the activities going on.

When the most important processes are understood by our consultants, we ask the best people to work with us in focus groups to determine how office technology or any other change can improve the process. We then quantify the improvement potential by any one of a number of accepted industrial engineering and interviewing practices.

PRIMARY ELEMENTS

1. Identify the Management Issues that Drive the Business

We first conduct Scoping Interviews with key executives to gain a sense for the mission of the enterprise, its history, future directions, strengths and weaknesses in its product offering, customer relations, profitability/cost pressures, etc. Scoping interviews provide us the insight into where management wants the organization to go and why.

In addition to providing management insight into the business mission of the organization and critical success factors, those senior executives must frequently direct us to managers that have the most knowledge of what the organization is trying to do to fulfill its mission. We call this set of interviews Directional Interviews because at their conclusion, we know precisely where management wants help and which categories of activities we will study. We know the names of most of the respected employees who perform the work, and probably know the majority of the work efforts that can benefit most from technology or any other changes.

2. Interview the Movers and Shakers

In addition to the key executives mentioned above, we will now complete the necessary interviews and flow diagrams with those individuals that are providing the how to meet your business executives. These are the individuals or class of individuals that are responsible on a daily basis to "make it happen."

3. Identify The Key Work Efforts in Your Organization

We will complete the list of key work efforts, (usually 60-100 for an organization) and obtain your management sign-off that these are indeed the critical activities to your enterprise.

4. Identify Improvement and/or Problem Areas Using a Holistic Approach

Peat Marwick is unique in its problem identification/solving methodology in that we employ a Holistic Approach to identifying productivity and quality improvement opportunities. When a problem or opportunity is identified, we test for solutions against twenty proven improvement techniques developed by The Information Systems, Industrial Engineering and Human Resource disciplines. Invariably, a combination of these techniques provides the optimum solution, not just office technology alone. (Figure 1)

Even though "80-20 OA" is primarily an office technology study, the Holistic Approach assures a multifunctional, multidisciplined view of each problem. We are not interested in mechanizing unimportant tasks and have no interest in pushing technology for technology's sake, or automating procedures that might best be improved by a host of other techniques outside the Information System's sphere of influence. We believe that white collar productivity can be improved by changing a combination of four things: the People, their Tools, their Processes, or the way they are Managed. Our Human Resource specialists provide major insights into the best solutions to change People. We utilized a combination of Information Systems and Productivity experts to change the Tools. Our Industrial Engineers provide the major expertise for Process changes. We do not believe that any other management consulting firm brings this type of scope to assisting its clients to capitalize on all opportunities and resolve all problems.

5. Quantify Effectiveness Gains

While productivity measurement is a major emphasis of the "80-20 OA" methodology, Peat Marwick's extensive experience in quantifying gains from office technology has repeatedly demonstrated that these devices are often best justified because of effectiveness arguments. The ability of the knowledge work to expand his or her creativity and thought processes has been noted to produce quantum gains in quality, customer service, response time, and an ability to tackle demands that were impossible to consider just months before.

Because Peat Marwick has taken an early lead in the development of measurement of quality and effectiveness; we will answer management's concerns expressed in the Scoping and Directional Interview phase with acceptable quantifications of the effectiveness improvements possible.

6. Quantify Efficiency/Productivity Gains

While senior executives have frequently been able to leverage effectiveness gains from office technology for major competitive advantage, the measurement of productivity gains is paramount if they are to justify their expenditures to budget-minded superiors.

Peat Marwick utilizes a variety of industrial engineering techniques to measure the before and after productivity impact of office technology. Our Operation Management practice area is headed by a past president of the 43,000 worldwide member Institute of Industrial Engineers (in this country, The AIIE American Institute of Industrial Engineers). We pioneer and utilize the latest thinking and techniques in the measurement and improvement of productivity. Indeed, our "80-20 OA" methodology is spear headed by individuals from our 150-person Operations Management practice and then supported by experts from our 650-person Information Systems practice and the 175 practitioners from our Human Resource/Organizational Development disciplines. Additionally, Peat Marwick's Cost Control accountants, coupled with our industry experts, know all the areas where tracking and monitoring of expenditures is critical and how to derive correct data from cost-reporting systems. All of these individuals recognize the importance of quantifying results.

Peat Marwick has invested in a variety of microprocessor and mainframe applications over the years to increase the creativity and decrease the time consumed for consultants to perform certain types of quantification activities. This not only provides a standard of high quality product for our clients, but also takes advantage of mechanization to decrease our fees.

Our structured FOCUS on Productivity and Quality (P's & Q's) process is designed to provide measurement data on both efficiency and effectiveness. FOCUS P's & Q's is best utilized when measuring white collar/knowledge workers where work efforts are not discrete in nature or are performed by groups.

For example, preparing periodic financial cost exhibits is frequently a discrete job of updating year-to-date figures with this period's results. It is clerical, routine, and might be automated by office technology.

Preparing the monthly report which analyzes the trends in the periodic data, possibly including extrapolations and graphic presentations of the data is a not-so-discrete work effort of a knowledge worker. Because the complexity and duration of the work effort "Analyze and Report on Periodic Financial Data" will vary based on the inferences in the data and the creativity of the analyst, it is important to gain insight into effectiveness issues as well as "time spent" issues.

Additionally, the work effort "Analyze and Report on Periodic Financial Data" will probably be performed primarily by a financial analyst (prior to the introduction of integrated office technology) with some analytical help from a supervisor and typing support from a clerk typist.

In the new integrated office, it could be that a secretary or administrator not previously involved in the process will guide some software through analytical routines. The analyst will check the data from a remote location where he/she is visiting to obtain a commitment to improve the situation (a task he/she never had time to do before), the supervisor will actually update existing text from the prior period, the administrator will prepare most of the exhibits and graphics using approved practices and conventions. The analyst might type the agenda and introductory overhead slides.

The unit cost of this operation will be driven down substantially by integrated office technology. The quality of data and quality of communications should improve significantly with more time for analysis, more powerful computer programs at work, and high-quality graphics. The effectiveness of the department might improve because analysis and presentations on the data will be more timely. In the past, with the little time allotted for this task, all that management might have learned was that there was a budget overrun. Now in the same time frame, management understands where the overrun occurred and a financial analyst has helped a management person plan action to get back on course.

Peat Marwick's FOCUS P's & Q's is a structured methodology, supported by microcomputer data manipulation, to quantify both the efficiency and effectiveness gains experienced from all the job shifting that can occur upon the introduction of technology to help the knowledge worker. Whether it is relatively simple office technology, sophisticated CAD, CAM, CIM, JIT, or organizational changes prompted by other phenomena, this methodology captures the impact of change on individual or group efforts.

In most instances, Peat Marwick conducts one or two secondary-level measurements which are used to confirm or challenge findings from the main measurement technique. Regardless which Peat Marwick productivity or quality improvement measurement technique is incorporated, the output of this phase is a quantification of the productivity gains available from the technology.

7. Relate Benefits to Management's Concerns

With the cooperation and feedback from a variety of levels within the client organization, Peat Marwick's management consultants will present the quantified effectiveness and efficiency gains to management. Because we are management consultants rather than strictly measurement engineers or systems designers, senior management welcomes our recommendations to leverage these gains to produce tactical, strategic, or competitive advantages.

Peat Marwick's presentations and written reports are designed to answer the senior executives concerns that the new technologies can address the critical success factors of the enterprise and offer productivity gains to pay for the technology. The format most often is a direct listing of:

a. management's concerns

b. the way technology addresses the concerns

c. effectiveness gains and opportunities

d. productivity/quality gains and the payoff from technology

In all the studies we have performed to date, management has decided to continue and even increase their investment in technology because of the business gains available. The dollars and cents mentioned in the productivity section of the presentation serves merely as "icing on the cake." Management has bought the technology for the business gains it offers, and is pleased that it also has a productivity payoff.

8. Design of the Integrated System

Once Peat Marwick has defined the business needs that technology can address, they can be converted to functional specifications for a systems design. An RFP or RFI can be written to see which hardware or software can best address the specification. Peat Marwick's Information Systems professionals can assist the client's systems department in these efforts or take the assignment in its entirety. Our systems experts are available to assist in any of the steps from design through implementation.

9. Business Training

One of the most frequently overlooked aspects of installing office technology is business training. Vendors offer some training in how to operate their hardware and software. These courses are most frequently inadequate for novice users of systems and are woefully inadequate in training an employee to relate the new technological power to his or her department's objectives, let alone the business mission of the enterprise.

Peat Marwick has witnessed installation after installation where major effectiveness and productivity gains have been lost because management has not invested properly in training to protect their significant investment in technology.

There are a variety of reasons why people fail to realize the importance of investing in business training: the vendor would like you to believe that you can roll his technology in with no additional cost: the systems community that orders the equipment is so far up the learning curve of systems that they do not realize that other employees need basic and intermediate training: the people who order and install equipment are examining clinical function specifications of key strokes, access times, and failure rates, and are not familiar at all with the business needs that created the specification, etc.

Peat Marwick strongly believes that training dollars should be set aside to assist employees in installing the technology so it best meets the business needs of the enterprise.

One technique to help the training, installation, and continued success of the technology, as new employees are introduced to the system, is to have a "Picture Book" produced. The Picture Book features an example of the most common outputs or products of a given department. The book references the old unit cost of the product and recommends which individuals or teams of managers, professionals, technicians, administrators, clerks or secretaries should work to produce that product. The Picture Book shows employees and how to manage to the newer reduced cost.

10. Implementation

Peat Marwick's team of Systems, Human Resource, Productivity, and Quality management consultants are available to assist in implementing the new technology. Our experience has shown that this type of assistance is particularly useful when there is a great deal of work shifting from one employee level to another, or where the introduction of technology prompts departmental reorganization/downsizing. The project teams in charge of the design of advanced office systems rarely have the depth of staff, breadth of skills, or commission from Management to assure that department managers change their policies and procedures to achieve the lower unit cost structure.

SUMMARY

The "80-20 OA" methodology has been proven to be an effective way for management to cost justify its investment in new technology.

Those companies that do not quantify the possible benefits available from technology almost inevitably do not harvest the benefits. Management must know the advantages in the first place, manage to objectives, and monitor the results.

Figures 2 and 3 represent some findings that are common to many studies Peat Marwick has completed.

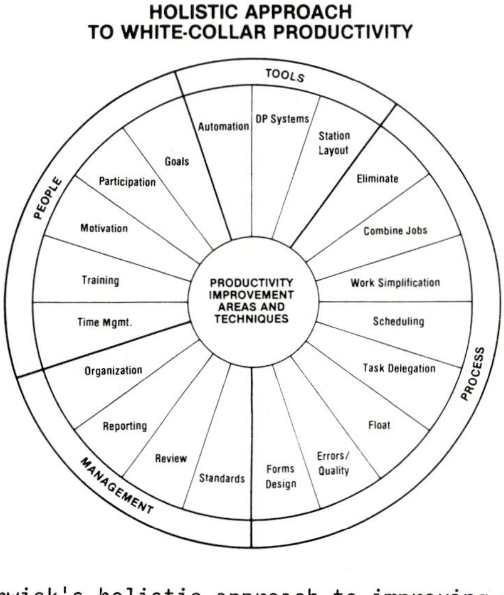

Fig. 1. Peat Marwick's holistic approach to improving productivity emphasizes the use of a proper blend of productivity improvement techniques.

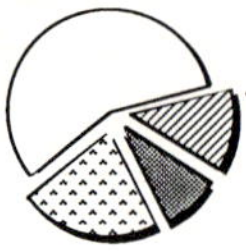

Better management planning and control and project design will reduce rework in the latter phases due to changes, yielding even lower final costs.

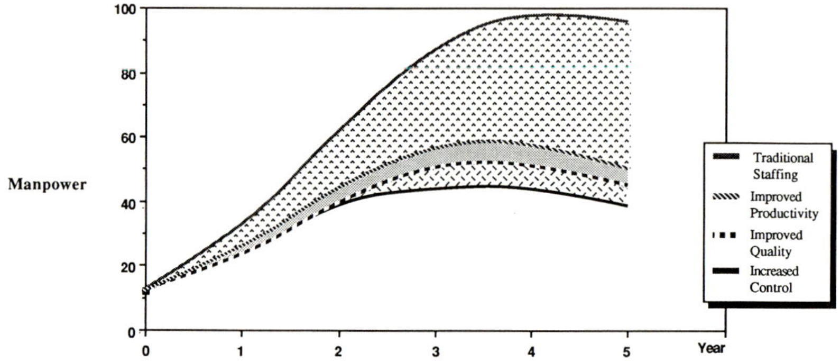

Traditional Staffing	30	62	88	100	95
Improved Productivity Way	24	44.6	57.2	58	49.6
Costs of Quality Savings	2.4	4.5	5.7	5.8	5.0
Improved Productivity &Quality Way	21.6	40.1	51.5	52.2	44.6
15% Less Design Changes	-	-	7.7	7.8	6.7
Improved Productivity, Quality and Control	21.6	40.1	43.8	44.4	37.9

Figure 2

Better design up front – avoid 15% of costs from design changes

Management can harvest the 24% higher through-put and quality-added in the design phases and cut rework due to changes in half.

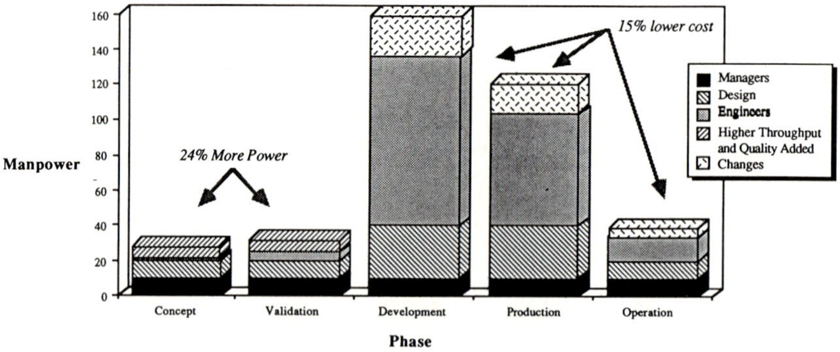

Managers	10	10	10	10	10
Design Engineers	10	10	30	30	10
Engineers	1	5	80	50	10
Engineers Saved	—	—	18	13.5	4.5
Cum Delta	—	—	18	31.5	35

Figure 3

Social, Ergonomic and Stress Aspects of Work with Computers,
edited by G. Salvendy, S.L. Sauter and J.J. Hurrell, Jr.
Elsevier Science Publishers B.V., Amsterdam, 1987 — Printed in The Netherlands

COGNITIVE PRESS IN COMPUTER-MEDIATED WORK

T.K. BIKSON
Behavioral Sciences Department, The RAND Corporation, P.O. Box 2138, Santa Monica CA 90406 (United States)

ABSTRACT

A framework is presented for conceptualizing the adaptation to computer-mediated tasks as a function of cognitive resources and demands in relation to established baseline performance. Information work is inherently cognitive since it involves generation, transformation and transmission of symbolic stimuli. Introduction of computers challenges extant response repertoires, but research suggests that white collar employees cope well with this initial stress. The incremental, cumulative and continuous changes to follow--termed press--can present greater difficulty. The framework suggests two avenues for facilitating adaptation: amplifying users' cognitive resources, and modulating task demands by shaping work technology around users' needs.

1 INTRODUCTION

Stress and adaptation are dynamic constructs that have been applied to explain human responses over a number of event domains--from changes in level of a perceptual stimulus (e.g., Helson, 1964) to large-scale social role changes (e.g., Kahn et al., 1964). Technological changes in the workplace and their consequences for human performance constitute another domain usefully studied by means of these concepts (Bell, Bikson, Rich and Wuchitech, 1976).

It is by now a commonplace observation that the U.S. is becoming an "information society," with ever larger proportions of the workforce manipulating symbols and images rather than objects and materials. The number of white collar employees, which exceeded 55 million in 1985, will top 65 million by 1995 (U.S. Department of Labor, 1983). These people are engaged in the generation, transformation or transmission of information (Bikson, 1987).

The key to the transition is the advance in computer-based technologies Bikson and Eveland, 1987). By 1995, it is likely there will be one computer workstation for every information worker (Roessner, 1984). The view taken here is that continuous innovation in information tools makes the white collar environment a "predisposing" one (Jaco, 1970). That is, it subjects its constituents to a high frequency of stress or press. New tools have this consequence by presenting task demands that fall outside well established work behavior repertoires, an inevitable concommitant of the innovation process.

These demands are inherently cognitive ones. This follows not just because appraising and responding to them theoretically require cognition (e.g., Lazarus, 1974; Osipow and Spokane, 1984) but also because the basic tasks of information work are themselves predominantly cognitive (cf. Karasek, 1981). This paper outlines a conceptual framework for understanding cognitive stress and press in the transition to computer-mediated work, applying it to to data gathered in recent RAND research (Bikson, 1987). Implications for facilitating adaptation to continued advance in electronic tools are discussed.

2 CONCEPTUAL FRAMEWORK: COGNITIVE STRESS, PRESS, AND ADAPTATION

2.1 Overview

The conceptual framework presented here was developed for a more general study of adaptation to rapid technological change (Bell, Bikson, Rich and Wuchitech, 1976). It draws on the research tradition initiated by Seyle (1956), emphasizing contributions from perception psychology (Helson, 1964) as well as from clinical and social-psychological research (e.g., McGrath, 1970; Lowenthal and Chiriboga, 1973). The figure below summarizes this composite approach.

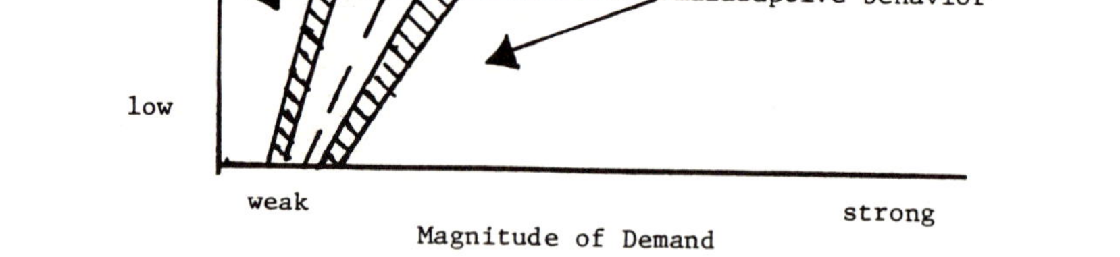

Fig. 1. Framework for conceptualizing adaptation to computer-mediated work.

Outcomes of interactions between people and information work environments are seen as a function of the level of task demands in relation to the task-relevant resources at hand, given cumulative baseline performance (adaptation level). Outcomes, roughly speaking, fall into one of three categories: the central (shaded) area represents greatest congruence between demands and resources, yielding adaptive task behavior and positive affect; striped areas on either side denote marginally adaptive behavior and tolerable affect; and the uncolored region represents maladaptive behavior and negative affect, a function of persistent divergence between task needs and capabilities.

2.2 Resources

In Figure 1, the ordinate represents the available balance of resources relevant to completion of a set of information tasks. Existing literature cites a great number of potentially relevant resources, ranging from genetic predisposition (Sharit and Salvendy, 1982) and personality traits (Brief and Atieh, 1986) to general health and social support (e.g., Williams, Ware and Donald, 1981). Such personal and social characteristics, however, are applicable to coping with all sorts of stress and have no special significance in white collar work.

Of particular relevance for adapting to electronic tools in information work are cognitive resources such as ability, experience, skill, training and learning support (cf. Turner, 1986). The flexibility of the tools provided and their conformability to the tasks at hand further condition adaptation (cf. Turner, 1984). The term "balance" calls attention to the mix of resources in work environments and emphasizes their complementarity (e.g., skill deficits can be compensated by stronger training). Individual variables are de-emphasized, giving more attention to organizational policy influences on task-related resources.

2.3 Demands

The magnitude of task demands (work-related stimuli that present an opportunity or a requirement for a response) appears on the abcissa in Figure 1. Work environments regularly present task demands that are met as a matter of course (see adaptation level, below). However, some task demands differ sufficiently from their predecessors to disrupt or threaten disruption of ongoing work work behavior. Two types of challenges to adaptation are proposed: stress and press.

Stress is the more recent and better-researched construct. By most accounts, stress results from a singular and visible change in the task environment eventuating in demands that fall outside the existing behavior repertoire (e.g., McGrath, 1970; Dohrenwend and Dohrenwend, 1974); current work methods do not generate an adequate response. Initial introduction of computers as information tools in an ongoing work context should--for most employees--be regarded as an instance. It constitutes a "critical work event," borrowing Osipow and Spokane's (1984) phrase (cf. "stressful life event," Holmes and Rahe, 1967). Stressors are taken to be normatively neutral. As occasions for unfreezing of current behavior patterns, such challenges might generate higher-level performance, lower-level performance, or simply different task performance on approximately the same level.

"Press" is a term used originally by Murray (1938) to denote elements in the environment that, together with a need, evoke a response. Like stressors, press are neutral in the sense that their positive or negative character is determined by functional consequences--how they operate to help or hinder the fulfilling of the need; for purposes of this discussion, task demands (getting work done) constitute the need. Unlike stressors, press are likely to be recurrent, chronic and/or cumulative, and individually scarcely noticeable. What is notable about press is not the effect of any one event but the number of events tending in the same direction (cf. aggregate effects of incremental changes in the environment, Studer and Barton, 1974). What makes such merely incremental shifts important is their uniform bias in a particular direction and their summation (Bell, Bikson, Rich and Wuchitech, 1976). In this way press can also eventuate in the disruption of previously viable task responses. Information overload (see Bikson, Quint and Johnson, 1984) is an instance of press. The same might be said of "underload."

2.4 Adaptation

People at work have learned response repertoires for dealing with a range of day-to-day task demands. The impact of an individual event is in large part determined by how similar it is to other environmental events. The broken diagonal line in Figure 1 represents the theoretical mean adaptation level for a person as a function of varying amounts of task-relevant demands and resources; the shaded area around it is a zone of normal deviation. Defined by approximate congruence between resources and demands made on them, activities within this region do not represent changes from baseline performance (cf. homeostasis, as in Osipow and Spokane, 1984).

Threshold-level events occur within the striped regions in Figure 1, at the bounds of adaptive behavior. Such events tax extant performance limits but available resources are sufficiently congruent with the demands to allow marginally adaptive responses to emerge. New behaviors, in turn, shift the adaptive performance zone to include areas previously outside the response repertoire and a new adaptation level results. In this way performance gains can be achieved as part of a developmental process--neither stress nor press is a necessary condition for positive change.

When task demands and available resources are seriously discrepant (in Figure 1, when their intersection lies outside the adaptation threshold), an adaptation problem arises; the existing response repertoire will not meet a recognized need. Such events induce "arousal," "disturbance," and the like (Berlyne, 1960, Lazarus, 1974), motivating attempts at problem-solving. Mastery of events that have caused a "hitch" in performance (Dohrenwend and Dohrenwend, 1974) is experienced with positive affect and boosts the adaptation level. Alternatively, failure to cope with stress or press results over the long term in impaired performance and negative affect (strain).

As Figure 1 illustrates, such maladaptive outcomes are predicted by either a deficiency or an excess of task resources in relation to demands. Deskilling, for example, would be regarded as a change in work environment resulting in an imbalance of the latter sort. This framework also suggests that preventive or ameliorative interventions can occur on either axis: adaptation to new work technologies can be facilitated by careful (re)design of tools, tasks and jobs as well as by an improved resource mix.

3 COMPUTERIZATION OF INFORMATION WORK: STRESS, THEN PRESS

3.1 Initial Stress

A wealth of literature on organizational innovation supports the thesis that the first introduction of a new tool is likely to be stressful in the sense just defined (Bikson, Gutek and Mankin, 1981). When it is a computer-based tool, the focus is usually on the physical workstation, which may well be a misappraisal. That is, the hardware interface is more like a vehicle or agent of change (Keen, 1984). The substance of the change--a change in how symbols are presented and handled--will be reflected in different ways of doing information tasks, bundling tasks into jobs, and distributing jobs among work groups (Bikson and Eveland, 1987).

Data gathered in a cross-sectional study of 55 work units in 26 different private-sector organizations implementing interactive information systems (Bikson, 1987) suggest that even familiar tasks present demands that exceed capabilities when new electronic tools are first introduced. We inquired about how organizations get basic task needs met during this period.

TABLE 1

WHAT INTERIM MEASURES ARE TAKEN TO HANDLE WORK WHILE USERS ARE LEARNING THE SYSTEM?

	Percent of Groups (n=55)
Send out some of the work	2
Hire temporary workers	6
People work longer hours while learning	19
People fall back on old methods	19
Work is redistributed within the group	21
No interim measures	31

Responses indicated that little attention is given to technological change as it involves task redesign and behavioral adjustment. However, the study also found that employees--even without additional interim resources--are generally willing and able to make the transition to an electronic information environment (Bikson, 1987). Many employees treat it as a positive and exciting challenge (e.g., Bikson, Stasz and Mankin, 1985; Stasz, Bikson and Shapiro, 1985); most report higher satisfaction (Gutek, Sasse and Bikson, 1986), suggesting successful adaptation to the initial stress.

3.2 <u>What About Press</u>?

The subsequent incorporation of computer systems into day-to-day information work is likely to generate press. An impact of small cumulative changes, press is harder to observe. And, if stress measurement presents problems, press must be even more difficult. Nonetheless, it is arguable that events like those grouped under "role ambiguity," "role conflict" and "role loss" are individually small but collectively syndromatic and more properly conceptualized as press than stress (Bell, Bikson, Rich and Wuchitech, 1976). There is considerable evidence of a similar nature for cognitive press in interactive information environments.

For example, employees surveyed in the large-scale field study of computer-based work (n=531) were found to perform more information tasks, and to do them faster and better than before this technology was installed (Gutek, Sasse and Bikson, 1986); their managers tended to concur. As a consequency, performance standards for work had increased in over half of the 55 units studied. By contrast, job descriptions changed in only a third of the sites and pay levels, in only a fifth. Employees in the main, then, are working harder/faster/better on cognitive tasks to keep up with the same job. As an editorial in Government Computer News (Perry, 1987) put it, now "the impossible takes longer." These appear to be instances of cumulative unidirectional effects of the sort here termed "press."

Accompanying these effects are changes in task variety and skill level. In the sample we studied (Bikson, 1987), a majority of employees and their managers reported that new and higher skills were involved. For some, the result was greater repetition (cf. Turner, 1984) and for others, greater variety. The difference, not surprisingly, is strongly associated with the status of an employee's work unit in the organization (Gutek, Sasse and Bikson, 1986). It is also a reflection of management policy, since organizations can choose whether to enrich or simply enlarge jobs (Bikson, Stasz and Mankin, 1985). Although employees evaluate their jobs and their tools more positively under conditions of greater task variety (Bikson and Gutek, 1983), the amount of cognitive press may not differ--rather, the difference may be in whether accumulating task demands exert press above or below adaptation levels.

Two additional contributors to press in today's electronic information environments should be noted. (1) While it is hard to project the emergence of specific tools at particular times, it is clear that there is no end in sight (Bikson and Eveland, 1986). Consequently skill obsolescence and skill replacement are not just a one-time threat to adaptation; information workers are confronting a demand for continuous change in task behaviors. (2) The need to meet that demand becomes increasingly important as computer systems become part of the infrastructure of white collar work, making the performance of information tasks depend quite directly on acceptance and mastery of the advancing technologies. The conclusion suggested is that most employees find the initial stress of conversion to computer use by and large manageable. It is the host of follow-on changes in information intensive work (cognitive press) that most endanger adaptation.

4 FACILITATING ADAPTATION TO COGNITIVE PRESS

The likelihood of long-term press in periods of rapidly advancing work technologies is corroborated by work unit managers (Bikson, 1987). When asked "how much change will there be in the nature of work for this group over the next 2-3 years," few managers thought the end was in sight. Over 70 percent believed moderate to major changes were yet to occur, even though most of the 55 work groups had been using computer-based tools for over a year. Given this scenario it is important to look at ways of facilitating adaptation to cognitive press. The framework presented here offers two approaches.

4.1 Enhancing Cognitive Resources

The most direct way to enhance workers' cognitive resources is to provide high quality training that enables them to understand and control the technology. An examination of training practices (Bikson and Gutek, 1984), however, suggests that organizations rarely offer the kind of learning that would enable users to take full advantage of their powerful, flexible new tools Rather, what training there is tends to be very concrete in orientation and to emphasize rote performance of sets of pre-defined steps.

In contrast, successful adaptation to changing tools requires a cognitive model of what they are. Providing users a model of how these tools function permits development of problem-solving strategies, allows for generalization and transfer to new, different or less-structured tasks, and makes the transition to the next generation of tools easier (Bikson and Eveland, 1986). Second, organizations should expect to provide long term learning support. White collar workers are better educated than ever before and should be able to make advanced use of information technology; providing only novice-level training does not permit them to deploy their strong task competencies and judgmental skills in conjunction with their tools. Finally, in addition to continuous learning, organizations should find ways of promoting development of local expertise and knowledge-sharing within work groups (Bikson, Stasz and Mankin 1985). Collaborative learning encourages "cognitive bootstrapping" and taps the collective knowledge resources of group members (Brown and Newman, 1985).

4.2 Alleviating Task Demands

Software design features have received much less attention than hardware properties in research related to effects of computer-mediated work, even though the software interface is most directly linked to the way information tasks are performed. Users' needs to alter their long-practiced work behaviors can be alleviated if the technology meets them half way--that is, if the software is well-adapted to the tasks at hand. Cross-sectional field research in varied task domains (Bikson, 1987) yields two software characteristics that are strongly associated with successful integration of information systems into ongoing work: functionality (or, goodness of fit between applications and job functions); and modifiability (or, extent to which the interface to these functions can be altered and controlled by users). Modifiability deserves emphasis because it allows the same system to serve highly individualized task behaviors and to change with users' changing skills and needs.

Job design features can also significantly influence cognitive press. As evidenced in a number of studies (Turner, 1984; Bikson, Stasz and Mankin, 1985; Bikson and Eveland, 1986), differences in the way tasks are grouped into jobs conditions their effects. For example, rapid pacing or repetition can better be accommodated if they characterize some but not all the tasks in a job. Job redesign thus allows organizations to alter demands that are not readily alleviated at the task level. By intervening at the subtask level (with software design) or the supratask level (with job (re)design), organizations can do much to modulate the cognitive press of computer-based information work.

4.3 Getting It Done

Preceeding paragraphs discuss alternative approaches for facilitating adaptation to electronic information tools, one addressing users' cognitive resources and the other addressing task and job demands. When managers of white collar work groups were asked about what roles organizations should take in helping employees adapt to technological change, they were undecided (see Table 2).

TABLE 2

WHAT IS THE ORGANIZATION'S RESPONSIBILITY?

Response	Percent of Groups
It's not the organization's problem	36.1
It's the organization's problem in the short run	30.5
There is a continuing role for the organization in this area	33.4

Although only a third of managers appear to look beyond the short-term stress of converting to computers, research on implementation suggests that management's commitment to support of technological change is an important ingredient in its long run success (Bikson, Gutek and Mankin, 1981). This does not mean that management should make all the decisions. On the contrary, user participation in decisionmaking is a consistent predictor of successful integration of electronic tools into ongoing work (Bikson, 1987). It is likely that their involvement in this process is the most direct way to link tool and job design to the substantive nature of the work performed (Bikson and Eveland, 1986).

5 COMMENTS

The intent of this presentation has been to take a longer view of human adaptation to new computer-based work tools than is represented by most research. Studies of interactive systems in varied white collar settings suggest that initial introduction of computers is the first in a series of changes in information-intensive work that will continue for years or decades. RAND research finds that coping with the first event is less difficult than many had predicted; managing the succession of events that follow is likely to be harder. The accumulation of cognitive press is less visible, and neither organizations nor individuals are well prepared to cope with it.

Guides to successful adaptation can be drawn from current implementation research. Recommendations include boosting cognitive resources and shaping task demands around human needs rather than technological capability. Potential interventions reviewed above generally involve empowering users (e.g., by making systems more intelligible to and modifiable by them, and by giving them key roles in implementation decisions). Recommendations also reinforce the group or collaborative nature of organized work (e.g., sharing knowledge resources, redistributing tasks). Such approaches to ameliorating cognitive press in technologized environments are consistent with more general research findings about significant mediators of stress and strain (autonomy and social support).

Finally, the presentation emphasizes implications for management policy. Organizations are able to exert considerable influence on the major determinants of stress and press in information work. In a time of rapid technological advance, organizations and individuals need to relinquish the homeostasis model and learn to manage change adaptively.

6 REFERENCES

Bell, D., Bikson, T.K., Rich, P. and Wuchitech, J., 1976. The Role of Education in Facilitating Adaptation to Technological Change: An Analytical Framework and Review of the Literature. The RAND Corporation, R-1821-NSF, Santa Monica CA.

Berlyne, D., 1960. Conflict, Arousal, and Curiosity. McGraw-Hill, New York.

Bikson, T.K., 1987. Understanding the Implementation of Office Technology. In: R. Kraut (Editor), Technology and the Transformation of White Collar Work, Hillsdale, NJ, Chapter 9: 155-176.

Bikson, T.K., and Eveland, J.D., 1986. New Office Technology: Planning for People, Pergamon Press, New York.

Bikson, T.K., and Gutek, B.A., 1984. Training in Automated Offices: An Empirical Study of Design Methods. In: J.W. Rijnsdorp and T.J. Plomp (Editors), Training for Tomorrow, Pergamon Press, New York, 129-145.

Bikson, T.K., Gutek, B.A., and Mankin, D.A., 1981. Implementation of Information Technology in Office Settings: Review of Relevant Literature. The RAND Corporation, P-6697.

Bikson, T.K., Quint, B. and Johnson, L.L., 1984. Scientific and Technical Information Transfer: Issues and Options. The RAND Corporation, R-2131-NSF, Santa Monica CA.
Bikson, T.K., Stasz, C., and Mankin, D.A., 1985. Computer-Mediated Work: Individual and Organizational Impact in One Corporate Headquarters. The RAND Corporation, R-3308-OTA.
Brief, A.P. and Atieh, J.M., 1986. Studying job stress: Are we making mountains out of molehills? National Academy of Management, Chicago.
Brown, J.S. and Newman, S.E., 1985. Issues in cognitive and social ergonomics: From our house to bauhouse. Human-Computer Interaction, 1: 359-391.
U.S. Department of Labor, 1983. Employment and Earnings, Vol. 30, No. 2. DOL, Washington DC.
Dohrenwend, B.S. and Dohrenwend, B.P., 1974. Stressful Life Events: Their Nature and Effects. Wiley-Interscience, New York.
Gutek, B.A., Sasse, S.H., and Bikson, T.K., 1986. The Fit Between Technology and Workgroup Structure: The Structural Contingency Approach and Office Automation, Proceedings of the Academy of Management, Chicago.
Helson, H., 1964. Adaptation Level Theory. Harper and Row, New York.
Holmes, T.H. and Rahe, R.H., 1967. The social readjustment rating scale. Journal of Psychosomatic Research, 11: 213-218.
Jaco, E.G., 1970. Mental Illness in Response to Stress. In: S. Levine and N. Scotch (Editors), Social Stress. Aldina, Chicago.
Kahn, R.L., Wolfe, D.N., Quinn, R., Snoek, J.D., and Rosenthal, R.A., 1964. Organizational Stress: Studies in Role Conflict and Ambiguity. Wiley, New York.
Karasek, R.A., 1981. Job Socialization and Job Strain: The Implications of Two Related Psychosocial Mechanisms for Job Design. In: B. Gardell and I. Johansson (Editors), Man and Working Life: A Social Science Contribution to Work Reform. Wiley, London.
Keen, P.W.G., 1984. The VDT as an Agent of Change. In: J.L. Bennett et al. (Editors), Visual Display Terminals: Usability Issues and Health Concerns. Prentice-Hall, Englewood Cliffs NJ.
Lazarus, R.S., 1974. Cognitive and Coping Processes in Emotion. In: B. Weiner (Editor), Cognitive Views of Human Motivation. Academic Press, New York.
Lowenthal, M. and Chiriboga, D., 1973. Social Stress and Adaptation: Toward a Life-Course Perspective. In: C. Eisdorfer and M.P. Lawton (Editors), The Psychology of Adult Development and Aging. American Psychological Association, Washington DC.
McGrath, J.E., 1970. A Conceptual Formulation for Research on Stress. In: J.E. McGrath (Editor), Social and Psychological Factors in Stress. Holt, Rinehart and Winston, New York.
Murray, H.A., 1938. Explorations in Personality. Oxford University Press, New York.
Osipow, S.H. and Spokane, A.R., 1984. Measuring Occupational Stress, Strain and Coping. In: S. Oskamp (Editor), Applied Social Psychological Annual, Vol. 5. Sage, Beverly Hills CA.
Perry, W.E., 1987. Professional stress: The impossible takes longer. Government Computer News, April 24: 31.
Roessner, J.D., 1984. Market Penetration of Office Automation Equipment: Trends and Forecasts. Georgia Institute of Technology, Atlanta GA (unpubl.), 56 pp.
Seyle, H., 1956. The Stress of Life. McGraw Hill, New York.
Sharit, J. and Salvendy, G., 1982. Occupational stress: Review and reappraisal. Human Factors, 24(2): 129-162.
Stasz, C., Bikson, T.K., and Shapiro, N.Z., 1986. Assessing the Forest Service's Implementation of an Agency-Wide Information System: An

Exploratory Study, The RAND Corporation, N-2463-USFS, Santa Monica CA.
Studer, R. and Barton, E., 1974. Aggregate effects of incremental changes in the social environment of the older person. Proceedings of the American Psychological Association, APA, Washington DC.
Turner, J.A., 1984. Computer mediated work: The interplay between technology and structured jobs. Communications of the ACM, 17(12): 1210-1217.
Turner, J.A., 1986. Integrated Information Systems Work Design and Ergonomic Issuess. In: J. Sethi, R.S. Schuler and D. Caro (Editors), Strategic Management of Technostress in an Information Society (in press), 24 pp.
Williams, A.W., Ware, J.W. and Donald, C.A., 1981. A model of mental health, life events and social supports applicable to general populations. Journal of Health and Social Behavior, 22(3): 141-166.

Social, Ergonomic and Stress Aspects of Work with Computers,
edited by G. Salvendy, S.L. Sauter and J.J. Hurrell, Jr.
Elsevier Science Publishers B.V., Amsterdam, 1987 — Printed in The Netherlands

Working Conditions and Health Complaints of Women Office Workers

Barbara G. F. Cohen

Research Psychologist

Division of Biomedical and Behavioral Sciences
National Institute for Occupational Safety and Health
Cincinnati, Ohio 45226

Chaya S. Piotrkowski

Associate Professor of Psychology

St. Johns University
Grand Central and Utopia Pkwys
Jamaica, New York 11439

Kevin E. Coray

Director, Information Management Staff

Office of Human Resource Information Management
Health and Human Services
Washington, D. C. 20201

ABSTRACT

The present study, conducted by the National Institute for Occupational Safety and Health, examines the relationship between the working conditions and the health complaints of women office workers in both public and private sector organizations. Female secretaries, clerks, and clerical information processors volunteered to fill out self-administered questionnaires of which 625 usable forms were returned for an overall response rate of 71.7%. While the questionnaires gathered data about working conditions, health problems, coping behaviors and family life, this paper reports only about working conditions and one index of health complaints.

Specific working conditions tapped by the questionnaire items alluded to physical environment, skill utilization, job autonomy, workload, interpersonal relations and the organizational climate. The hypothesis underlying the analysis is that better working conditions are related to less frequent health complaints among female office workers. This hypothesis was supported. Among physical conditions found to be related to more frequent health complaints were reports of poor air quality, noise, and inadequate lighting. Associated with fewer health complaints were measures of lesser tension and positive aspects of organizational climate such as supportive management, lack of coworker tension, being treated with dignity at work and absence of sexual harassment. It was concluded that health complaints may be related to a host of factors associated with the organizational and interpersonal context of office work as well as job design and related ergonomic concerns.

INTRODUCTION

Since the introduction of office automation, occupational health concerns among clerical workers have focused on visual, musculoskeletal, and stress problems associated with computerized office jobs (e.g. Smith et al., 1981; Sauter et al., 1983; Grandjean et al., 1983; Arndt, 1983; Dainoff, 1986; Bammer, 1986). Research examining working conditions and health complaints in a newspaper facility (Smith et al., 1981) indicated a pattern in which clerical workers using video display terminals reported the most physical symptoms and the highest stress level among three groups that were surveyed. Second highest scores on health complaints and on psychological stress scales were attained by clerical workers using traditional office equipment rather than VDTs, demonstrating the stressful nature of office work. More recently, Wall (1985), exploring various aspects of severe musculoskeletal problems in office workers, concluded that the origin of the problems were multicausal--biomechanical, individual, organizational and social.

In the present study, 625 female office workers were surveyed to investigate working conditions that may be related to health complaints. Specifically examined are worker perceptions of their physical environment, opportunities for skill utilization, job autonomy, workload, interpersonal relations and the organizational climate. The hypothesis underlying the analysis is that better working conditions are related to less frequent health complaints among female office workers. In this preliminary analysis, only simple correlations are reported.

METHODS

Data Collection:

The target population was comprised of secretaries, clerks, and clerical information processors. Self-administered questionnaires were distributed to women in these occupational groups at four office sites (two public and two private sector establishments). A total of 625 usable forms were returned for an overall response rate of 71.7%. In absolute numbers, the sample was split evenly between public and private sector respondents.

Sample Characteristics

Respondents ranged in age from 17 through 67 years (x = 36.6, SD = 12.1). Most were Caucasian (71.7%). Approximately 40% had high school diplomas and almost 60% had at least some college education. Almost half of all respondents were married, about one-third were single and a smaller proportion were separated, divorced or widowed. Almost one-third had children living at home.

Measures

While the questionnaire administered to the sample of office workers gathered data about a wide variety of working conditions, health problems, coping behaviors and family life, this paper reports only on the relationship between working conditions and one index of health complaints.

Some job conditions were measured with established self-report scales. Only measures that were internally consistent (Cronbach's alpha >.65) were retained. These included five Work Environment Scales, WES, (Moos et al., 1974) and several additional scales: i.e. Variance in Workload, Role Conflict, Underutilization of Abilities (Caplan et al., 1975) as well as Non-support from Boss (Haynes and Feinleib, 1980). Other items were factor analyzed to create other internally consistent measures. Second-order factor analysis of all the aforementioned scales was used to further group the indices of working conditions. This process resulted in the identification of six factors reflecting key dimensions of the work situation (See Table 1). This report is limited to those individual measures with high loadings on each of these factors.

A health complaints index was created from the results of a factor analysis of items from a health check-list contained in the questionnaire. This check-list has been used in several prior NIOSH investigations (e.g. Smith et al., 1981; Colligan et al., 1979). Three resulting factors (musculoskeletal symptoms, visual problems, head/back problems) corresponded with the list of symptoms generally reported by VDT operators surveyed in other studies (e.g. Smith et al., 1981; Dainoff et al.,

1982; Laubli, 1980; Gunnarsson and Soderberg, 1983; Gunnarsson and Ostberg; 1977; Hunting et al., 1980; Elias et al., 1980; Coe et al., 1980). Items with high factor loadings on these three factors were combined to form an overall health complaints index (18 items) which reflect the symptom pattern revealed in the literature. Coefficient alpha for this health complaints index is .91. In previous analyses (Cohen and Coray, 1986) it was found that VDT clerical operators reported complaints on this index more frequently than other office workers.

This health complaints index was not found to be significantly correlated with age nor race nor marital status. Correlations with education ($r = -.13$, $p < .01$) and the presence of children at home ($r = .15$, $p < .001$) are significant but low. The possibility that persons with increased health complaints were more likely to participate in the study was assessed by comparing the frequency of health complaints in a site with a high response rate (92%) to a site with a lower response rate (66%). Respondents in the latter site did not report significantly more frequent complaints suggesting minimal self-selection bias.

RESULTS

Consistent with other research (e.g. Smith et al., 1981; Colligan et al., 1979) this group of employed women reported several symptoms but occurring on an infrequent basis. While scores could range from 1 (Never) to 4 (Constantly) on each item of the checklist, the average score for the items on the health complaints index was 1.7 (SD =.47). Correlations between the health complaints index and scales tapping the key dimensions of the work situation are presented in Table 1. Table 1 indicates the strength and direction of the associations between working conditions and health complaints. The signs of all correlations are consistent with the hypothesis: Better working conditions are associated with less frequent health complaints.

DISCUSSION

Because even small correlations can be highly significant in large samples, the focus is on percentage of variance accounted for in health complaints, rather than significance levels per se. Using $r = .20$ as a lower

Table 1

Correlations (Two-tailed) Between the Health Complaints Index and Self-Report Measures of Working Conditions.

Working Condition Scales	r's	Nature of Relationship Between Scale and Health Index
		Increased health complaints are associated with reports of:
1 PHYSICAL CONDITIONS		
Air Quality	.26 a	being bothered more by poor air quality
Lighting	-.29 a	less adequate lighting
Noisiness	.22 a	being bothered more by noise
Space and Privacy	.10 b	less space and privacy
2 WORKLOAD		
Role Conflict	.25 a	more frequent role conflict
Work Pressure (WES)	.21 a	greater pressure orientation
Variance in Workload	-.20 a	more frequent variance increases
Underwork	.04	less frequent underload
Workload	-.28 a	more workload
3 INTERPERSONAL TENSION		
Stress from Co-workers	.22 a	greater source of stress
Sexual Harassment	.28 a	more frequent problems
4 ORGANIZATIONAL CLIMATE		
Clarity (WES)	-.17 a	less clarity in organizational roles
Innovation (WES)	-.11 b	less encouragement for new ideas
Non-support from Boss	-.18 a	less supportive boss
Staff-support (WES)	-.23 a	less supportive management
Task Orientation (WES)	-.16 a	less emphasis on tasks
Treated with Dignity	-.23 a	less fairness and respect
People Get Along	-.19 a	less co-worker support
Acceptance	-.25 a	less peer acceptance and less personal environment
Work Recognition	.18 a	less appreciation for work
5 SKILL UTILIZATION		
Remain in Current job	-.04	less inclined to stay
Underutilization of Abilities	-.13 b	less opportunity to use skills
Skills Used	.02	skills not used
6 JOB CONTROL		
Less Task Autonomy	.18 a	less latitude for change
Rule Orientation	.17 a	greater organizational emphasis

NOTE: Actual N's range from 558-615
$a = p < .001$
$b = p < .01$

limit, some interesting trends are found. Physical working conditions such as inadequate lighting, poor air quality and noise are found to be related to more frequent health complaints. Ergonomic and environmental measurements taken in work areas for a subsample of the survey respondents will allow further examination of these findings.

However, physical conditions were not the only factors related to increased health complaints. Work overload (but not underload) was also associated with increased complaints. Equally interesting, measures of the interpersonal context were related to health complaints. Perceived stress from coworkers as well as sexual harassment were related to increased health complaints; human relations aspects of organizational climate (e.g. staff support, being treated with dignity at work, and acceptance) were related to fewer health complaints.

These findings are consistent with the conclusion drawn by Cohen et al., (1981). Noting the slight correlation between time on VDTs and number of health complaints in a sample of office workers, Cohen et al. concluded that job content and organizational factors must be considered in understanding the health complaints of VDT operators and other office workers. In other words, the organizational and interpersonal context of office work, as well as job design, may be as important to well-being as ergonomic factors.

While each individual correlation between health complaints and working conditions was modest in size, in future reports, the authors will consider additive and interactive effects among the various measures of working conditions as they may improve the prediction of health complaints. This pattern of health complaints may be associated not simply with VDT operation per se, but a host of factors associated with the stresses of office work in general and the special organizational and job design features associated with the new technology.

References

Arndt, R. 1983. Working posture and musculoskeletal problems of video display terminal operators-Review and appraisal, American Industrial Hygiene Association Journal, 44: 437-446.

Bammer, G. and Blignault, I. 1986. Psychosocial aspects of repetition strain injuries, Proceedings of Australian National University Research on RSI.

Caplan, R., Cobb, S., French, J.R.P., Van Harrison, D. and Pinneau, S. 1975. Job Demands and Worker Health, National Institute for Occupational Safety and Health, Cincinnati, No. 75-160.

Coe, J., Cuttle, K., McClellon, W., Warden, M. and Turner, P. 1980. Visual Display Units, New Zealand Department of Health Report No. W/1/80, Wellington, New Zealand.

Cohen, B.G.F., and Coray, K. E. 1986. The relationship of select ergonomic and personal factors with health complaints of female office workers. American Psychological Association Annual Conference, Washington, D.C.

Cohen, B.G.F., Smith, M.J., Stammerjohn, L.W. 1981. Psychosocial Factors contributing to job stress of clerical VDT operators, In G. Salvendy and M. Smith (Eds.) Machine Pacing and Occupational Stress, London, Taylor and Francis: 337-345.

Colligan, M.J. and Murphy L.R. 1979. Mass Psychogenic Illness in Organizational Settings: An Overview, Journal of Occupational Psychology, 52: 77-90.

Dainoff, M., Fraser, L. and Taylor, B.J. 1982. Visual, musculoskeletal and performance differences between good and poor VDT workstations. Proceedings of the Human Factors Society 26th Annual Meeting, Seattle.

Dainoff, M. and Dainoff, M. 1986. People and Productivity: A Manager's Guide to Ergonomics in the Electronic Office. Toronto: Holt and Reinhart Publishing Co.

Elias, R., Cail, F., Tisserand, M. and Christman, M. 1980. Investigations in operators working with CRT display; relationships between task content and psychophysiological alterations, in E. Grandjean and E. Vigliani (Eds.) Ergonomic Aspects of Visual Display Terminals, London: Taylor & Francis, Ltd.: 211-218.

Grandjean, E. Hunting, W. and Piderman, M. 1983. VDT Workstation design: preferred settings and their effects, Human Factors, 25: 161-175.

Gunnarsson, E. and Ostberg, O. 1977. The Physical and Psychological Working Environment in a Terminal Based Computer Storage and Retrieval System. (Stockholm, Sweden: National Board of Occupational Safety and Health Report 35).

Gunnarsson, E. and Soderberg, I. 1983. Eye strain resulting from VDT work at the Swedish Telecommunications Administration, Applied Ergonomics, 14: 61-69.

Hunting,W., Laubli, T., Grandjean, E. 1980 Constrained postures and VDU operators, in E. Grandjean and E. Vigliani (Eds.) Ergonomic Aspects of Visual Display Terminals, London, Taylor and Francis, Ltd: 175-184.

Laubli,T. Hunting, W. and Grandjean, E. 1980. Visual Impairments related to environmental conditions in VDU operators. In E. Grandjean and E. Vigliani (Eds.) Ergonomic Aspects of Visual Display Terminals, London: Taylor and Francis, Ltd.: 85-94.

Moos, H., Insel, P., and Humphrey, B. 1974. Family, Work and Group Environment Scales Manual, California: Consulting Psychologists Press, Inc.

Sauter, S.L., Gottlieb, M.S. and Jones, K.C., Dodson, V. and Rohrer, D. 1983. Job and health implications of VDT use; initial results of the Wisconsin NIOSH study, Communications of the Association for Computing Machinery, 26: 284-294.

Smith, M.J., Stammerjohn, L., Cohen, B.G.F. and Happ, A. 1981. An investigation of health complaints and job stress in video display operations, Human Factors, 23: 387-400.

Wall, J. 1985. Investigation of attribution of cause, exposure factors and psychosocial stress parameters as they relate to musculoskeletal complaints in keyboard operators, Doctoral thesis, Dept. of Psychology, Australian National University.

AUTHOR INDEX